Praise for High D
DVD Handb

"The HD Handbook is essential for anyone interested in getting the most out of next gen discs. The authors' even-handed approach balances breadth and depth of information perfectly, making sense of these complex new formats. An indispensable reference."

— **Joe Rice**
Sr. Technical Director
MX Production Services

"The...definitive handbook on HD DVD and Blu-ray. Whether you are new to DVD and next gen technologies, or a seasoned professional, you will find up-to-date information, clearly illustrated examples, and real-world knowledge that will be invaluable to you. The book covers technical specs, authoring, compression, project planning, quality control, graphical design and much more. It truly is the bible for everyone involved in high definition DVD".

— **Michael Ory**
Managing Director, Blink Digital NY
Ascent Media Group, LLC

"This book provides insight into first-hand experiences launching both new formats ... that makes it an invaluable resource and a must-read for everyone in this business – professional and newbie."

— **Michael Zink**
Director, Advanced Technology
Technicolor

"This is a great book for getting an early foundation of understanding for both competing high definition optical disc formats. Inside, you'll find the information you need presented in a concise format from some of the most authoritative sources in the industry."

— **Eric Carson**
DCA, Inc.

"Written by industry-leading experts, the High Definition DVD Handbook offers a broad examination of the HD DVD and Blu-ray formats, providing a balanced analysis of their strengths and weaknesses. Combined with tips and tricks garnered from years of production expertise in Hollywood's top authoring facilities, this book is sure to become the reference for HD authors."

— **Daniel Robertson**
drinteractive
Victoria, Australia

High Definition DVD Handbook

Producing for HD DVD and Blu-ray Disc

About the Authors

Mark R. Johnson is the CEO of Javelin Ventures LLC, a software company pioneering the development of new tools for DVD, HD DVD and Blu-ray Disc production. Formerly Vice President of Research & Development at Technicolor Creative Services, Mark played an active role in the creation of both the HD DVD-Video and Blu-ray Disc specifications. He co-authored *DVD Demystified*, Third Edition with Jim Taylor and Charles Crawford, released in November, 2005. As an award winning DVD author, Mark has helped to set the standards for DVD production with his breakthroughs in advanced applications of the DVD specifications. Mark lives in Pasadena, CA.

Charles G. Crawford is the Co-Founder of Television Production Services, Inc. and Heritage Series, LLC, production companies specializing in traditional and interactive production and title development. He has been involved with DVD technology since 1998 when he wrote the Operations and Reference Manuals for Matsushita/Panasonic's award winning DVD Authoring System. He co-authored *DVD Demystified*, Third Edition with Jim Taylor and Mark R. Johnson. Chuck has been involved in video production for over 35 years at the local, network, and international levels, and has received three national EMMY awards for his technical and directorial expertise, plus numerous other industry awards for his production skills. Chuck lives in Washington, DC.

Christen M. Armbrust, is a leader in the development and implementation of new digital media. In 2005, he joined Technicolor Creative Services, a Thomson Company in Burbank, CA to define, develop and implement their new encoding, authoring and mastering processes for the next generation discs - HD DVD and Blu-ray Disc. Chris is the Founder of Marin Digital, a company which focuses on "new" technology and has produced over 1100 DVD titles. Chris has delivered white papers and has spoken at SMPTE, NAB/BEA, Replitech Europe, DVDA and other DVD related conferences. Chris commutes to Burbank and lives in Sausalito, CA.

High Definition DVD Handbook
Producing for HD DVD and Blu-ray Disc

Mark R. Johnson
Charles G. Crawford
Christen M. Armbrust

New York   Chicago   San Francisco   Lisbon   London   Madrid
Mexico City   Milan   New Delhi   San Juan   Seoul
Singapore   Sydney   Toronto

The McGraw Hill Companies

Cataloging-in-Production Data is on file with the Library of Congress

McGraw-Hill books are available at special quantity discounts to use as premiums and sales promotions, or for use in corporate training programs. For more information, please write to the Director of Special Sales, Professional Publishing, McGraw-Hill, Two Penn Plaza, New York, NY 10121-2298. Or contact your local bookstore.

1 2 3 4 5 6 7 8 9 0 DOC/DOC 0 1 2 1 0 9 8 7

ISBN-13: P/N 978-0-07-149585-1 of set
            978-0-07-148585-2
ISBN-10: P/N 0-07-149585-1 of set
            0-07-148585-6

This book is printed on acid-free paper.

Dedications

To my dearest friends and family, who pulled together at the most difficult of times to make this book possible.

— Mark Johnson

To my wife and partner, Samantha Cheng, for her zealous focus on the goal and her unflagging patience during the journey.

— Chuck Crawford

To my wife, Grace, whose loving support, understanding and encouragement gave me the inspiration to keep going throughout the effort.

— Chris Armbrust

Acknowledgments

Like any production, be it a book or television show, it is without a doubt a team effort. We gratefully and respectfully acknowledge those with whom we have been so fortunate to have as part of our team.

Thank you to Jim Taylor for inspiring us to move forward in writing this book. To Steve Chapman, our Publisher for his patience, Nathan J. Best, Graphics and Layout, Elisabeth G. Baker, Copy Editor and Ari Zagnit, for always returning our calls.

A special thank you to our technical reviewers without whom this book would have no value - Sam Archer, Jess Bowers, Randy Berg, Eric Carson, Jeffrey Cheng, Andrew Clinick, Kevin Collins, Roger Dressler, Don Eklund, Lee Evans, William F. Foote, Matthias Frener, Deiter Hametner, Gunthar Hartwig, Phil Henkel, Gary Hung, Jim Lin, Ole Lutjens, Phillip Maness, Hideki Mimura, Gail Myles, Mike Ory, Nicolai Otto, Jeff Ouyang, Charles Potter, Dulce Ponceleon, Joe Rice, Daniel Robertson, Andrew Rosen, Mike Schmit, Robert Seidel, Stacey Spears, Phil Starner, Peter Torr, Yasufumi Tsumagari, Steve Venezia, Mike Ward and Michael Zink.

Mahalo to our Ohana who thought they were going to paradise to rest and relax and ended up reading and editing chapters instead - Alyson Follenius, Julie King, Chip Reif and Mopsi Villereal.

— Samantha Cheng, Managing Editor

Foreword

As I started writing this foreword I paused to consider how it might read after five years, and I realized things would be quite different then. As of 2007, high definition video is still very new. Even though high definition video was conceived in the eighties and born in the nineties, it was not until after the turn of the century that it began to transform mainstream video, and not until 2006 that mass-market disc formats Blu-ray and HD DVD began to deliver full 1080p resolution directly to consumers. The "format war" is in full swing and no one knows how it will turn out, although some of us have long predicted détente in the form of dual-format players and even dual-format discs. It's unclear how quickly the transition from SD to HD displays will happen. So in spite of the fact that everyone knows what HD is and has seen plenty of eye-catching HD, it's still relatively new, untested and misunderstood. Thus this book. This book's authors have a wealth of real-world experience in what it takes to make HD work, from shooting to encoding to editing to authoring to final disc. This is not a theoretical book about formats and specifications, it's a handbook, soon to be dog-eared from regular use, that should be indispensable for anyone authoring high definition discs. As this book makes abundantly clear, those DVD-Video discs that we used to think were so tough to make were a walk in the park compared to the new high-definition disc formats. In this book a few pioneers of HD authoring have paused long enough to pull the arrows out of their backs, figure out how the darn things got there, and jot down some helpful advice for those following behind. And to top it off it's not a boring read – it's wry, not dry. And the jokes will still be funny after five years.

— Jim Taylor
Senior Vice President & General Manager
Advanced Technology Group,
Sonic Solutions
Chairman, DVD Association
Author of *DVD Demystified*

Table of Contents

Table of Figures and Tables

Preface

This book was first conceived following the successful but challenging completion of *DVD Demystified*, Third Edition. The HD DVD and Blu-ray Disc formats were nearing their market debut, and it seemed the time was ripe for an in-depth look at how to produce content for these sparkling little discs. With all of our closest friends and colleagues (and several thousand pages of freshly bound specifications) in mind, we sought to create a resource that would distill the confusing myriad of details into a useful collection of facts, figures and helpful tips.

Little did we realize how fluid the specifications for both HD DVD and Blu-ray would continue to be throughout the year that this book was in production. Early success in understanding these formats gave way to specification revisions, corrections, and updated implementations. Players and discs continued appearing on the market at the same time as the tools used to create them. In an amazing dance only chaos math could describe, player manufacturers, software tool developers, content producers and disc manufacturers all worked together (and sometimes in spite of each other) to create a working symphony of amazing technologies. With the occasional bent note here and misplaced chord there, the industry successfully launched not one, but two blue laser formats around the world.

Now for the first time, we can see the tangible substance of what was for years relegated to philosophical and often heated debates. Now, as the realities of price and performance mix with the concerns of daily production, it's clear what makes a specification into a standard. Not to overlook the tremendous effort that hundreds of industry visionaries and pioneers put into them, but the specifications alone amount to a substantial pile of paper and ink. We now have the privilege of witnessing an industry giving birth to standards that put into practice that which has been described by the specifications.

If we've done our job well, you will find this book a helpful tool as you make your way through these exciting new formats. There's far too much to cover in a single text, but we hope this handbook will provide you with an essential understanding to help you realize your own creative vision!

— Mark R. Johnson
January 2007

Chapter 1
Introduction

Two formats, both alike in enterprise,
in fair featureland, where players stage their game.
For arcane grudge revives a rending tumult,
as ego 'n' greed beget churlish foes to blame.[1]

Hoping to build on the highly successful standard DVD formats, content producers and copyright holders ambitiously sought agreement on a next generation disc format, in consult with leading electronics and computer manufacturers. The movie studios and entertainment producers were looking to technology to provide an enhanced presentation environment where their libraries could be repurposed using the features and benefits of high definition widescreen display in enhanced aural settings. Yet, what has emerged from the pursuit of their goal is the introduction of two incompatible, yet eerily alike, optical disc technologies, HD DVD and Blu-ray Disc™, aka BD.

Both disc formats utilize blue laser optics to embed and retrieve data from discs, sized to match CDs and DVDs. Both formats provide an enhanced user experience, with interactive menu and presentation tools. And, both formats incorporate data encryption techniques that protect and preserve the rights of the content providers.

One may wonder why we have two formats if, in fact, they perform in similar fashion. What could possibly have created the current dilemma? The seed for the dissension was sown in the long ago, but never forgotten, contest between competing videotape technologies, Betamax versus VHS. Coupled with long simmering antagonism over patent and royalty rights, major electronics manufacturers staked out territories in the newly developing field of blue laser technology. Sony chose to pursue greater capacity, while Toshiba sought compatibility with the burgeoning standard DVD technology. Further differences developed in the areas of content protection and disc construction. Each side sought allies for their approach, further entrenching their positions. There were attempts at format unification, but neither side has been willing to compromise for the greater good.

The marketplace may decide a winner between these formats or may bypass both, turning instead to other emerging innovations, such as, video-on-demand, subscription hard drives, non-realtime download, or other network streaming technologies. Or, both formats may survive albeit with limited growth and moderate acceptance.

No matter the outcome of the marketplace reception, the disc authoring community, striving to capitalize on the next frontier, has been saddled with tools for these not too dissimilar high definition optical disc formats. Suffice it to say that the business would be easier if there was one format for the industry. Alas.

---

[1]with apologies to William Shakespeare (1564-1616), Prologue to "Romeo and Juliet", a tragedy (1596)

1

About this Book

High Definition DVD Handbook: Producing for HD DVD and Blu-ray Disc endeavors to address the processes, problems and peculiarities of both high definition disc formats. The authors' goal is to provide a thorough foundation and an exhaustive understanding for conceiving, creating, and completing productions that take advantage of the features, functions and foibles of optical discs authored to meet the specifications of the next generation discs, HD DVD and Blu-ray Disc.

First off, we are going to shorthand the title of this book to *HD Authoring Handbook*. This will allow us to save on ink and paper.

This *HD Authoring Handbook* is a technical handbook, intended to provide a reader with all of the information needed to conquer the tasks of authoring and programming for the next generation optical discs. Unlike standard DVD, with off-the-shelf toolkits and software applications, the next-generation disc flavors require that an individual attempting to author an HD DVD or BD project has both an understanding of the disc technology as well as familiarity with Java™ and JavaScript™ computer language programming.

Two of the authors of this book co-authored, with Jim Taylor, *DVD Demystified Third Edition*, and this *HD Authoring Handbook* is a companion to that publication. As such, we will be drawing on the information in *DVD Demystified Third Edition* to aid in the presentation of this book. Although this handbook will, on occasion, duplicate some of the content in *DVD Demystified Third Edition*, we urge our readers to consult that publication for a more complete exposure to and grounding in DVD technology.

This book is divided into parts, which are divided into chapters, which are subdivided into sections. Following the final chapter, is *Appendix A*, which provides information about the disc that accompanies this book. Please see the following section, *About the Disc*, for more information on Appendix A. And, following Appendix A is a *Glossary* of esoteric, obscure, and mundane words and terms to aid those who know they saw a word somewhere but cannot remember what it means or where they saw it.

By the way, for those of you expecting or hoping that the authors of this book will choose, let alone endorse, one format over the other, fuggedaboutit. Forces far greater than we, having invested vast amounts of time, effort, and money trying to bring about format unification, have ceased their efforts. Far be it for us to step in where others no longer tread. Nosirree.

Part I – Fundamentals and Technologies

The new disc formats require an understanding of the technologies that can be utilized in the production for and the presentation of the next generation optical discs. This part of the book provides information on the technologies that a production team should be familiar with while developing the content for a disc project, from concept to playout.

Chapter 2, Lights...Camera...ACTION!, takes a look at the format choices for acquiring footage and some of the issues confronted when deciding to use either film or video tools on a production. Successfully capturing high definition imagery requires an understanding of

1

picture resolutions, frame rates, data rates and digital technologies. Producers face a range of digital options and the choices made during the acquisition stage of a project will impact the end product. The adage of "we can fix it in post production" no longer applies. Knowing what happens to the image data when shooting the footage will help to alleviate problems during the subsequent production steps.

Chapter 3, Pictures and Sound, discusses some of the issues and conditions encountered when interacting with the next generation HD technologies — playing, viewing or recording. For example, how does the latest in connection technology, HDMI (High Definition Multimedia Interface), affect the crop of players and displays in the marketplace. And, given the two HD disc formats, what are the settop player and computer configurations needed for the latest discs. Meanwhile, there remains a huge installed base of legacy equipment that is not capable of achieving the full level of functionality that the new disc formats and players may support.

Chapter 4, Coding Technologies, exposes the strengths and weaknesses of the expanded encoding choices for the next generation discs — video: MPEG-2, MPEG-4 AVC, or SMPTE® VC-1; audio: Dolby® Digital Plus, Dolby® TrueHD, DTS-HD™, or DTS-HD™ Lossless. This chapter also delves into the additional audio options for streaming content — mp3, aacPlus version 2, Windows Media® Audio, and more. Information is also provided on what differentiates the encode choices and what must a producer know to make an informed decision for the best presentation of their content. This chapter will familiarize the reader with these new data schemes and will help explain where the tangible differences are to help make the judgment call of which technologies to use.

Chapter 5, Persistent Storage, provides information on one of the most unique features incorporated in the specifications for the new disc formats — onboard player memory storage at a minimum of 64 kilobytes. Possible applications utilizing this capability include storing user preferences, playlists, and configuration information, as well as, downloading material such as trailers, bonus materials or updates supplementing the content of the disc currently loaded in the player.

Chapter 6, Network/Streaming Technologies, explores what may be the Holy Grail for next generation discs, merging disc content with data accessed on the Internet. Now you can create content on the web and either download that content to the player or stream it from the web without the need to have the enduser configure a computer (or in the corporate environment, get IT approval to set it up). Network or streaming technology makes it possible to create interactive applications that take advantage of the massive local storage of the HD DVD or Blu-ray Disc combined with the fresh content and information from a network.

Part II – Formats

This part discusses the features, the functional differences, and the technologies needed to create and display the next generation optical disc formats.

Chapter 7, The HD DVD Format, introduces the two primary modes for delivering HD DVD content that will play in PCs and consumer electronic devices — *Standard Content* and

1

Advanced Content. Standard Content is essentially the high definition version of today's DVD with a few new bells and whistles, while Advanced Content is an entirely new way to create content for discs. This chapter will look at the key differences between these two approaches and will provide some guidelines for selecting the content approach that is right for your project.

Chapter 8, The Blu-ray Disc (BD) Format, presents the two different modes for putting Blu-ray content on a disc — *HD Movie Mode (HDMV)* and *Blu-ray Java (BD-J)*. Unlike HD DVD, the Blu-ray modes diverge significantly from the traditional approaches used for current DVDs. Key differences between the two BD modes help determine which approach may be most effective for a given project.

Chapter 9, Compression and Authoring Tools, contains information on the choices available in the selection of the appropriate tools for a project, which are allied with the selection of a format mode for a project. For example, if a project is relatively straightforward with video playback and little need for or use of fancy interactive graphics, then the tools selected would be those optimized for this project type, allowing you to finish in minimal time with minimal effort. In this chapter, we take a look at the wide range of tools that are either on the market or expected shortly, and we assess the most effective tools for each type of project.

Part III – Production

The art of constructing a next generation format disc requires thorough preparation and touches on a plethora of considerations. This part of the book traverses the slopes, peaks and valleys of the production process, from concept to completion. Rather than continuing to wax poetic, we'll cite a few of the many challenges that need to be overcome while producing a high definition disc project.

A major hurdle to be addressed in a production setup is that the file sizes can be HUGE compared to DVD! A project starting with uncompressed 1080p video and eight channels of 24 bit 96 kHz sampled audio will result in terabytes of data. So you'll need storage capacity to match. And, that's if you're working on one project at a time!

Also, the time required for compression varies from realtime for hardware encoders to a range of 4 times to 80 times realtime for software encoders. And, these are time estimates for one-pass encoding.

Further, the target image for a full disc can be from fifteen to fifty Gigabytes. The first generation of Blu-ray or HD DVD recorders will write to disc at 2x speed, thus taking from 70 to 140 minutes just to write the disc, and double those times when adding a verify pass.

This section of our book serves as a guide through each stage of production, from the simple to the sublime. (We couldn't help ourselves with that one.)

Chapter 10, Setting Up for Production, breaks down the initial stage of the production process into three general categories — Single-user Consumer Production, Small-scale Prosumer/Corporate Production, and Large-scale Feature Production. This chapter will examine what is needed to enter HD DVD and/or BD production, whether it's putting a home movie collection on a few large-capacity discs or executing multiple simultaneous feature film to disc productions. Workflow is examined for each production category, giving special attention to the hardware/software tools, network flexibility needs, and budget considerations for each category. Granted, things may change a great deal over the next few years as software and hardware becomes available at increasingly lower cost, but there may be a few tricks you can use to get in early without a heavy investment.

Chapter 11, Project Design and Planning, addresses how to set up a project. The capabilities of the next generation discs are much more extensive than DVD and planning for a project is much more complex. The functionality requirements of the project and the skill sets of the production team, along with the tools available, dictates the amount of planning required for a project. Next generation disc production introduces never-before-encountered influences on project planning and scheduling. Programming test periods and software code development are but two of the new factors that may affect a project timeline. Reworking or tweaking a title must be planned for in both the workflow and in the overall schedule. Given the complexity and sophistication of the new formats, it will be much easier to miss a deadline when producing a next generation disc. Adequate flexibility is paramount when designing and executing a project.

Chapter 12, Graphic Design, illuminates the tools and methods available for creating the graphics elements of a title. Graphics design will be determined by the authoring tool used on a project, and is closely intertwined with the authoring/programming efforts for the title. You will need to optimize what can be programmed and how to create the assets to expedite the programing effort. Element naming conventions and hand-off formats will need to be defined, so that the graphics design and creation process proceeds apace of the project timeline.

Chapter 13, Compression, explores the video and audio compression tools and methods available for creating HD DVD and BD content. Whether you have selected MPEG-2, MPEG-4 AVC or VC-1 to encode the project video, there are a number of things that can be done to ensure that you achieve the highest possible quality that the tool will allow. In this chapter, we'll discuss the different types of coding parameters, such as GOP size, scene detection, motion search options, and de-blocking filters, that each method offers and how these can be used to improve the quality and efficiency of the video encoding. And, as with video compression, the new audio compression methods supported by the next-generation disc formats offer another new set of tools and parameters. We'll take a look at the different processes and demystify each option, showing which ones will be most effective for a project. In addition, we'll look at how to approach several common problem conditions, including how to keep the compression from taking so long!

Chapter 14, Authoring/Programming, describes the step-by-step processes necessary to create a next generation format shiny disc. This chapter provides details on arguably the most critical phase of a project. If the instruction set for title playout is wrong, then the disc is a failure. Information is provided on how to create the four disc format flavors — HD DVD Standard Content, HD DVD Advanced Content (HDi), Blu-ray HD Movie mode (HDMV),

1

and Blu-ray Java™ (BD-J). Along the way, we'll look at a few simple programs and identify some of the more common problems to watch out for and to avoid. With clear, working examples, by the time you're done, you should come out of this chapter with a solid understanding of how to author a next generation disc without driving yourself bonkers.

Chapter 15, Quality Control, outlines the tools and techniques needed for assessing the performance of a title. These new format discs more closely resemble a software project than a DVD project. Tools are needed to perform code inspection, to verify multiplexed data streams, as well as to confirm the functionality of a title. Among the concerns that will need to be addressed is the fact that players (hardware or software) differ in how they implement the HD specifications. Coordination with testing facilities will help to insure that the disc performs as designed. You will need to make your disc immune to the anomalies of the different implementations.

Chapter 16, So, You Think You're Done, examines the final stages of a project. Duplication or replication procedures are determined pursuant to the disc format selection, with a limited number of facilities available to perform those tasks. Additionally, there are considerations of packaging and labelling, covering the finishing touches when preparing the discs for distribution and merchandising.

Part IV – Content Protection and Management

This final part of the handbook elaborates on the intricacies and idiosyncracies of the protection schemes developed for use with HD DVD and BD.

Chapter 17, Advanced Access Content System (AACS), offers insight into the problem of deciding whether or not to include AACS protection on a disc and, more importantly, how? Whether you are creating an HD DVD or a Blu-ray Disc, AACS is a robust, feature-rich, but very complicated system for protecting content. It makes it possible to use content on disc, in persistent storage, or over a network. And, AACS provides a mechanism called *Managed Copy* that allows the content producer to determine how and when a viewer is allowed to make copies of the content from disc or elsewhere. Unfortunately, there are also a great number of restrictions and challenges associated with the use of AACS, and just getting it to work properly can be quite a challenge! In this chapter, we'll look at when AACS should be used and how to get it onto a disc in its various forms.

Chapter 18, Blu-ray Disc's BD+, for those creating Blu-ray titles, expounds on the additional level of content protection that is provided in the form of BD+. Among other things, BD+ makes it possible to include special security software on a disc that can help to control how that disc may be accessed, as well as to provide additional features such as embedded watermarking or piracy countermeasures. The actual process of creating BD+ security software (known as *content code*) goes well beyond the scope of this book, but we will take a look at the process that would be used to activate the features of BD+ and to integrate it with the disc programming. In addition, we'll provide some pointers for where to find more information about Self Protecting Digital Content (SPDC) and how to increase the security of your discs.

About the Disc

This *HD Authoring Handbook* is accompanied by a standard DVD. The disc is composed of DVD-ROM components only and will not play in a DVD-Video player. When the disc is viewed via a computer DVD-ROM drive, please select the file **AboutThisDisc.pdf**, to view the disc contents in a folder/file structure.

The disc contains folders and files that provide additional tools and assets for the reader of this book. Included on the disc are calculators for determining disc size and audio volume, along with an editor for creating HD DVD (*.aca) Archive files. Instruction sheets and tutorials have been prepared for the various elements in the Adobe® portable document file (.pdf) format, and are included in the respective folders of the disc. We also gathered data from a variety of manufacturers and suppliers detailing their latest high definition applications and products.

As an added bonus, we have included a series of professionally produced sound files that may be used for menu and element sound effects. These sounds are free for use but you must reference Willie Chu of New York City with disc and/or jacket credit in a form similar to the following — *Interactive audio sound effects courtesy of Willie Chu NYC.*

Please be aware that elements on the disc may require unique minimum system capabilities. Information about system requirements is provided in *Appendix A* and in the **AboutThisDisc.pdf** file.

As this book went to press, we were still pulling the disc assets together from an array of sources, so it is not possible to list specific contents in this section of the book. We apologize for that dilemma, and can only request that you accept the conundrum as unavoidable.

Standards and Conventions

The authors have adopted a standard approach for the presentation of several terms, conditions, and items when used in this handbook. Not the least of which is that we will use the article "an" before "HD DVD". We recognize that this might not conform to traditional grammar, but we feel that it is easier to read and say "an HD DVD". We think that the other way, "a HD DVD", is just too much of a mind-stumble and, when encountered, causes one to pause uncomfortably, resulting in a momentary loss of concentration and a realization that sometimes the English language is a little screwy. Conversely, we will follow the standard practice of "a" before "BD", too.

The following icons, symbols and typographical conventions are used throughout this *HD Authoring Handbook* —

①   A number in a circle indicates a step-by-step process that is to be followed.

❶   A number in a filled circle indicates a simple numbered list.

▪   A gray square denotes a simple list, bulleted item or items.

■   A black square denotes alternative methods to perform tasks or actions.

●   A circle denotes associated actions, elements, or commands.

1

NOTE
This symbol indicates that the information contained herein is an important fact or comment.

TIP
This symbol denotes that the information provided herein may be a useful consideration.

WARNING
This symbol indicates an alert to an action that you may want to avoid and/or the information stated herein is critical to know.

Other Conventions

Spelling There are many words that may be spelt in more than one way. Actually, we just cheated there. We were going to use "spelled", but wanted to keep your attention. That was a cheap trick, not because spelt is not a word, it is. It was a cheap trick because it does not mean spelled.[2]

The word *disc*, with a c, is used to indicate optical discs, whereas magnetic disks are spelled with a k. To maintain the usage rule, magneto-optical discs, which combine the formats, also use a c as the discs are read by laser. No less an authority than the *New York Times* updated their manual in 1999 to follow the, by then, long since accepted industry practice of c for a laser read disc. Those who continue to misuse k when c is de rigueur should be viewed with some skepticism especially if pontificating.

HD The term *HD* may be used to mean high definition or high density. In this book, we will use the term as shorthand for high definition.

iHD or HDi and HD DVD AC or HDAC When used in a sentence, iHD is HD DVD AC but HDi is not HDAC. Oh, my, one yearns for an "acronym free zone". Addressing the second pair first, HDAC is the shortcut for HD DVD Advanced Content. Used sparingly, HDAC is a quicker way of saying HD DVD AC. But the terms iHD and HDi are not interchangeable. The term *iHD* was casually adopted during the early development of next generation shiny discs to mean "interactive high definition". The legality of usage was overlooked, however, and the Internet High Definition standards organization adopted "iHD" along with the web address www.iHD.org, while the International Healthcare Distribution firm seized www.IHD.com. So, Microsoft shifted the letters around and filed a trademark for "HDi". Now, technically, HDi is a registered mark for Microsoft's implementation of their tools for the next generation disc formats. Magnanimously muddying the waters, Microsoft is casually allowing widespread use of the term. We will strive to avoid either term, but will claim ignorance if we fail.

---

[2] Spelt is a noun, and is a hardy wheat grown mostly in Europe for livestock feed.

PC and **OS** The term *PC* is used for "personal computer" and *OS* means "Operating System". When it is necessary to make a distinction as to hardware or OS, brand names such as Windows®, Mac OS®, Linux, Apple®, et cetera, will be indicated.

Title The word *title* takes on two primary meanings in the optical disc realm. Generally, title may be used to indicate the total contents of a disc, for example, the accompanying disc to this book is a DVD title. And, specifically, title is the largest data unit of a disc.

Widescreen The term *widescreen*, as used in this handbook, is used to indicate an aspect ratio of 1.78 (16:9).

Authoring The word *authoring* has two primary meanings when used in disc production. In a general sense, authoring may be used to describe the overall production effort, as in, authoring a title takes a long time. In a specific sense, authoring refers to the process of creating the navigation and programming command structure required for a disc.

Bits, Bytes, Units and Notations

The evolution to next generation optical disc technologies does, on one hand, simplify the frame size factors for displays, although we still have to deal with the two frequency factors of 50 Hz and 60 Hz. On the other hand, the evolution does not solve the confusion that exists when mixing computer storage measurements with data computation measurements. The language of numbers when defining disc capacities borders on the incomprehensible. When you are working with bits and bytes, millions and billions take on different meanings when based on powers of two (binary) rather than powers of ten (decimal). This dilemma quickly became evident to those working with DVDs, where a 4.7 billion byte disc has only 4.37 gigabytes of capacity.

Since their introduction, computers use the binary numbering system which is based on powers of two — two, four, eight, sixteen, etc. Computers adopted this format, which uses only two digits, zero and one, because each binary digit can be represented by a single digital switch set to either on or off (zero or one). Each switch is called a bit[3], and eight bits comprise a byte. Binary circuits that detect the difference between the two states, on or off, are less complex and less expensive than circuits that detect the difference between ten states (ten digits, zero through nine).

In the early stages of computer development, values were relatively small, and the ranks of computer users were filled mostly by engineers who readily understood the binary method of computation. As values increased, terms began to blur. For example, two to the tenth power is 1024, while ten to the third power is 1000, yet both of these values cavalierly were described by the same prefix, kilo-. The next prefix, mega-, was used to describe both two to the twentieth power, which is 1,048,576, and ten to the sixth power, which is 1,000,000. Still, because it was mostly engineers who worked with these values, the inequalities were disregarded.

---

[3]By the way, this'll put a smile on your face. Bit is a contraction of "binary digit". Yeah, right.

However, as numbers increase, the disparity between values changes from problematic to significant. Whereas, at the kilobyte level the difference between decimal and binary values is 2.4%, the difference at the gigabyte level is 7.4% — 1,000,000,000(10^9) or 1,073,741,824(2^{30}). Thus, when extrapolated to reflect the actual capacity of a disc or device, determining the binary value becomes very significant.

Comprehending this numbering maze takes an even more bizarre turn when one considers that capacities expressed in decimal terms use larger numbers than capacities expressed in binary terms, even though the binary value reflects the true capacity of the media or device. And, marketing types seize on the larger number when promoting a product or device, ignoring the true capacity of the media or of the computer drive. This hijacking of the decimal terms to promote that larger is bigger (yes, that's a play on words), merely contributes to the distortion and confusion. As more devices become available, the blurring of thousands, millions, billions with kilo-, mega-, and giga- grows apace.

Attempts at standardization have heightened awareness of the issue, but clarity remains elusive. The engineering genie is out of the bottle and the marketing weenies are hiding the stopper.

In 1999, the International Electrotechnical Commission (IEC) introduced new prefixes for the binary increments, supplanting the common uses of the established prefixes promulgated by Systeme International d'Unites (SI), the international standards organization for setting measurement notations such as millimeters and kilograms (see Table 1.1).

Table 1.1 Measurement Prefixes

| SI | | | IEC | | | Usage |
Symbol Prefix		Common Use	Symbol	Prefix	Computer Use	Difference
k or K	kilo	[k] 1,000 (10^3)	Ki	kibi	[K] 1024 (2^{10})	2.4%
M	mega	1,000,000 (10^6)	Mi	mebi	1,048,576 (2^{20})	4.9%
G	giga	1,000,000,000 (10^9)	Gi	gibi	1,073,741,824 (2^{30})	7.4%
T	tera	1,000,000,000,000 (10^{12})	Ti	tebi	1,099,511,627,776 (2^{40})	10%
P	peta	1,000,000,000,000,000 (10^{15})	Pi	pebi	1,125,899,906,842,624 (2^{50})	12.6%

The IEC prefixes combine with byte to form words — kibibyte, mebibyte, gibibyte, tebibyte, and, our favorite, pebibyte. These terms may catch on, or attempts to use them may cause further confusion. Rest assured, some wags somewhere have most certainly already worked up mercilessly trite jokes about "kibbles and bits". Fortunately, though, we have not heard any of them.

However, old habits are hard to break, and we will continue to see the dual uses of the SI prefixes applied to both decimal and binary values. Given that marketing influences will remain dominant for sometime to come, it is vital to understand how to translate billions of bytes in decimal values to binary gigabyte values. A simple formula may be utilized for this purpose — y times 10^9 divided by 2^{30}, where y is the prefix of the decimal value (see Table 1.2).

Table 1.2 Media Gigabyte Conversions

Media Type	Decimal Size Claim	Actual Binary Value
DVD (SL)	4.7 billion bytes	4.37 gigabytes
DVD (DL)	8.5 billion bytes	7.9 gigabytes
HD DVD (SL)	15 billion bytes	13.96 gigabytes
HD DVD (DL)	30 billion bytes	27.93 gigabytes
BD (SL)	25 billion bytes	23.28 gigabytes
BD (DL)	50 billion bytes	46.56 gigabytes

Note: SL is the abbreviation for single layer, and DL is the abbreviation for dual layer.

This handbook uses 1024 as the basis for byte rate measurements, with notations of KB/s (kilobytes per second) and MB/s (megabytes per second). For data transmission values, which are generally measured in thousands and millions of bits per second, this handbook will use 1000 as the basis for bit rates, with notations of kbps (thousand bits per second) and Mbps (million bits per second). Please see Table 1.3 for a listing of standard notations.

Table 1.3 Notations Used in This Book

Notation	Definition	Value	Alternate Notations	Example
b	bit	(1)		
kbps	thousand bits per second	10^3	Kbps, kb/s, Kb/s	56 kbps modem
Mbps	million bits per second	10^6	mbps, mb/s, Mb/s	11.08 Mbps DVD data rate
B	byte	(8 bits)		
KB	kilobytes	2^{10}	Kbytes, KiB	2 KB per DVD sector
KB/s	kilobytes per second	2^{10}	KiB/s	150 KB/s CD-ROM data rate
MB	megabytes	2^{20}	Mbytes, MiB	650 MB in CD-ROM
M bytes	million bytes	10^6	Mbytes, MB	682 M bytes in CD-ROM
MB/s	megabytes per second	2^{20}	MiB/s	1.32 MB/s DVD data rate
GB	gigabytes	2^{30}	Gbytes, GiB	4.37 GB in a DVD
G bytes	billion bytes	10^9	Gbytes, GB	4.7 G bytes in a DVD
TB	terabytes	2^{40}	Tbytes, TiB	
T bytes	trillion bytes	10^{15}	Tbytes, TB	

Chapter 2
Lights, Camera...Action!

On television or watched in a movie theater, where is the production going to play? That's a primary question that should be answered at the start of a project. Yet, the line separating those two venues is no longer distinct, with tools now available that can be applied to either endeavor. Advances in technology have blurred, to the point of erasure, the previously unique paths that had been followed when crafting images for presentation in a theater or on television.

Tools available to movie makers and content producers span a myriad of technology frontiers — electronic cinema, digital intermediate, advanced video compression, and variable frame rate videography, to name a few. However, there are traps and travails with the new tools that can throw a project off its tracks, potentially and tragically rendering the effort futile and incomplete.

The rapid development of breakthrough technologies is only surpassed by the relentless onslaught of marketing jargon announcing the new tools. But, rather than clarifying new avenues and approaches made possible by the debuting equipment or newly developed software, the overwhelming hype frequently obscures realities that will be encountered during the production process.

What happens when you combine footage from a film source with footage from a video source? What obstacles need to be overcome when dredging archives for historical scenes that will then be married to computer generated animations? And, what needs to be done when applying the latest video compression process to an elaborately composited, multilayer montage?

Those questions are but a few in the multitude of queries and concerns that are raised as we move on to the production of high definition format discs and video presentations. As much as we would like to think we can provide answers to these and other questions, we regrettably acknowledge that is nigh on impossible given the breadth of issues. We will, however, attempt to furnish context and information that contribute to an understanding of how to utilize and to apply recently introduced tools and technologies. Some might say that advances in technology have crossed the Rubicon, while others might say that advances in technology have crossed the River Styx.

Film or Tape or ? ?

Fortunately, the sustainable legacy of film tools and techniques provides a solid and reliable path to accomplishing a superior visual presentation. Augmented by tried and true audio tools, successful productions can be and are created without muddying the production assets by utilizing hybrid techniques that draw on video and computers. But the storytelling craft can be dramatically enhanced through the use of video and computers. Ahh, there's the rub then.

2

It is incumbent that producers and developers learn the how, the what, and the why of the tools and the techniques that are brought to bear on their production. Budget considerations are, nowadays, relatively balanced whether the principle photography is done on film or in high definition video. Footage acquired on film can be transferred to a digital workstation for editing and color control, while HD video footage can be transferred to the same workstation. The workstation performs all of the requisite tasks for completion and can then execute a "filmout" that creates a film print of the finished production. The various combinations of the acquisition, editing, and release stages result in a leveling of the ultimate costs of a project. Knowing what each tool and each step can accomplish, a producer can achieve a higher level of production while maintaining budget discipline.

Some producers are eschewing film acquisition entirely, creating their projects with video and computer tools exclusively, and only going to film for the release of theatrical prints. The advances in electronic cinematography have created an entirely new workflow on the stage and in the edit room. But, it is with this "no film" approach that trouble lurks under every rock in the quarry.

The flexibility that a video camera provides is inversely proportional to the constraints generated by the resolution that the camera may output. The latest generations of video cameras do qualify as high definition video sources. However, with the exception of the large factor video cameras, each of the smaller video cameras perform elaborate, albeit elegant, data manipulation when coding visual information to storage. Whether it is to videotape, hard drive, disc or memory card, the data encoded by these cameras is the result of sophisticated realtime data manipulation in conjunction with varying numbers of sensor elements. And, yes, this manipulation might include different types and levels of data compression. Armed with the knowledge of what exactly is being saved, a production team can use the smaller video cameras, taking advantage of their reduced encumbrances and greater freedom. Software and hardware tool processes can be applied to the footage, insuring that the captured images seamlessly integrate with the other elements of a production. Failure to recognize how and what data is saved and stored by video cameras may result in issues such as, mismatched frame rates, image distortion, and/or loss of synchronization between audio and video.

Television Systems: A Primer

It is important to understand how the world's television systems were developed. Not to get the reader bogged down in minutiae, but without a basic knowledge of how and why US television operates differently from other countries, it is very difficult to explain, let alone understand, the variety of frame rates and video resolutions that are now available with high definition television.

There are three standard definition television (SDTV) systems in the world. The NTSC standard is used in Canada, Mexico, the U.S, and Japan. NTSC is shorthand for National Television Systems Committee. The PAL standard and its variations PAL-M and PAL-N are used in most of Europe, as well as China, India, South America, Africa, and Australia. PAL is shorthand for Phase Alternation with Line (yes, technically, that should be "PAwL", but only the uppercase letters are used in the acronym and, besides, "PAwL" is tough to pronounce). A third standard, known as SECAM, is in use in Russia and France. SECAM is the acronym for Système Électronique pour Couleur avec Mémoire. The English translation of SECAM is Sequential Color and Memory. Production for SECAM is accomplished using PAL format equipment. These television standards are adapted from the electrical standards of the respective countries.

Long before the invention of television, the US standard for electricity was developed so that manufacturers would be able to make products that would work when plugged in to a common electrical source. The US standard was set at 110/120 volts, with the system changing polarity from positive to negative and back again, 60 times a second. This sequence is referred to as *alternating current*, with each positive/negative polarity change termed a *cycle*. In Europe, the standard was set at 220/240 volts and 50 cycles per second.

When first introduced, television broadcasts were in black and white only, and the standard set by the NTSC mandated the rate of 30 frames per second with two fields comprising a frame, for 60 fields per second. NTSC uses 525 scan lines to compose a frame, with the odd lines displayed as field one and the even lines displayed as field two. The fields interlace on presentation to compose a frame. Whereas, the PAL standard adopted a rate of 25 frames per second, also comprised of two fields, for 50 fields per second, and uses 625 scan lines to compose a frame.

The scan line counts are the product of a string of small integer factors that, at the time, could only be reliably supported by vacuum tube divider circuits — 525 is 7x5x5x3, and 625 is 5x5x5x5. These divider circuits derive the field rate from the power line rate. These factors combined to give monochrome NTSC 525/60 television a line rate of 30x(7x5x5x3), or exactly 15.750 kHz. Monochrome PAL 625/50 television has a line rate of 25x(5x5x5x5), or exactly 15.625 kHz. With the introduction of color for television, the NTSC number takes on immense significance, whereas the PAL signal structure was mostly unaffected.

In 1953, when adding color to the monochrome signal, a second NTSC determined that it would be appropriate to choose a color subcarrier in the region of 3.6 MHz. This subcarrier would imbed the color information within the television signal. But there was a technical concern with adding the color information without disturbing the sound subcarrier element of the signal. Too close a relationship in the subcarrier frequencies would result in distortion that would become visible in the luminance component of the signal.

Remember that the responsibility for setting broadcast standards resides with the Federal Communications Commission (FCC). If the FCC had altered the sound subcarrier frequency to be increased by the fraction 1001/1000 (as recommended by the second NTSC) — that is, increased by 4.5 kHz to about 4.5045 MHz — then the color subcarrier in NTSC could have been exactly 3.583125 MHz, the line and the field rates would have been unchanged, and we would have retained exactly 30 frames per second!

But, true to form, alas, the FCC refused to alter the sound subcarrier, thus allowing then existing black and white television receivers to remain in use. Instead, the FCC chose to reduce both the line rate and the field rate by the fraction 1001/1000, effectively dropping those rates to about 15.734 kHz and 59.94 Hz, respectively. Thus the 59.94 rate was born![1] And the snowball begins to roll downhill!

---

[1]Actually, the number 59.94 is rounded off, as the fraction results in an incomplete number that repeats the integers in the sequence 59.940059940059940059400599400599400...forever.

Unfortunately the field rate of 60 divided by 1.001, rounded to 59.94 for convenience, means that 60 fields consumes slightly more than one second, so 30 frames no longer agrees with one second of clock time. As a side note, this precipitated the invention of dropframe timecode that, ostensibly, alleviated this disparity.

Regrettably, the shortsighted, although pragmatic, FCC decision helped to bring about the frame and video rate complexity that exists today. This is further amplified by the continuing convergence of video and computer technologies and terminologies, let alone the emergence of high definition standards and formats. As noted in chapter one, a billion might mean something other than a thousand million.

It is imperative that the ultimate display arena is established prior to the start of production — in theaters or on television. The tools and techniques that are employed for creating the project assets must be compatible and geared to meeting the requirements for the final presentation.

Frame Rates

With the exceptions of cinema and computer display, we are in a universe of NTSC and PAL television systems. Even with cinema, though, that may be the initial display, only to be rapidly followed by conversion to video for alternate market exploitation. Thus, no matter the acquisition, television boundaries will be encountered by virtually every production.

The explosion in frame rate flexibility of video acquisition cameras has been generated by the desire to match film acquisition settings, so that a video camera can achieve the "cine-look" of film. Alas, that is a highly subjective goal. One person's "cine-" is another person's "(fill in your own word here)". At the upper reaches of the technology, that is to say, the more expensive end of the spectrum, the "cine-look" attained by a video camera is indistinguishable from a film source. And, there are tools that can be used that will enhance footage from a lower-end camera that may render the final product compatible with a satisfactory "film-like" look.

But, when using video tools in place of film, extreme care must be exercised in all aspects of image preparation — lighting, set design, cast choices, costumes, makeup, time of day, and phase of the moon. It must be understood that video tools have limitations in extreme conditions that may preclude their use. Further, when using video in place of film, the "silk purse" approach to post production no longer applies. If the image acquired by the source is less than desired, no amount of manipulation after the 'get' can save the picture.

But, what exactly is "frame rate"? Is it a component of acquisition or of display, or both? You're gonna love us for this one...it's both. And, more.

Frame rate in acquisition establishes the speed with which action is transcribed to media, capturing moments in time either as two fields or as a frame. Frame rate in display establishes the frequency with which frames are presented for viewing. Frame rate is expressed in the number of frames per second, or, by the number of frames in Hertz. What's "Hertz"? Something to confuse the crap out of us, that's what. (Okay, that's a mini-rant. Hertz is a unit of frequency equivalent to cycles per second, as in one cycle is one Hertz. Alternating current of 60 cycles per second can be called 60 Hertz. Hertz is abbreviated "Hz".)

Additionally, there are two styles of frame orientation — interlace and progressive. Interlace is the method of interweaving the two fields that comprise a frame. Progressive is the method of contiguously presenting all of the lines that comprise a frame, as in a snapshot. Interlace is represented as "i" and progressive is represented as "p", when describing the frame style.

There are three primary frame rate standards for television and movies —

- **60i:** 60 fields, interlaced, composing 30 frames; (PAL) 50 fields, interlaced, composing 25 frames. For NTSC, this is really 59.94 fields, interlaced, composing 29.97 frames per second.
- **30p:** 30 frames, progressive, for video cameras, producing an image without interlaced field artifacts. Again, for NTSC, this is really 29.97 frames per second.
- **24p:** 24 frames, progressive, for film and capable video cameras.

Depending on the camera, there are any number of frame rates that may be used for the acquisition rate. With a film camera, the standard acquisition rate is either 24 or 25 frames per second(fps), depending on the region of display. Film cameras can be run off-speed, either sped up or slowed down, to accomplish various special effects and/or to capture scenes under unique conditions.

Since their invention, NTSC color video cameras have captured images at the rate of 59.94 fields per second. Nowadays, high definition capable video cameras can be set to record either 59.94 fields per second or 59.94 frames per second, as well as rates from 2 frames per second to 60 frames per second. Unfortunately, we identify all of these rates with the abbreviation "fps".

Very often, the goal is to give video images a "cine-look", so the opportunity to shoot footage at a rate equivalent to the film rate is highly prized. However, 59.94 fields per second is equal to shooting 29.97 frames per second. We have to get from 29.97 to 24, and you cannot get there from there. But you can get to 23.976[2], which is sometimes rounded to 23.98. Employing the frame rate of 23.98 also allows for an easy expansion to the 29.97 or 59.94 rates by the use of what are called "pulldown" techniques, wherein fields or frames are duplicated to meet the higher rate. This allows for the integration of "cine-look" video with other video assets, as well as the presentation of the footage via television. (Although that might defeat the quest for "cine-", wouldn't it? But, we digress...)

Further, with sophisticated circuitry, sensors, and integrated chips, some video cameras may be set to record 24 frames per second, also known as "true 24". 24 fps is not compatible with an NTSC display, but video shot at 24 frames per second can readily integrate with film originals that are shot at 24 frames per second. It is this fluidity of medium that accommodates the extensive use of computer generated imagery and augmentation that is evident in motion pictures today.

As we have said, if you are producing for theatrical presentation exclusively you can establish 24 fps as the production criteria, whether shooting film or video. The follow-on production steps would all respect the 24 fps rate, and all will remain as intended.

---

[2]23.976 is shorthand for the result of dividing 24 by 1.001. The actual quotient is the incomplete number 23.976002397600239760023976...forever, akin to the result that gives us 59.94.

2

But, "true 24" and "23.98" are not the same. In fact, 23.98 can become something called *progressive segmented frame* (PsF), wherein the odd lines from the frame are segregated from the even lines for the frame. This is not the same as interlace fields because the progressive frame segments are not presented sequentially. Rather, the two segments are recombined prior to presentation as a single frame. This technique allows for easier integration with interlace technologies.

Beyond the imaging source is the recording device for saving and storing the images acquired. The recording rate for devices may be set independently from the frame rate, depending on the technology employed. Once again, it is the ultimate display criteria that should govern the settings for recording. The 24 fps rate may be used when the footage is destined for cinema, whereas a rate of 29.97 Hz or 59.94 Hz should be used when footage will appear on television. Put another way, when the intent is to shoot cine-like video, set the camera to image 23.98 fps and the recording device to match NTSC display criteria.

And what about frame rate for display, as in the frequency that images are presented for viewing? The methods for presentation vary between platforms — projectors, television receivers, and displays. A projector runs at a speed of 24 frames per second, but will show each frame twice, thereby displaying the film at the rate of 48 Hz. This action is done to minimize the perception of flicker between frame images.

Television receivers present the image standard for their region, PAL or NTSC, as either 25 or 29.97 frames per second, but given the interlaced field construction of the frames, the display rate is either 50 or 59.94 Hz.

Technology has provided new ways to display images that are not bound to television standards and do not use a cathode-ray tube (CRT) for presentation. These displays are capable of presenting standard television, high definition television, and images from a variety of sources — disc players, hard drives, computers, et cetera. The circuitry employed by these technologies provides vastly improved control over the data being displayed. And, we are now able to view images with display rates that are multiples of either television or film frame rates — 48, 60, 72 Hz and higher. These increased frame display rates can significantly improve the perceived clarity, depth, and color characteristics of a presentation, depending on the breadth of features included in the display.

High Definition Image Resolutions

Image resolution is determined by several factors — image size, frame style, color or chroma sampling, and data bit rate or depth. When multiplied by frame rate, we get the data rate required for streaming images to a display and/or to storage. High definition and beyond picture sizes generate exceptionally large numbers, both in terms of data rate and cumulative data for storage.

The high definition digital television formats defined by the ATSC (Advanced Television Systems Committee)[3] are —

---

[3]The ATSC standards are intended to replace the NTSC system and are being adopted by many other countries besides the US. The ATSC has defined systems for both standard and high definition television, but only the high definition formats are included here.

- 1280 pixels per line × 720 lines, in 16:9 widescreen aspect, with progressive frames, at frame rates of 23.976, 24, 29.97, 30, 59.94 and 60 per second
- 1920 pixels per line × 1080 lines, in 16:9 widescreen aspect, with either —
  - interlace frames, at frame rates of 29.97 (59.94 fields) and 30 (60 fields) frames per second, or
  - progressive frames, at frame rates of 23.976, 24, 29.97 and 30 frames per second

Shorthand is applied to these standards, resulting in the terms 720p, 1080i, and 1080p. Adding frame rates to those terms, creates terms like 720/24p or 1080/60i. Although, in the case of 720, the "p" is unnecessary as all of the frame rates for 720 lines are progressive.

Also, please note that the much ballyhooed image rates of 1080/60p and 1080/59.94p are not included in the ATSC high definition standards. Another component of the ATSC standards is the use of MPEG-2 compression in the transmission of data for the ATSC formats. All of the ATSC image formats are compressed somewhat to accommodate the bandwidth allocation for television channel transmission. With the conversion to digital television, the channel bandwidth allocation is only 19.39 megabits per second for over-the-air transmission. The high data rates generated by images at 1080/60p (59.94) would result in extremely poor image quality when compressed, transmitted, uncompressed and displayed.

The ATSC format standards notwithstanding, the next generation optical disc formats identify high definition as —

- picture sizes, in 16:9 widescreen aspect — 1920 pixels per line × 1080 lines, 1440 pixels per line × 1080 lines, 1280 pixels per line × 720 lines
- display frame rates —
  - for 1080, frames rates of 23.976p, 25p, 29.97p, 50i, 59.94i per second
  - for 720, at frame rates 23.976p, 25p, 50p, 59.94p per second

The HD disc frame rate of 23.976 for both 1080 and 720 includes a 2:3 pulldown instruction in order to display at 29.97 frames per second. Depending on the connection type and the display type, this frame rate may be either interlace or progressive.

Typically, HD discs will have content that is encoded at either 1080/23.976p for film-based content or 1080/59.94i for video-based content. And, please note that marketing jargon will refer to these numbers as 1080/24p and 1080/60i. Isn't that special, eh?

Whereas ATSC incorporates MPEG-2 compression in meeting the delivery and display requirements for television broadcast, the new disc formats can use either MPEG-2 or MPEG-4 AVC or VC-1 for compression, and are not restricted by the digital television channel bandwidth limitation. These expanded compression choices for HD discs allow for vastly improved data rates, at higher compression levels. And, the HD DVD maximum data rate for video playback is 29.40 mb/s, while the Blu-ray Disc maximum data rate for video playback is 40.0 mb/s. Thus, the seeds for confusion are sown when comparing data rates for broadcast high definition with next generation HD discs.

2

Chroma Subsampling and Bit Depth

Another component of image data is the sampling that is performed on an image. This is referred to as chroma subsampling, and is generally expressed as a three number ratio, such as, 4:4:4 or 4:2:2 or 4:2:0, among others. These numbers reflect the sampling rate and method that is applied to the luminance and chrominance components of an image. The theory behind this sampling process is that the human eye is more sensitive to changes in brightness (luminance) than changes in color (chrominance), and there is no perceived loss when sampling color details at a lower rate. The signal is divided into a luminance element (Y') and two color difference elements that are derived as red minus luminance (C_r) and blue minus luminance (C_b)[5]. But, the three numbers are not directly related to the three signal elements and instead reflect the methodology of the sample (otherwise, the 4:2:0 variant would have no data for one of the color channels...doh!).

The terms 4:4:4 and 4:2:2 were developed for standard definition digital video. Nowadays, the sampling rates for high definition video are 22:22:22 and 22:11:11, respectively, but casual use of the SD terms has been adopted for the HD rates.

Technically, sampling is performed using rates that are multiples of 3.375 MHz, and 4 times 3.375 is 13.5 MHz and 22 times 3.375 is 74.25 MHz. Thus, 4:4:4 means that the signal elements are each sampled at 13.5 MHz (74.25 MHz for HD). Sampling at 4:4:4 provides close to real-life color representation and is considered to be lossless. Taking advantage of the different sensitivity to color perception, 4:2:2 samples the luminance at 13.5 MHz (74.25 MHz), but the color difference signals are each sampled at 6.75 MHz (37.125 MHz) or one half the frequency of 4:4:4. The ratio 4:2:0 means that the luminance is still sampled at 13.5 MHz, but the color difference channels are subsampled by a factor of 2 both horizontally and vertically, in a two-line grouping structure. The 4:2:0 sampling scheme is the method in use for HD DVD and BD, as well as standard DVD.

Bit depth reflects the number of bits used to store information about each sample. The higher the bit depth means that more bits per pixel are used for an image, and the more pixels used for an image then the larger the image file. 8 bits provide up to 256 color gradations from black to white, while 10 bits provide up to 1024 color gradations from black to white. And, there are 12 bit schemes that, in conjunction with 4:4:4, provides for the best image quality currently available.

High Definition Data Streams

The goal is capturing data digitally at the highest possible quality given the current capabilities of technology. Translating these image format sizes and frame rates to continuous data streams requires advanced degrees in mathematics or an association with some extremely knowledgeable and quick partners. In our case, we have a couple of authors who can rattle off data rate computations on request (see Table 2.1).

---

[5]The terms Y', C_b, and C_r denote digital encoded component video signals. The terms Y, R-Y, and B-Y are used to denote analog component signals. Technically, these terms are not interchangeable, although tech talk frequently crosses the term boundaries.

Table 2.1 Image Data Stream Examples

Format	Frames per sec	Frame Size	Number of Pixels	Bits per Pixel	Data Rates Mb/s	MB/s
Sampling Rate: 4:4:4 RGB 12 bit						
1080/60i	29.97	1920 × 1080	2,073,600	36	2,237.2	266.7
1080/24p	23.976	1920 × 1080	2,073,600	36	1,789.8	213.4
720/60p	59.94	1280 × 720	921,600	36	1,988.7	237.1
720/24p	23.976	1280 × 720	921,600	36	795.5	94.8
Sampling Rate: 4:2:2 YUV 10 bit						
1080/60i	29.97	1920 × 1080	2,073,600	20	1,242.9	148.2
1080/24p	23.976	1920 × 1080	2,073,600	20	994.3	118.5
720/60p	59.94	1280 × 720	921,600	20	1,104.8	131.7
720/24p	23.976	1280 × 720	921,600	20	441.9	52.7
Sampling Rate: 4:2:0 YUV 8 bit						
1080/60i	29.97	1920 × 1080	2,073,600	12	745.7	88.9
1080/24p	23.976	1920 × 1080	2,073,600	12	596.6	71.1
720/60p	59.94	1280 × 720	921,600	12	662.9	79.0
720/24p	23.976	1280 × 720	921,600	12	265.2	31.6

Table 2.1 provides a sampling of the data rates that are generated when working with high definition data streams. What is important to note is that even before compression is executed on an image data stream, the amount of data can be reduced dramatically by choosing a different sampling scheme, and/or by resampling the image data to meet the requirements of HD DVD and BD playout.

There is a concerted effort underway to address the issue of producing HD images in the 1080 line format with 60 progressive frames per second. Termed 1080/60p, the hope is that cameras will be capable of acquiring images in that image format and productions will be able to take advantage of the perceived superior resolution. But, the quest is conveniently overlooking the hard fact of exactly how much data can be transmitted via an HD Serial Digital Interface (HD SDI) link, the standard for connecting cameras to recording devices, monitors, et cetera. The maximum data rate for a single HD SDI link is 1.485 Gbit/s, including audio. As you can see in Table 2.1, the data rate for 1080/60 interlaced video, sampled at 4:4:4 12 bit, is 2.24 Gbit/s, and the data rate at 4:2:2 YUV 10 bit is 1.24 Gbit/s. Which means that HD SDI cannot accommodate 4:4:4 12 bit data, but can accommodate 4:2:2 10 bit data. When these data rates are doubled to accommodate 60p, it becomes obvious that some other interconnection is required.

Standards have been established for a dual-link HD SDI structure which ties two HD SDI links together in parallel, which is intended to accommodate 4:4:4 sampling or 1080/60p. And, another standard was adopted in June 2006 for a 2.97 Gbit/s interface that uses a single cable, which may be used to replace the dual-link approach.

2

That, then, is the conundrum. Even if cameras and sensors are created that are capable of 1080/60p acquisition, the production infrastructure will have to be changed. New tools would need to be developed, let alone the massive tasks of updating editing and workstation interfaces to accommodate the realtime needs when presenting images at these gargantuan data rates. All of which are considerable undertakings.

And, yes, these 1080/60p data rates could be dramatically reduced with compression, but wouldn't that be self-defeating? We'll leave it to others to decide.

Film Formats

Ahh, good ol' film. Point and shoot...get it in the can...send it to the lab...print it, project it and look at it. Sure made for an easy to understand process. And, film is, arguably, still the best medium for acquiring images, providing the greatest flexibility for manipulation after acquisition while retaining the camera original footage in pristine condition. But, unless the production budget is unlimited, maintaining a film process to release is an expensive proposition. Especially when tools and techniques have been developed that are time-saving, money-saving and more sophisticated than many a special effects maven ever envisioned.

The tools for finishing a film production have crossed the technology frontier to computers and workstations. The transfer via telecine to hard drive arrays, removable storage or videotape has made the transition to non-film finishing services relatively efficient and, certainly, more affordable. Techniques in the digital intermediate (DI) realm have replaced opticals and film laboratories, providing for extensive integration of film, video and computer graphics elements. We have moved far beyond the days of expensive work prints and answer prints, let alone waiting for elements to return from the optical house.

Primarily, the film sizes of 35 mm and 70 mm are used for feature films. To a lesser extent, 16 mm and Super 16 mm may also be used for feature work, but are more likely to be used in production for television or documentaries. 35 mm film produces resolutions that may be described as 2K and 4K (see Table 2.2). Although Table 2.2 refers to pixel values, the film is not composed of pixels and lines. The pixel values are applied during an ingest or transfer process with a scanner or telecine. The data rates noted in Table 2.2 may be transcoded to other values depending on the process used to ingest or transfer the film material.

Table 2.2 Film Format Examples with RGB 10 bit Data Rates and Resolutions

Film Format	Aspect Ratio	Camera Aperture (mm)	Frame Size (pixels)	Frame File size	24 fps Data Rate
16mm	1.37:1	10.26 × 7.49	1712 × 1240	7.5 MB	182.2 MB/s
Super 16 mm	1.66:1	12.52 × 7.41	2048 × 1240	9.1 MB	218 MB/s
35 mm, "Academy" 2K	1.37:1	21.95 × 16.00	1828 × 1332	8.7 MB	209 MB/s
35 mm, "Academy" 4K			3656 × 2664	34.8 MB	836 MB/s
35 mm, 2x anamorphic 2K	2.40:1	21.31 × 18.16	2048 × 872	6.4 MB	153 MB/s
35 mm, 2x anamorphic 4K			3656 × 1556	20.3 MB	488 MB/s
Super 35 mm 2K	1.33:1	24.89 × 18.67	2048 × 1556	11.4 MB	274 MB/s
Super 35 mm 4K			4096 × 3112	45.6 MB	1,094 MB/s

The term 35 mm refers to the literal width of the film stock, not the image area. Using various combinations of camera aperture settings, camera lenses, and projection aperture settings and lenses, 35 mm film provides for image areas that occasion a wide variety of names, uses and aspect ratios. Additionally, Super 35 mm film uses the same film stock as 35 mm, however Super 35 usurps the optical sound track area for additional image negative area (see Figure 2.1).

Figure 2.1 Relative Framing Samples for 35 mm Film

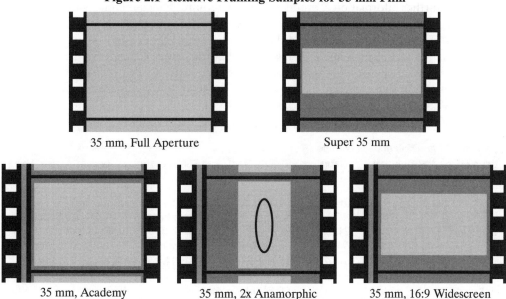

35 mm, Full Aperture   Super 35 mm

35 mm, Academy   35 mm, 2x Anamorphic   35 mm, 16:9 Widescreen

This imaging flexibility inherent to large gauge film is very difficult to translate to digital video cameras. Where film allows for adjusting the image area by using a portion of the film negative, most video cameras do not have larger-than-needed image sensors or "pickup" areas. Only the very high-end video cameras have sensors that provide for larger imaging resolutions than what may be used for the ultimate release format. Couple this film flexibility with the ease and convenience of telecine and digital editing workstations, it will be highly unlikely for film to ever be replaced by digital video cinematography. The mediums will continue to refine their differences and similarities.

Video Formats

The adoption of high definition standards by television broadcasters has opened the floodgates of a withering and bewildering array of HD capable cameras, camera/recorders, and recorders. Manufacturers are striving to keep up with the rapidly evolving demands of the production and broadcast communities, while struggling to maintain sophisticated research and development in an era of dramatically lower revenue streams with dwindling profits.

HD video cameras have crossed into the feature movie realm, utilizing unprecedented

2

imaging capabilities that can make digital video indistinguishable from film, or so it is hoped. But, just as it is hyper-critical that extreme care and unrelenting attention to detail are the hallmarks of high-quality film production, so is it true in the video production world.

Table 2.3 provides an overview of the breadth of HD video tools currently or soon to be available, as of early 2007. When perusing the table information, it is important that the reader does not try to make detailed item for item comparisons. The specifics of a product are too numerous to present in an overview spreadsheet, and the data here is meant only as a spur to seek further information. Careful evaluation requires hands-on time with a device for test shooting and recording before proceeding to production or purchase.

The table is a "work-in-progress" and there may be incomplete entries due to an inability to ascertain the particular item or items. The table data was gleaned from a variety of sources, as many of the product developers are reluctant to divulge their methodologies. Perhaps, out of fear that someone may come along someday and try to create a comparative database. Ah, well, that day is here.

Table 2.3 HD Video Camera and Recorder Attributes

Model / Resolution	Native Resolution	Frame Rates	Sampling/ Bit Rate	Data Rate Compression	Imaging Sensor Type	No. / Size
Large Factor Cameras						
Arriflex D20/ 1920 × 1080 and higher	HD 16:9 at 2880 × 1620, Film 4:3 at 3018 × 2200	1 to 60 fps, incl 23.976, 24, ,25, 29.97, 30	YUV 4:2:2 10 bit via single HD SDI, RGB 4:4:4 10 bit via dual HD SDI	Internal data bus up to 10 Gb/s none	Custom CMOS with Bayer mask[a]	One/ Super 35mm, 24.9 mm × 18.7 mm
DALSA Origin/ 4096 × 2048	4096 × 2048, 16 bits total per pixel	0 to 36 fps, variable	RGB 4:4:4 16 bit, each frame output in DPX file format	400 MB/s none	Custom CCD with Bayer mask	One/ 35 mm × 17 mm
Grass Valley Viper/ 1920 × 1080 1280 × 720	16:9 and 2.37:1 aspect ratios	23.98p, 24p, 25p, 29.97p, 50i, 59.94i, & 720/50p and 59.94p	$Y'C_bC_r$ 4:2:2 22 bit via HD SDI, RGB 4:4:4 10 bit or 22 bit via dual HD SDI	1.5 Gb/s none	Patented Frame Transfer(FT) CCDs	Three/ 2/3", 9.2 Mpixel
Kinetta/ 1920 × 1080	1920 × 1080/ Cineon and DPX files	1 to 60 fps @ 1080p	YUV 4:2:2 10 bit via single HD SDI, RGB 4:4:4 10 bit via dual HD SDI	———		One/
Panavision Genesis/ 1920 × 1080	12.4 Mpixel, 16:9 aspect ratio	1 to 50 fps	RGB 4:4:4 10 bit via dual HD SDI	——— none	RGB CCD	One/ Super 35mm, 23.6 mm × 13.3 mm
Red Digital RED ONE/ 4520 × 2540	4900 × 2580	1 to 60 fps for 2540p, 4K, 2K, 1080p, 720p/ 1 to 120 fps for 2K, 720p	RGB 4:4:4 10 bit YUV 4:2:2 10 bit	——— none	Mysterium™ custom CMOS with Bayer mask	One/ Super 35mm, 12 Mpixel

continues

Table 2.3 HD Video Camera and Recorder Attributes (continued)

Model / Resolution	Native Resolution	Frame Rates	Sampling/ Bit Rate	Data Rate Compression	Imaging Sensor Type	No. / Size
Large Factor Cameras (*continued*)						
Silicon Imaging SI-1920HDVR/ 1920 × 1080 1280 × 720	1920 × 1080	1080 @ 24p, 25p, 30p/ 720 @ variable, 12 to 72 fps	Cineform RAW™ 10 bit sampling	96 Mb/s Cineform VBR Wavelet 5:1	CMOS with custom Bayer codec	One/ 2/3", 16:9 1920 × 1080
Camera/Recorders						
Canon XL H1/ HDV 1920 × 1080	1440 × 1080	23.976i, 29.97i, 59.94i, 25i, 50i	4:2:2 10 bit for HD SDI output, 4:2:0 8 bit HDV	25 Mb/s MPEG-2	CCD with pixel shift	Three/ 1/3", 1440 × 1080
JVC GY-HD250U/ HDV 1280 × 720[b]	1280 × 720	24p, 25p, 30p, 50p, 60p	4:2:0 8 bit HDV	25 Mb/s MPEG-2	Interline-transfer CCD	Three/ 1/3", 1.1 Mpixel
Grass Valley Infinity/ 1920 × 1080 1280 × 720	1920 × 1080	1080 @ 50i, 59.94i/ 720 @ 50p, 60p	JPEG2000 at 4:2:2 10 bit, MPEG-2 at 4:2:0 8 bit	110 Mb/s JPEG2000 or MPEG-2	CCD ?	Three/ 2/3"
Panasonic AG-HVX200/ 1920 × 1080 1280 × 720	1440 × 810 effective resolution	1080 @ 23.98p, 25p, 29.97p, 30p, 50i, 59.94i/ 720@ increments from 12 to 60p	4:2:2 8 bit	100 Mb/s DCT 6.7:1	CCD with spatial offset	Three/ 1/3", 960 × 540
Panasonic AJ-HDX900/ 1920 × 1080 1280 × 720	1280 × 720	1080 @ 23.98p, 25p, 29.97p, 50i, 59.94i/ 720@ 23.98p, 25p, 29.97p, 50p, 59.94p	4:2:2 14 bit	100 Mb/s DCT 6.3:1	CCD	Three/ 2/3",
Panasonic AJ-HPX2000[c]/ 1920 × 1080 1280 × 720	1280 × 720	1080 @ 23.98p, 25p, 29.97p, 50i, 59.94i/ 720@ 23.98p, 25p, 29.97p, 50p, 59.94p	4:2:2 14 bit	100 Mb/s DCT or MPEG-4 AVC[c] 6.7:1	CCD	Three/ 2/3"
Panasonic VariCam AJ-HDC27H/ 1920 × 1080[d] 1280 × 720	1280 × 720	4p to 60p fps	4:2:2 12 bit	100 Mb/s DCT 6.7:1	CCD	Three/ 2/3", 1280 × 720 1.1 Mpixel

continues

Table 2.3 HD Video Camera and Recorder Attributes (continued)

Model / Resolution	Native Resolution	Frame Rates	Sampling/ Bit Rate	Data Rate Compression	Imaging Sensor Type	Imaging Sensor No. / Size
Camera/Recorders (continued)						
Sony HVR-Z1/ HDV 1920×1080	1440×1080	50i[e], 59.94i[e]	4:2:0 8 bit HDV	25 Mb/s MPEG-2	CCD with pixel shift	Three/ 1/3", 1.1 Mpixels
Sony CineAlta F23[f]/ HDCAM 1920×1080	1920×1080	23.98p, 24p, 25p, 29.97p, 50p, 59.94p, 50i, 59.94i	RGB 4:4:4 10 bit at 1p to 30p fps, YUV 4:2:2 10 bit at 1p to 60p fps	_____	Progressive CCD	Three/ 2/3", 2.2 Mpixels
Sony CineAlta HDW-F900R/ HDCAM 1920×1080	1920×1080	23.98p, 24p, 25p, 29.97p, 50i, 59.94i	$Y'P_bP_r$ 4:2:2 12 bit	144 Mb/s DCT	FIT CCD	Three/ 2/3", 2.2 Mpixels
Sony CineAlta HDCU-F950[g]/ 1920×1080	1920×1080	23.98p, 24p, 25p, 29.97p, 50p, 59.94p, 50i, 59.94i	YUV 4:2:2 10 bit via single HD SDI, RGB 4:4:4 10 bit via dual HD SDI	_____	FIT CCD	Three/ 2/3", 2.2 Mpixels
Sony PDW-F350/ XDCAM HD/ 1920×1080	1440×1080	23.98p, 25p, 29.97p, 50i, 59.94i	$Y'P_bP_r$ 4:2:2 12 bit	35 Mb/s MPEG-2 MP@HL	Interline Transfer CCD	Three/ 1/2", 1.5 Mpixel
Recorders						
HD D5 / 1920×1080	1920×1080	23.98p, 24p, 25p, 29.97p, 50i, 59.94i & 720/59.94p	4:2:2 / 10 bit	258 Mb/s[h] Intraframe 5:1		
DVCPRO HD / 1920×1080	1440×1080	50i, 59.94i, 60i & 720 @ 59.94pm 60p	4:2:2 / 8 bit	100 Mb/s DCT 6.7:1		
HDCAM/ 1920×1080	1440×1080	23.98p, 24p, 29.97p, 30p, 50i, 59.94i	3:1:1 / 8 bit	115.6 Mb/s DCT 4.3:1		
HDCAM SR SRW-5000/ 1920×1080	1920×1080	23.98p, 24p, 29.97p, 30p, 50i, 59.94i & 720/59.94p	4:2:2 10bit via single HD SDI, 4:4:4 10bit via dual HD SDI	440 Mbps[i] MPEG-4@SP 4:2:2@2.7:1 4:4:4@4.2:1		

[a]The D20's custom CMOS sensor is specified to capture images at up to 150 fps and provides raw data at a bit depth of 12 bits.

[b]The GY-HD250U can execute a "cross convert" of the HDV 1280x720 image to create a 1080/60i output.

continues

Table 2.3 HD Video Camera and Recorder Attributes (continued)

cAJ-HPX2000 is slated for release Spring 2007, with the AVC optional board available by Summer 2007.

dVaricam output must converted by a Panasonic frame converter, AJ-UFC1800 or AJ-FRC27, or other 3rd party product to achieve 1080/24p image resolution.

eThe HVR-Z1 utilizes the 50i and 59.94i frame rates to create Cineframe 24, 25 and 30 modes.

fF23 CineAlta is expected to ship Spring 2007. The F23 supports top or rear docking of the SRW-1 HD recorder.

gThe HDC-F950 may not, technically, be a camcorder as it requires a connection to a video recorder — either the SRW-5000 or SRW-1 HD video recorders for 4:4:4 recording or HDCAM recorders for 4:2:2 recording. But, record start/stop is initiated from the camera, so is it a camcorder?

hHD D5 uses various data rates, from 258 to 323 Mbps, depending on the format/frame rate and the number of audio channels being recorded. For example, 258 Mbps supports 1080/23.98p with 8 audio channels, while 323 Mbps supports 1080/59.94i and 720/59.94p with 8 audio channels.

i HDCAM SR is also capable of 880 Mbps recording at 2x speed, sampling either one 4:4:4 10 bit stream or two 4:2:2 10 bit streams.

Buried in the details of the cameras and recorders are dark little secrets that reveal the true capabilities of the products. For instance, as noted earlier in this chapter the data rate of a high definition 1920×1080/24p data stream, sampled as RGB 4:4:4 12 bit, is 213.4 MB/s. When a data rate of 100 Mb/s is given for a particular product, that means there is already a great deal of data manipulation and reduction performed on the signal BEFORE any compression and recording is executed by the particular device. It is this disparity between the literal data stream and the actual data recording that results is image artifacts and potential image quality loss.

Just as a software transcoding from RGB 4:4:4 12 bit to YUV 4:2:0 may reduce a data stream to manageable levels for next generation discs, lowering the data rates of cameras and recorders can still provide usable images at manipulable file sizes. However, any changes to the data must be carefully evaluated to insure that the ultimate use of the image is not compromised. Welcome to the "digital age"!

Chapter 3
Pictures and Sound

As the next generation disc formats begin to take advantage of the enhanced features inherent in their specifications, consumer electronics manufacturers are striving to keep up. Delayed introductions, less than promised deliveries, and incompatible connections have mangled the marketplace reception of new formats and new devices. The first wave of players, in either format, was quickly diagnosed as suffering from premature debut. The HD DVD player from Toshiba needed an immediate firmware update and the BD player from Samsung had a less than stellar decoder chip that caused video distortion. Although correctable, these are but two instances that do not bode well for a general acceptance of HD format discs. Couple these player snafus with the misunderstandings and misperceptions regarding high definition technologies in general and it is small wonder that, save for early-adopters, the market continues to stay away from either HD disc format.

As 2006 drew to a close, we saw the long delayed introduction of the Sony Playstation™ 3, with its Blu-ray Disc™ player component. Preceded by the splashy premiere of the Microsoft Xbox 360™ HD DVD player add-on, the gaming war was finally fully joined. Except that the long-anticipated battle of the titans was generally ignored by endusers, who have taken to the Nintendo Wii™, instead. So, neither disc format has gained much of an advantage via game console sales, perhaps demonstrating that gamers are not movie enthusiasts.

In what might be considered a success, sales of HD-capable televisions and displays are growing each year by very large numbers. However, is it a success if the viewer is unable to watch HD disc productions because their ball-busting behemoth cannot communicate with the player? Content protection challenges have created a maze of setups and standards that are difficult to fathom even for the most experienced installation technicians. Which itself may be a looming requirement, that of needing a professional installer to properly make all of the requisite connections in order to watch HD TV!

In fact, there is a serious "HD disconnect" occurring, wherein buyers have purchased HD-ready displays but are choosing to not watch high definition content. There may be more than a few reasons for this dichotomy, some of which are —

- the digital HD set is merely replacing an older analog model,
- the viewer is unwilling to spend more money for HD content packages from either cable or satellite providers,
- viewers feel that there is no real improvement in a high definition picture over a digital picture, or
- viewers think they are automatically watching HD content because the show being viewed was introduced as being "Available in HD".

This apparent HD disconnect is most likely also influencing purchasing decisions about next generation players and discs. Beyond the reluctance to choose one format over another, the inability to discern an advantage or an improved viewing experience is weighing heavily against the adoption of the HD disc formats.

3

High Definition Multimedia Interface (HDMI)

The struggle to protect intellectual property is the primary cause of the difficulties being faced by player and by display manufacturers. The desire to view the best possible image generates the demand for high resolution, high definition displays. But, the need to protect the content from illicit duplication has created the unforgiving connection standards that have been developed, which are in addition to the AACS and BD+ scenarios for the HD disc content. A chief concern is that an HD display could be used as a conduit for recording the image being presented. Or, a player could have its output siphoned off to a recording device. So a standard was needed that protected the literal connection between the image source and the image display.

Expanding on the HDCP (High-bandwidth Digital Content Protection) process that was developed for the Digital Video Interface (DVI), manufacturers are now instituting the High Definition Multimedia Interface (HDMI) procedures and connection standards.

HDMI supports television and computer video formats, plus multichannel audio configurations, via a single connection cable. First promulgated in 2002 as HDMI version 1.0, the technology has matured to version 1.3 as of June, 2006. Each successive iteration provided for ever-increasing data throughput, growing from 4.9 Gbit/s with version 1.0 to 10.2 Gbit/s with version 1.3. With each new version of standards, devices adopted the then current data rate. Unfortunately this has lead to the situation evidenced currently where legacy devices that only supported the lower data rate are being connected to newer devices that recognize and/or output the higher rates. This mismatch of data capability may create the condition where the HDMI "handshake" between the devices cannot take place. What occurs subsequent to this connection refusal is the transmission of a less than optimum image format with an attendant diminution of video and audio quality.

The backwards compatibility of the HDMI 1.3 connection should allow for devices with the earlier HDMI 1.1 and 1.2 versions to be recognized. These earlier versions support all of the mandatory audio and video formats of both next generation disc formats. Once again, though, here is where confusion may be introduced in the connection equation. For example, a display may not have the capability to automatically recognize the signal being sent to it by a player and does not reset for the new source. The display settings need to be manually adjusted before a proper connection can be sanctioned. This dilemma could be further complicated by the capabilities of the player. The enduser is then faced with the task of accurately setting both devices in order to establish the best connection.

But, without the protection standards that have been developed for content and for transmission, the movie studios and content producers would not participate in the high definition presentation arena. So, the difficulty of working with these protection schemes has become something of a necessary evil.

HD Disc Players

During 2006, next generation disc players were slowly introduced for each HD disc format. The slow rollout was partially a result of the timeline for the then still-developing standards, but was primarily necessitated by a severely limited manufacturing capacity for the blue laser diodes at the heart of the technology. The restricted supply of the laser diodes precipitated the extreme delay in the introduction of the Sony Playstation 3, as the desired distribution numbers could not have been maintained if the device were introduced earlier.

As 2006 came to a close, the long-circulating rumors of a dual-format player finally proved to be true. LG Electronics announced their intention to bring to market early in 2007 a single player capable of playing both the HD DVD and the Blu-ray Disc formats. As this book went to press there were no other announcements that others were following LG's lead. As to the viability of a dual-format player, it is certainly a welcome event as it should help alleviate the stagnant adoption rates for the HD disc formats, but the pricing may remain a hindrance initially. The LG player is expected to cost approximately $1200, which is about the same cost as buying two individual format players, although it is not known what the eventual "street price" may be for the dual format player.

Also announced before the end of 2006, Broadcom Corporation has created a single chip solution that can be the engine for a universal player. This development expands the options for manufacturers, as well as for other home entertainment and networking applications.

And, the Warner Brothers announcement of their patented "Total HD" approach to disc construction preceded the 2007 CES convention in Las Vegas. The Total HD disc will be two-sided with HD DVD content on one side and BD content on the other side. The costs for manufacturing this dual-format disc were not announced before this book went to press.

Settop Players

Settop players are available for either HD disc format. The specifications for the HD DVD players are set and the players for that format will not change much over time. However, the same cannot be said for Blu-ray Disc players. BD player functionality is subdivided into two groups[1] —

- **Profile 1(v1.0)** players do not support network connectivity, enhanced uses of persistent storage, and some other advanced features (picture-in-picture and secondary audio, among them). As of November 1, 2007, Profile 1(v1.0) will be supplanted by **Profile 1(v1.1)** which mandates the inclusion of all of the features not yet implemented except network connectivity. (Still with us, so far?)
- **Profile 2** players support network connectivity, as well as the other mandated features.

Early on, it was felt by the Blu-ray Forum that authoring techniques when the format was first introduced would only be capable of a limited feature set that did not include network access or use of the format's other envisioned advanced features. As a result, the initial players could be less than fully featured and needed to only support playback of the first generation of titles. This decision also helped to keep the price of the BD players lower, although with an inaugural price of over $1000US, it's hard to say just exactly how much impact the lessening of features really contributed to the eventual price. Be that as it may, a result of this

---

[1]There is a third (fourth?) BD player threshold, termed Profile 3, which is intended for audio-only play and does not require video decoding or the inclusion of BD-J programming.

multi-stage player profile decision may likely be the disillusionment of the early buyers as later titles are released that draw upon the network, the onboard storage, and the other advanced features of the format.

It is important to recognize that after June, 2007 there will still be two BD player profiles, those with network connectivity and those without. Profile 1(v1.1) players will not include the networking feature and may be called "BD-Video" players, while Profile 2 players will include the networking feature and may be called "BD-Live" players.

Further splintering the BD users group is the inclusion of a Blu-ray Disc player in the Sony Playstation 3(PS3). Although late to market, it is expected that the PS3 will enjoy rapid sales growth, even with the competition from other game consoles noted earlier. But the PS3 player is not required to support many of the features of a standard BD format player, such as, upscaling standard DVDs to HD image resolutions, nor providing analog multichannel audio output of 5.1 or 7.1 surround configurations.

Meanwhile, the HD DVD front has advanced to a second generation of players that support HDMI 1.3 and are capable of 1080p playout support (BD had this as an initial feature). And, Toshiba has been joined by Thomson and RCA in the production of settop HD DVD players[2]. The introduction of the HD DVD add-on for the Microsoft Xbox 360, rounds out the HD DVD player options. Although here, too, the gaming console player element for the format does not support all of the features of a standalone player. The Xbox 360 HD DVD player does not support HDMI and does not have any digital connectivity. So, the Xbox 360 gamers may watch movies only via analog connections, if they choose.

There is a VGA option for the Xbox 360 player that may be used with the HD DVD drive. The unique ability of this VGA adapter is that the VGA connection can pass a 1080p signal, while the component video connections cannot. Any improved performance of this VGA adapter is totally dependent on the VGA input capabilities of the connected display, however. Some of the big-screen 1080p displays may not react very well to a VGA input.

HD Disc Recorders

As the HD disc formats mature, HD disc recorders are becoming available for each format[3]. These recordable drives are now available for installation in laptop or desktop computers, and may be used in standalone HD disc recorders, too. The introduction plans for the standalone devices are a little muddled as representatives from Sony, Panasonic, and Toshiba, as recently as Ceatec in October of 2006, announced that their HD disc recorders would be available in Japan but not in the US.

Other manufacturers may introduce multiformat recording devices and will have little compunction to restrain themselves from going after the US domestic market. For instance, Ben-Q has unveiled a "trio" optical writer, with three types of lasers that support multiple media formats — CD, DVD and BD. And Okoro Media Systems has introduced digital

[2]The BD player front has expanded to include players made by Samsung, Sony, and Panasonic, among others.

[3]Sony launched a Blu-ray Disc recorder in 2003 in Japan, the BDZ-S77, but it recorded on cartridge-type media and would not play feature films.

entertainment systems built with a Blu-ray recording drive that can function as a media center PC or the central device in a home theater. These HD devices come with very large storage drives and rudimentary disc burning tools. Toshiba's first HD DVD recorder comes with a one terabyte hard drive, for example.

There are extensive plans to introduce HD recordable drives for computers worldwide. These are becoming available in both the internal and the external drive flavors, using recordable and re-recordable (R/RE) media. Additionally, both single and dual layer discs can be utilized. These capabilities apply to both HD DVD and Blu-ray Disc, so the primary difference between the recordable formats will be capacity. As has been previously announced, Blu-ray Disc dual layer capacity is 50 GBs while HD DVD dual layer capacity is 30 GBs. Although, both formats continue to develop additional layering capabilities and there may well be further announcements early in 2007 about expanded disc capacities.

It is difficult to pin down the minimum requirements for a computer that incorporates a high definition optical drive, whether a laptop or a desktop. But we'll take a stab at it —

- a really fast processor (really, really fast and more than one), and
- a whole lotta RAM, and
- a really late model display/graphics accelerator board, and
- a whole lot more RAM, and
- a very large hard drive (maybe more than one here, too), and
- a whole lot more RAM!

Seriously, the specifics of what constitutes a computer that can support and smoothly reproduce high definition optical disc content are elusive and very difficult to accurately assess. The data rate and throughput demands that a 1080p image at 24 frames per second puts on a computer are massive and relentless. Couple those demands with the typical operating environment of a computer with its power needs, display needs, media player needs, application needs, and that's a whole lotta needs! Thus the computing power required for HD OD (high definition optical disc) play, let alone record, remains an area for experimentation and imagination. As a reference, here are the specifications for the Sony VAIO™ laptop, model VGN-AR270 —

- Intel® Core™ 2 Duo Processor T7600 @ 2.33 GHz
- Two 200 GB hard drives
- 2 GB PC2-4200 DDR2, 533 MHz SDRAM
- NVIDIA® GeForce® Go 7600GT graphics card with 256 MB VRAM
- 17" WUXGA display
- Blu-ray Disc R/RE optical drive

In January of 2007, Toshiba announced the introduction of their latest HD DVD read/write drive, the SD-H903A. Although the HD DVD burner was not announced as a component of a computer, it is relatively safe to say that a minimum configuration for HD DVD is similar to that for Blu-ray Disc. For further perspective, here are the specifications for the Toshiba Qosmio™ G30 laptop with an HD DVD drive —

- Intel Core Duo T2600 @ 2.16 GHz
- Two 120 GB hard drives

- 2 GB DDR2 533 MHz SDRAM
- NVIDIA GeForce Go 7600 graphics card with 256 MB VRAM
- 17" WXGA display
- HD DVD optical drive

Additional Player Considerations

Format Specification Adherence

The authoring specifications that are being used to compile the contents of HD discs are slowly becoming more well understood. As this understanding evolves, programming techniques are being refined to more closely match the intentions of the format specifications. Similarly, as their operations mature, manufacturers are building devices that more closely meet the requirements for the next generation players. The challenge that results from this dual-track progress is that the earlier released discs that play on the earlier players may not work on later models or, conversely (perversely?), the latest discs will only play on the latest players. At best, this can be summed up as "growing pains". At worst, this could result in disc and/or player recalls, as was the case with the release of the *Speed* title on Blu-ray Disc before the end of 2006. The producing studio, Fox Home Entertainment, offered replacement discs for *Speed* as there was a problem with the discs when played on Samsung BD players.

Audio Formats

The HD disc formats offer greatly expanded selections of audio formats, compared with what was available on standard DVD. That sentence may be a little difficult to read given the extensive use of plurals. Yet, we have to deal with that because each disc format, HD DVD and BD, offers unique audio format combinations. These offerings are subdivided between mandatory and optional selections —

For Blu-ray Disc, the mandatory formats are:
- Dolby® Digital (AC-3), up to 5.1 channels
- DTS Digital Surround®, up to 5.1 channels
- Linear PCM, up to 5.1 channels

For Blu-ray Disc, the optional formats are:
- Dolby® Digital Plus, up to 7.1 channels
- Dolby® True HD, up to 7.1 channels
- DTS HD™, up to 7.1 channels

For HD DVD, the mandatory formats are:
- Dolby Digital (AC-3), up to 5.1 channels
- Dolby Digital Plus, up to 7.1 channels
- Dolby True HD, up to 2.0 channels
- DTS Digital Surround, up to 5.1 channels
- Linear PCM, up to 5.1 channels

For HD DVD, the optional formats are:

- Dolby True HD, 7.1 channels
- DTS Digital Surround, up to 6.1 channels
- DTS-HD Master, up to 7.1 channels

Certainly a litany that lends itself to confusion. Oh, before we forget, it is only necessary to include one of the mandatory audio formats on an HD disc. It is totally unnecessary to have all of the mandatory audio formats represented on a single disc. Just in case anyone was confused about that, too.

Nonetheless, these audio formats present multiple challenges both to disc authors and to player manufacturers. While the HD disc authoring team really only needs to include one mandatory audio format stream, player builders must include decoder circuitry for all mandatory audio formats. The challenge for manufacturers only begins there. Even the mandatory formats may have audio stream core mixes, breakouts, or mixdowns, as well as a number of channel and sample rate configurations. Deciding how to support the formats and to what depth that support may go, is a major concern to manufacturers. And, by extension, given the extensive options that exist, even within the mandatory formats, consumers are being confronted by a dizzying array of player features and capabilities.

HDTV Displays

In recent years, there has been a dramatic expansion in the area of new display technologies. We have gone from a world dominated by *cathode ray tubes* (CRTs) to one in which it will soon be difficult to find a CRT. With advances in these new display technologies, it will be important to understand how the next-generation disc formats will be affected.

What's Wrong with CRT?

If you go down to your local consumer electronics store, you will see a very interesting trend — the cathode ray tube (CRT) television set has virtually disappeared, and almost all of the high definition displays use some form of new technology that is not CRT. One of the leading reasons for this shift is size. It is very difficult to manufacture a glass tube CRT in very large sizes. Given that the CRT is a type of vacuum tube technology, the tube itself must be able to structurally withstand the pressure of containing a vacuum without imploding. As the tube grows larger, this becomes more difficult without any sort of internal support. And, it's a lot of glass, making it expensive and HEAVY.

Does that mean CRT is going away? Probably, but not completely. It's true that a 500+ pound CRT television will have a difficult time competing against the convenience of a relatively light, wall-mounted plasma screen or LCD. Even in situations where taking up a large space is not a problem, the convenience of a lightweight system may outweigh (pun intended) the CRT heavyweight. However, it would be a shame to lose CRTs. For one thing, there has been over 60 years of research applied to developing the color display technologies that CRT uses, including phosphors, screen masks and color dot distribution to name a few.

3

Types of New Displays

Although the CRT is fading from the landscape, it should not be dismissed too quickly. In order to really see the continuing benefit of CRTs, put a properly calibrated CRT television next to any of the newer display types and just try to get them to look as good as the CRT. The still-evident limitations of these other display types are revealed when viewed next to a CRT. But, that comparison posture is slowly losing credibility as advances in the rapidly developing display technologies continue to chip away at the CRT's pedestal. Below are summaries of some of the current and new display technologies.

■ **Digital light processing (DLP)** This type of display is based on the *digital micromirror device* (DMD), a silicon chip whose surface is covered by a matrix of tiny mirrors. Through electrical signals, these mirrors can toggle from an "off" position to an "on" position and back. Basically, each micromirror represents a pixel; by reflecting a light source off the mirror, one can project a black and white image. Flickering the mirror at variable high rates creates the impression of varying amounts of light, displaying a grayscale. Passing the image through a rotating color wheel that is properly synchronized with the mirror flashes generates a color display. This is the basis for most consumer rear-projection DLP televisions, which can offer a bright, crisp picture. However, depending on the color wheels used and the sophistication of the display logic, DLP displays tend to have difficulty reproducing some of the darker color ranges. Likewise, being a rear-projection device, there may be some *light bleed*, which can wash out the image, making black areas appear gray. Some DLP displays may exhibit color shift problems, exaggerated sharpness, and excessive brightness. The latter two characteristics, in particular, can overemphasize encoding artifacts in video.

■ **Plasma display device (PDP)** Plasma displays work by applying a charge to a small gas-filled cell. The gas becomes ionized and, in turn, interacts with a phosphor on the surface of the cell, which glows a given color. Because plasma displays work with phosphors, they are able to utilize much of the research that has gone into phosphor research for CRTs and reproduce more natural colors. The earlier problem of burn-in or after-image silhouetting has been reduced considerably in newer plasma models. However, plasma displays can consume large amounts of power compared to other types of displays, and some models still tend to suffer from bleed-over effects in which the plasma from one cell bleeds over to other cells.

Early plasma models also suffered from a dissipation of luminence within a relatively short time frame. Nowadays, though, new models are expected to provide as much as 60,000 hours of operation before the display reaches half of its original luminescence. That's equal to more than six and one-half years of continuous operation!

■ **Liquid-crystal display (LCD)** Liquid crystal displays operate by sandwiching color filters and polarizing filters on either side of a liquid crystal cell matrix, and placing a white backlight behind the sandwich. The backlight remains lit while the display is active. The liquid crystal in each cell, by default, stays "curled" up such that it blocks the polarized illumination from the backlight. When a charge is applied to a cell, the crystal "unwinds," allowing the illumination from the backlight to escape through the color filter. One of the most noticeable problems with early model LCDs was the limited viewing angle, although with the latest updates to these technologies, the viewing angle has been extended dramatically (up to as much as 178 degrees). Though, increased viewing angle tends to lead to increases in *light* *bleed* — white light that escapes to wash out the picture and raise black levels.

There remains the issue of the "unwinding" process, which takes time at a molecular level. Here, too, technological advances have reduced the time it takes to fully refresh an LCD display from a time of approximately 24 ms in earlier LCD models to time approaching that of 16.7 ms and even faster. Comparatively, a progressively displayed frame of film lasts for about 40 ms, but interlaced video at 59.94 fields per second requires an update to the screen every 16.7 ms.

■ **Liquid crystal on silicon (LCoS)** Similar to DLP rear projection displays, this type of television operates by projecting colored light through a translucent matrix of liquid crystals, in which each cell can be individually made opaque, transparent or somewhere in between. Because LCoS uses liquid crystal, these displays tend to suffer from the same performance issues as LCD and often require three separate light sources (or one light source with three pathways through three separate color filters) in order to provide full color reproduction.

Other New Display Technologies

Three other display technologies have been recently introduced and warrant considerable attention and evaluation. These are —

■ **D-ILA (Digital Direct Drive Image Light Amplifier)** Developed by JVC, D-ILA stuffs 2048×1536 pixels on a 1.3" chip, for greater than 1080 HD resolution.

■ **SXRD (Silicon X-tal Reflective Display)** Developed by Sony, SXRD builds on LCos technology, with dramatically improved color reproduction characteristics and an expanded contrast ratio that has been measured as high as 13,300:1.

■ **SED (Surface-conduction Electron-emitter Display)** jointly developed by Toshiba and Canon, SED boasts a 50,000:1 contrast ratio, and a 1 ms response time, built in a thin, flat panel form factor.

HD Video Displays Meet HD Discs

As the next-generation disc formats enter the marketplace, they will meet the two high definition video display formats — 720p (1280×720 resolution at 59.94 progressive frames per second) and 1080i (1920×1080 resolution at 29.97 interlaced frames per second). These are the two HD video formats that have been adopted by the television broadcasting industry. What is particularly challenging is determining what HD displays do to the video internally before it is displayed.

A considerable number of early model high-definition displays have a native resolution that falls within a range from 1280×720 up to 1600×1050. It is only the most recently produced displays that are being built with the 1920×1080 image resolution threshold. So, the issue of legacy devices will remain for an appreciable period of time.

Perhaps the greatest concern is that a display may state that it supports 720p, 1080i and 1080p video formats, but they rarely indicate what scaling, frame rate conversion, deinterlacing, or other signal processing these formats may undergo prior to being displayed. With the introduction of HD disc players and the HDMI technology, there is even further cause for concern that endusers may be erroneously prevented from viewing HD content because

the devices do not have the ability to recognize the diverse settings that may exist with older displays. Finding a good match between HD disc players and display devices may be just a matter of luck for some time to come.

Chapter 4
Coding Technologies

In the days of DVD, the questions of which coding technologies to use for your audio and video were relatively simple — for audio, the only compressed audio that was guaranteed to be supported by all players was Dolby® Digital (AC-3)[1]. Likewise, for full resolution video you only had the choice of MPEG-2. Sure, there were other options like Linear PCM, MPEG and DTS™ for audio or MPEG-1 for video, some of which were mandatory and others optional for players to support, but the choice was pretty simple to make since these other *codecs* (coder/decoders) were either not guaranteed to work on every player or did not offer the full resolution and feature set that was usually needed.

Well, those days are gone. Not only do you have to choose between two disc formats that do the same thing (HD DVD and BD), but you must now choose among a bevy of different audio and video codecs that purportedly offer essentially the same features and capabilities. What's more, the options tend to differ slightly between the two disc formats, as do the criteria you use to select them. For instance, a Blu-ray Disc™ offers a higher bandwidth for reading content from the disc, so audio options like multichannel Linear PCM that use very high bandwidth may still be reasonable for a project. In contrast, PCM would tend to be a poor choice on HD DVD, which has almost 40% less bandwidth.

So, why were so many choices included in the format specifications this time around? To answer this question, you have to understand how the format specifications were created and the motivations that exist therein.

Creation of a Format

To understand the dilemma we face with the new formats, it is important to understand how these formats were first created and, subsequently, the motivations behind the participants. In the cases of both HD DVD and Blu-ray, the application format specifications, which define how one can include audio, video and interactive content on a disc, are created by a group of industry representatives that come primarily from either content production (film, television or music studios), information technology (computer software and hardware manufacturers), or the consumer electronics industry (player manufacturers). However, there are competing motives involved as these different representatives come together. On the one hand, there is the common goal to build a format that will win the hearts and minds (and dollars) of the consumer marketplace, picking up where the DVD format is now slowing down. On the other hand, each representative is attempting to gain political or financial

---

[1]At the start of DVD, even Dolby Digital was optional for players in territories that used the PALvideo format, such as Europe. It was made mandatory in a subsequent update to the DVD-Video specifications.

4

advantage in the process. Most of the time, these motives work hand in hand since a successful consumer format will generally bring good business to its creators. However, there are times when these motives are clearly at odds.

In the DVD Forum, the *DVD Specifications for High Definition Video* were created by a group referred to as Working Group 1 (WG1). Its counterpart in the Blu-ray Disc Association is named Technical Experts Group 2 (TEG-2), which created the *System Description, Blu-ray Disc Read-Only Format, Part 3: Audio Visual Basic Specifications*. (Now that's a mouthful!) In each case, fifty to seventy companies gathered together once or twice each month over the course of several years to propose, debate and ultimately approve technologies for inclusion in the specifications. In most cases, the main proponent of a given technology was the technology's owner and patent holder. After all, patents & licensing are big businesses for these companies. If one managed to get one's technology adopted in the specification, especially as a mandatory requirement, then that meant one had an opportunity to collect royalty payments from player manufacturers, disc manufacturers, and sometimes the content owners themselves. Even at just a few cents per player, or a fraction of a cent per disc, this can add up to millions of dollars over the course of a few years. Likewise, *cross-licensing* deals between technology owners were common, in which companies would trade their technologies' license with each other in order to avoid the costs of royalties, thus allowing them to be more price competitive in the industry.

Besides the monetary drive behind format participation, political motives were involved. For example, many of the founding members of the DVD Forum, based primarily in Japan and in the United States, generally held the opinion that the profitable lifespan of the DVD format was shortened by the aggressive entry of foreign manufacturers who were able to undercut prices by allegedly not paying licensing royalties. Therefore, some companies sought to take a protectionist stance in the creation of the new formats such that it would be difficult for manufacturers, other than core members, to build discs and players that conform to the specifications. In other words, they intentionally sought to make the specifications difficult to understand so that competitors would have a hard time making players that work.

The result of all this eco-political scheming is a set of specifications that are both complicated and expensive.

Audio Coding Technologies

HD DVD and Blu-ray offer several new audio codecs, including Dolby® Digital Plus, Dolby TrueHD and DTS-HD High Resolution Audio™ and Master Audio™ that come as an addition to the Dolby Digital AC-3 and DTS codecs commonly used for DVD. Several optional low bandwidth audio codecs have also been provided for streaming content, such as mp3, aacPlus version 2, Windows Media® Audio (WMA), and more. However, not all of these new codecs are quite what they seem, and may not always offer the improvements their names imply.

The first criterion often used to determine which audio codec to use on a project is playability. In other words, will the audio you put on the disc be heard by your customers? Playability can only be guaranteed for audio codecs that are defined as mandatory in the specifications, unless you can guarantee which player(s) will be used, such as for a kiosk or other permanent installation. The second criterion most commonly applied is size. Which codecs will deliver the desired audio experience and still fit within the space available on the disc, as well as stay within the bandwidth limitations? Third is quality. Which of the available codecs will deliver the highest quality experience? And finally, the last criterion tends to be ease of use and availability, which can sometimes outweigh other criteria since it includes cost implications.

In the following sections, we will take a look at each of these criteria and how the different audio options fit. This information should help you make educated selections for your own projects.

Mandatory vs. Optional Codecs

Several times now, we have mentioned *mandatory* and *optional* codecs, but you may be asking yourself, "What does that really mean?" Simply that the HD DVD-Video and Blu-ray Disc specifications each define what a player must support at a minimum (mandatory), as well as what a player can support if the manufacturer chooses (optional). The combination provides room for competition in the marketplace, while also allowing for a wider range of applications. For example, a portable player is typically designed to be small, lightweight, and inexpensive. As such, one is typically listening to audio on a portable player solely through headphones. Therefore, it would not make sense to mandate that such players support full eight channel surround sound audio, since that would just add more cost and weight, and yet the user would not really benefit from any of the extra functionality. On the other hand, one would certainly want a high-end home theater system to offer the highest possible quantity and quality of audio channels, even if it means a larger, more expensive product.

In some cases, the specification of mandatory or optional is very clear cut, but not so in others. For instance, Table 4.1 shows that nearly all of the primary audio codecs available for HD DVD are listed as mandatory. However, all it takes is a brief glance at the long list of footnotes to realize that many of these codecs are not as mandatory as one might expect. In almost every case, players are actually only mandated to support a restricted subset of the audio stream's capabilities.

This approach of mandating only portions of a codec's capabilities can be good and bad. The good part is that you can use any of these mandated codecs and know that your customers will actually hear the audio. The bad part is that what they hear may not be quite what you had hoped. For example, DTS-HD Master Audio provides a means to include a lossless 8-channel audio stream on the disc with data rates up to Mbps, resulting in an extremely high quality surround sound experience. Players, though, are only required to decode the 5.1-channel 1.509 Mbps compressed DTS core portion of the audio stream and can completely ignore the rest. Not to say that this is not still a high quality experience, but it may not be quite the experience that was intended. These restrictions are often even more severe for Blu-ray Disc (see Table 5.2).

Table 4.1 HD DVD-Video Codec Support for Primary Audio
[Square brackets indicate optional levels of support.]

Codec	Mandatory	Sampling Frequency	Max. Data Rate	Max. Channels
Linear PCM[a]	Yes	48/96/[192] kHz	13.5 Mbps	6
Dolby Digital[b]	Yes	48 kHz	448 kbps	5.1
Dolby Digital Plus[c]	Yes	48 kHz	3.024 Mbps	5.1/[7.1]
Dolby TrueHD[d]	Yes	48/96/[192] kHz	18 Mbps	2/[6]/[8]
MPEG-1 Audio[e]	No	48 kHz	384 kbps	2
MPEG-2 Audio[e]	No	48 kHz	912 kbps	5.1/[7.1]
DTS-HD High Resolution[f]	Yes	48/[96] kHz	1.509/[3.0] Mbps	6.1/[7.1]
DTS-HD Master Audio[f,g]	No	48/96/192 kHz	18.432 Mbps	7.1

[a]Linear PCM sample frequency of 192 kHz is only allowed for 2-channel audio

[b]Dolby Digital is a subset of Dolby Digital Plus and is therefore supported as mandatory in HD DVD-Video. However, Dolby Digital streams must be authored as Dolby Digital Plus and do have the shown restrictions.

[c]Players that only support 2 or 6 channel output will only decode the 5.1-ch portion of a 7.1-ch Dolby Digital Plus stream

[d]Dolby TrueHD™ streams must include a 2-ch sub-stream. If more than 6 channels are present, they must also include a 6-ch sub-stream. Only decoding of the 2-ch substream is mandatory. The 192 kHz sampling frequency is only allowed for 2-ch streams.

[e]Although MPEG Audio is not mandatory for players that support 60 Hz video, it is mandatory for players that support 50 Hz video systems.

[f]DTS-HD streams contain a DTS core component (48kHz, 1.509 Mbps, 5.1 channels) plus zero or more extensions to provide increased sampling rates, channels, and accuracy (including lossless compression). Only decoding of the core component is required.

[g]DTS-HD Master Audio may include up to 7.1 channels at 48 kHz or 96 kHz sampling frequencies, but only 2.0 channels at 192 kHz.

Table 4.2 BDMV Codec Support for Primary Audio
[Square brackets indicate optional levels of support.]

Codec	Mandatory	Sampling Frequency	Max. Data Rate	Max. Channels
Linear PCM[a]	Yes	48/96/192 kHz	27.6 Mbps	8
Dolby Digital	Yes	48 kHz	640 kbps	5.1
Dolby Digital Plus[b,c]	Yes	48 kHz	0.640/[4.736][g] Mbps	5.1/[7.1]
Dolby TrueHD[a,b,d]	Yes	48/96/192 kHz	0.640/[18.64] Mbps	2/6/8
DTS	Yes	48 kHz	1.509 Mbps	5.1
DTS-HD High Resolution[e,f]	Yes	48/96 kHz	1.509/[6.0] Mbps	6.1/[7.1]
DTS-HD Master Audio[a,e,f]	No	48/96/192 kHz	24.5 Mbps	7.1

[a]A sampling frequency of 192 kHz is only allowed for 2, 4, and 6-channel audio that is losslessly encoded.

[b]Players are only required to decode the Dolby Digital (48 kHz, 640 kbps, 5.1-ch) portion of the audio stream.

[c]Dolby Digital Plus streams are composed of a Dolby Digital substream (up to 640 kbps) and an additional Dolby Digital Plus substream (up to 1,024 kbps). The data rate of 4.736 Mbps is intended for future applications using additional DD+ substreams.

continues

Table 4.2 BDMV Codec Support for Primary Audio *continued*

dDolby TrueHD streams must be accompanied by a Dolby Digital bitstream, and must include a 2-ch substream. A 6-ch substream is also required if more than 6 channels are present.

ePlayers are only required to decode the DTS core (48 kHz, 1.509 Mbps, 5.1-ch) plus one extension (either 6.1-ch or 96 kHz) of the audio stream.

fDTS-HD streams contain a DTS core component (48k Hz, 1.509 Mbps, 5.1 channels) plus additional extensions to provide increased sampling rates, channels, and accuracy (including lossless compression).

gThe BDMV specification currently limits a DD+ bitstream to a maximum of one DD substream and one DD+ substream to carry up to 7.1 channels. The maximum data rate for this stream configuration is limited to 1.664 Mbps (640 kbps for DD substream, and 1024 kbps for the DD+ substream). The data rate of 4.736 Mbps is intended for future applications using additional DD+ substreams.

When is Mandatory Not Mandatory?

Although disc specifications may mandate a given codec, that does not guarantee that the codec will be playable by the user to the full extent allowed by the specification. For instance, the BDMV specifications mandate support for DTS audio, a codec that has been used for years on DVD. However, not all surround sound audio receivers support DTS decoding. The number is extremely small, at this point some legacy audio receivers with digital inputs do not know how to decode DTS audio. For example, if your customer has their BD player connected to a pre-1997 A/V receiver via a digital connection (e.g., S/PDIF), they may not be able to hear the DTS audio you put on the disc even though support for the codec is mandated by the BDMV specifications. The user will, however, be able to hear the audio if they connect the analog audio output from the player to their receiver since the player is, in fact, required to decode that audio (though possibly only in stereo).

Choosing a Primary Audio Codec

Both of the disc formats distinguish between *primary* and *secondary* audio. Primary audio is generally audio content that is included on the disc itself and provides the primary audio experience that accompanies the video features of the disc. For example, a movie's soundtrack would be considered primary audio. Secondary audio is generally considered to be supplemental audio content that is often delivered via alternative means, such as, from the persistent storage in the player or streamed over a network connection. In both cases, the audio content is generally considered *long-form* content, meaning that it lasts for more than just a few seconds, in contrast with a button sound effect or other *drips*.

Some common uses of audio on a disc include movie soundtracks, alternative language soundtracks, audio commentaries, musical scores, and so on. For DVD, all audio was considered primary audio since the DVD player could only decode a single audio stream at one time. That is no longer the case with HD DVD and BD, which support decoding and mixing multiple audio sources simultaneously. For instance, it is now possible to have a director's commentary decoded and mixed in time with the film's soundtrack during playback. In this way, the soundtrack remains primary audio but the director's commentary is now handled as secondary audio. It is not required that a commentary be handled this way but it does offer certain advantages regarding efficient use of disc space and other new features.

Primary audio is usually the highest quality audio in a project. Because it is played from

4

disc, it is able to make use of the large available capacity and high bandwidth to deliver higher definition and more channels. Primary audio also does not have nearly as many restrictions on it as compared to secondary audio. As a result, when choosing a codec primary audio content, one typically would lean toward codecs with higher channel counts (i.e., surround sound rather than stereo) and higher data rates (less compression). However, there is no point in going beyond the capabilities of the source audio itself. For instance, if the original source audio is only 2-channel stereo, then there's not much point encoding it as 5.1 surround sound. Likewise, if the sample frequency of the source audio is only 48 kHz, then there is really no benefit of encoding it at a higher sample frequency, like 192 kHz. In both cases, you would be better off with a lower data rate codec and use the extra space to improve video quality.

Tables 4.1 and 4.2 each list the codecs that HD DVD-Video and BDMV allow for primary audio, as well as the restrictions each format puts on those codecs, respectively. As you can see, there are a lot of similarities between the two disc formats, but also some key differences, which may be important when determining which format is most appropriate for your content. For example, HD DVD mandates a larger number of audio codecs for players, including the Dolby TrueHD lossless audio compression format. The BDMV specifications only mandate that a player decode Linear PCM (uncompressed audio), Dolby Digital, and DTS. All of the other codecs on a BD player are required to have either AC-3 or DTS as a core component so that even a minimal player will be able to decode some audio content, even though it may be at a lower quality. On the other hand, a closer look will also reveal that Blu-ray generally supports higher maximum data rates for audio streams than does HD DVD. As a result, a Blu-ray disc might use multi-channel Linear PCM audio and avoid audio compression altogether. Both of these issues must be considered by anyone creating discs in which audio quality is an important factor.

NOTE DTS-HD can include a lossless extension (XLL) for providing lossless audio compression, which is referred to as DTS-HD Master Audio. The mechanism is different, but the result is similar to that offered by Meridian Lossless Packing used in Dolby TrueHD. Both HD DVD-Video and BDMV support DTS-HD Master Audio as optional such that only the 1.509 Mbps "lossy" DTS core component is required to be decoded by the player.

There are two principal causes for the differences in audio codec support between the two disc formats, legacy support and political influence. Evidence of legacy support can be seen by the fact that HD DVD-Video continues to support MPEG Audio as a mandatory option in players that support the 50 Hz video system (e.g., players sold in territories that use the PAL and SECAM video formats). During the discussion of audio formats for HD DVD, there was little or no comment made on this issue. In fact, the ongoing support for MPEG Audio was primarily a legacy carried over from the original DVD-Video specifications. Quite a lot of the HD DVD-Video specifications bear a striking resemblance to those of DVD-Video for the same reason. Blu-ray, however, had no prior legacy to uphold other than

BDAV, the video recording specification for Blu-ray that was released a few years earlier.

Political influence, though, had an enormous impact on codec selection differences between HD DVD and BD. When the Blu-ray specifications started, they were created by a group referred to as the Blu-ray Disc Founders (BDF), which was composed almost solely of consumer electronics (CE) manufacturers. Motion picture studios and information technology (IT) companies were not allowed to join the organization until years later when most of the specifications were already complete. The only studio that can claim any impact on the BDMV specifications in the early stage is Sony Pictures, through their cohorts in Sony Corporation's consumer electronics division, which was a founding member. As a result, many of the early decisions regarding BDMV were heavily biased toward issues that the CE manufacturers considered important, such as keeping player costs down and limiting manufacturing burden. At that point, many of the BDF members shared the view that "the studios are the problem" when it comes to format definition.

Evidence of this can be seen in the audio codec selection for BDMV in which no player is actually required to support anything beyond Linear PCM, legacy Dolby Digital, and legacy DTS. The reason for this is that the patents on Dolby Digital and DTS are due to expire in the next few years, releasing manufacturers from having to continue to pay those royalties. However, newer codecs like Dolby Digital Plus and DTS-HD incorporate new patented technologies that are not due to expire for many years to come, and therefore would continue to be subject to royalties throughout the lifespan of the disc formats.

Unlike Blu-ray, the HD DVD-Video specification was the initiative of a group that consisted of a fairly even mix of CE companies, IT companies and motion picture studios. In fact, with many of the CE companies favoring the BD format, this opened the door to the IT companies and studios to have much more influence over the specifications. For instance, HD DVD was the first to adopt the SMPTE VC-1 video codec, based on Microsoft *Windows Media Video 9* technology, and later went on to adopt the joint Disney/Microsoft *HDi* approach for implementing Advanced Content. (Oddly enough, Disney went on to be a major studio supporter of Blu-ray Disc, which had turned down HDi in favor of a Java-based approach.) Likewise, through pressure from the studios and recording companies, the HD DVD-Video format includes mandatory support for the Dolby Digital Plus and Dolby TrueHD audio codecs.

All political wrangling aside, the end results between the two disc formats is not terribly dissimilar. In fact, the player implementations that we are likely to see on the market will probably be much more alike than different. This is largely becuase licensing fees and chip implementations for the baseline codecs (e.g., Dolby Digital and DTS) will generally include the more advanced features of Dolby Digital Plus, DTS-HD, and potentially the lossless codecs, as well.

Terms like these make it difficult for a player not to support even the optional features of these codecs, so how do you choose which to use?

Linear Pulse Code Modulation (PCM)

Linear PCM is the "gold standard" against which all of the other codecs are compared. This is becuase it represents uncompressed digital audio. In fact, to say that Linear PCM is

uncompressed is a bit of a misnomer since, in fact, the audio data it represents is based on quantized samples taken at regular periods across a given time interval, therefore some information is lost in the process, depending on the bit depth and sampling frequency compared to the original analog source. (You may recall having seen the word *quantization* applied to video compression, as well.) However, the 16-, 20- and 24-bit sample sizes and 48 to 192 kHz sampling frequencies are generally considered to capture all of the fidelity that the human ear is capable of hearing. So, we come back to the statement that it is uncompressed. In any case, it is enough to say that Linear PCM is the starting point from which all of the other audio codecs begin. Any divergence from this original set of Linear PCM samples would, therefore, be considered error, noise, or distortion.

So, if Linear PCM is the gold standard, then why not just use it? The simple answer is size. Capturing Linear PCM audio involves taking a 16- to 24-bit sample of each audio channel at anywhere from 48,000 to 192,000 times each second. For stereo audio, that corresponds with 1.536 to 9.216 *million bits per second* (Mbps). For standard surround sound audio, the numbers jump up to 4.608 to 27.648 Mbps. To put things in perspective, most standard definition video on DVD is compressed to just 4.5 Mbps and the total maximum bandwidth allowed on a DVD for audio and video combined is just 10.08 Mbps! Of course, we are not just talking about DVD anymore. HD DVD supports a maximum bandwidth of approximately three times that of DVD, and the maximum bandwidth for Blu-ray is almost five over four times that of DVD. Between the higher bandwidth and greater disc capacity of these new formats, Linear PCM audio can be seriously considered for both HD DVD and Blu-ray Disc applications. In fact, many of the first Blu-ray movie titles released in 2006 included Linear PCM surround sound audio tracks as a standard feature (though these discs had few other features since there was no space left for them on the single-layer 25 Gbyte discs used at that time).

Nonetheless, Linear PCM files are still incredibly large, especially when compared to compressed counterparts from Dolby and DTS. For applications in which the highest audio quality is required, Linear PCM should be considered. However, for other applications in which audio is not the central concern, but perhaps disc space or other features may be of greater importance, then you should consider one of the compressed audio options described below.

Dolby Digital (AC-3)

The most commonly used audio format on DVD, Dolby Digital supports channel configurations from single channel mono to stereo to full 5.1 surround. In fact, Dolby Digital can even support up to 6.1 channels by matrixing an additional rear center channel into the left and right surround channels, which may be reproduced with either one or two rear speakers. This is marketed as *Dolby Digital Surround EX™*. Used on nearly every DVD title released to the consumer market, Dolby Digital has consistently offered good audio quality for home listeners with minimal impact on disc space or bandwidth. In addition, Dolby Digital is among the most widely supported compressed audio formats for home theater environments with proven playability on virtually every legacy A/V receiver that supports digital audio inputs.

Not only is Dolby Digital widely supported in playback environments, it is also very well supported with regard to encoders and processing tools that are available. Being so well established in the industry, it is difficult to find an audio facility that doesn't already have Dolby Digital encoding capabilities. Also, because Dolby Digital is an exact subset of Dolby

Digital Plus (DD+), a standard Dolby Digital stream may be used anywhere that a DD+ stream is required. For instance, the HD DVD-Video specifications specifically support Dolby Digital, but require such streams to be identified as DD+ when authored.

NOTE   While Dolby Digital Surround EX™ offers 6.1 channels based on the original Dolby Digital codec, this additional channel is not discrete. It is formed by manipulating the audio content in the rear left and rear right surround channels, which can sometimes have an audible effect on those channels. Dolby Digital Plus (DD+), on the other hand, is able to offer 6.1 and 7.1 channel configurations in which the additional channels are digitally discrete. In fact, Dolby Digital Plus can offer much more than just 7.1 channels, but the HD DVD-Video and BDMV specifications limit support to this level.

Dolby Digital Plus (DD+)

Dolby Digital Plus, also known as Enhanced AC-3, is an extension of Dolby Digital that adds a number of new features to the already popular codec. In particular, DD+ provides new spectral coding techniques that help to improve audio reproduction at very low data rates (such as for streaming applications), while also adding a mechanism for supporting more discrete channels and higher data rates.

Unfortunately, Dolby Digital Plus comes with a few problems, as well. First, legacy A/V receivers that support Dolby Digital via digital connections (e.g., S/PDIF) will not natively support the decoding of Dolby Digital Plus. Fortunately, Dolby Digital Plus can address this by down converting to legacy Dolby Digital at a high data rate (640 kbps) to preserve as much of the DD+ quality as possible. If the Dolby Digital Plus stream to be converted contains more than 5.1 channels of audio, the channels beyond 5.1 are discarded durning the conversion process, but due to the structure of the DD+ bitstream, the content from those channels is retained in the converted Dolby Digital stream by means of a downmix process that occurs in the Dolby Digital Pus encoder during the production process. However, only the 5.1 down mix of streams originally with 6.1 or 7.1 channels will be heard.

Second, most of the extra features of Dolby Digital Plus have the greatest effect at very low encoding rates rather than at the data rates one would typically expect on an HD DVD or Blu-ray disc. As Figure 4.1 shows, DD+ is able to retain its level of perceptual quality at lower data rates, but in the moderate and higher data rate ranges that are more consistent with what would be expected to be used on an HD DVD or BD disc, the difference in quality between AC-3 and DD+ are actually quite small. Likewise the curve flattens out as the data rate continues to increase. So, even though DD+ can support data rates as high as 3.0 Mbps and 4.7 Mbps on HD DVD and BD, respectively, such high rates would offer little perceived improvement to the audio experience in a 5.1-channel surround environment. (The best use of the extra data rate is for adding additional channels, though not many soundtracks are currently mixed for 6.1 or 7.1 channel surround.)

The final strike against Dolby Digital Plus is the lack of availability of encoding and decoding tools. Being a new codec, it will take some time before it is widely adopted among tool

4

manufacturers other than Dolby Laboratories. Therefore, it may be more difficult early on, to find facilities that are already equipped and ready to begin DD+ encoding. Likewise, those who are equipped may still have a bit of a learning curve to get through before they are truly able to use that equipment effectively. This, like the legacy issue, will improve over time.

Dolby TrueHD

For those who have followed DVD-Audio, Dolby TrueHD will be a familiar technology under a new name. Dolby TrueHD is, essentially, the marketing name Dolby Laboratories has begun using for Meridian Lossless Packing (MLP), a lossless compression scheme developed by Meridian Audio, Ltd., but now licensed by Dolby Labs. MLP was the codec of choice preferred by the recording industry for providing lossless audio compression on formats such as DVD-Audio. Able to deliver the exact same quality as uncompressed Linear PCM audio, but usually at much better compression rates, Dolby TrueHD provides an excellent alternative in cases where disc space or bandwidth are a concern, especially on the HD DVD format in which Dolby TrueHD support is mandated.

NOTE   One important advantage that losslessly compressed codecs like Dolby TrueHD and DTS-HD Master Audio (defined in the following section) have over Linear PCM is the ability to include a lossless downmix in the stream. For example, an 8-channel Linear PCM stream contains eight discrete audio channels that are all expected to be heard by the listener. However, if your sound system only has six or even just two speakers, you will not be able to hear the full presentation. In fact, with only two speakers, you may find you are missing out on quite a bit! A Dolby TrueHD and/or a DTS-HD Master Audio 8-channel stream, however, is required to also include both a 6-channel and a 2-channel downmix, each of which is also losslessly encoded. This has almost no impact on the data rate and file size, but provides for a greatly improved listening experience no matter how many speakers you have.

There are several challenges associated with the use of the Dolby TrueHD codec, however. The first is that HD DVD only mandates 2-channel decoding of Dolby TrueHD streams. Fortunately, the codec provides an effective means for including lossless 2-channel, 6-channel and 8-channel mixes all in the same audio stream with very little impact on data rate, as shown in Figure 4.2, presented at the end of this chapter. The second problem, however, is that legacy digital connections such as S/PDIF do not have the bandwidth necessary to deliver uncompressed surround sound data to a receiver. At best, only decoded stereo outputs can be delivered via legacy digital audio connections. In the past, DVD-Audio players, which also supported MLP, got around this limitation by simply decoding to analog and delivering multichannel analog signals to the home theater systems to which they were connected. This is also now being addressed by new HDMI connections that are able to support multichannel uncompressed audio as well as their increased support for high definition video.

The final difficulty with Dolby TrueHD, however, is that it is a *variable bit rate* (VBR) coding scheme. That means the data rate of the stream fluctuates over time depending on the complexity of the audio being compressed. This can be a problem when the variable bit rate audio is combined with variable bit rate video, which can lead to unexpected multiplexing issues due to peak data rate overruns. In areas with intricate high frequency detail, which is difficult to encode, the data rate of a stream may reach levels as high as an uncompressed Linear PCM stream, while in other areas it may be dramatically lower. Although the Dolby TrueHD encoder uses a complex scheme to minimize peak data rates, some peaks may be unavoidable. The operator cannot simply specify data rate limits as one would do for variable bit rate video because doing so would require changing the audio data itself, making it no longer lossless. Because Dolby TrueHD is a lossless compression method (a key difference from VBR video encoding), it can only ensure a lossless encode by being allowed to raise the data rate to encode all of the audio perfectly. What's more, the operator cannot predict what the range of required data rates will be until *after* the stream has been encoded since it is entirely content dependent. This can turn the disc production process upside down if you do not know how much bandwidth will be available for your video until after you are done encoding all of your audio!

WARNING   Even an audio master with very faint, virtually inaudible high frequency noise can result in a significant increase in the data rate of a Dolby TrueHD lossless stream. This can sometimes cause confusion, especially when the source audio seems very simple and clean. Simply filtering out the high frequency noise can result in a dramatic reduction of data rate in difficult areas, which are often the areas where such a filtering operation would be least noticeable. It is, of course, no longer a lossless production since you are removing data to improve the compression rate.

DTS & DTS-HD

One of the more celebrated features of both HD DVD and Blu-ray Disc is the addition of DTS audio as a mandatory requirement in players. Previously optional in DVD, DTS has come a long way toward widespread adoption in the consumer market. As a result, both CE manufacturers and studios easily agreed that its support in the new breed of players would be a welcome feature among customers.

Both DTS and DTS-HD are built upon the same Coherent Acoustics technology, which defines a "core + extension" structure for audio coding (see Figure 4.3, presented at the end of this chapter). In this case, the core component is typically composed of either 768 kbps or 1.509 Mbps 5.1-channel 48 kHz encoded audio. The extension unit is typically one of XCH or X96 (defined below). For DTS audio, the coded audio is composed of a single *core substream* that contains the core audio data plus zero or one optional extension. For DTS-HD, however, at least one *extension substream* exists, which can contain one or two extension units. A legacy DTS encoder only pays attention to the core substream, while a new DTS-HD decoder will use both the core substream and the extension substream to provide a more enhanced experience. What is important to understand here is that the core substream is essentially equivalent to the DTS audio that has been used in the past on DVD titles. The

new extension substream in DTS-HD is what allows it to transcend even that high quality experience, providing additional channels (up to 7.1), lossless compression, increased sampling rates (96 kHz, or 192 kHz for lossless encoding), and additional data rates (up to 24.5 Mbps).

NOTE Although the BDMV specifications support both legacy DTS and DTS-HD, the HD DVD-Video specifications support only the new DTS-HD audio codec. However, the digital audio output from an HD DVD-Video player contains a core DTS stream component that can be decoded by legacy DTS decoders. In other words, when you author an HD DVD title, you must use a DTS-HD audio stream rather than the old DTS streams that you may have used for DVD. However, legacy receivers will still be able to decode the core component of that stream.

The extensions supported by the legacy DTS codec include XCH and X96, defined below. The newer DTS-HD codec adds XXCH, XBR and XLL.

XCH

The Channel Extension, XCH, provides an additional discrete mono channel for output as a single or dual rear center channel, and is marketed under the name "DTS-ES®." Many legacy DTS decoders and A/V receivers support this extension when found in the core substream.

X96

The Sampling Frequency Extension, X96, also known as "Core+96k," provides a method for extending the sampling frequency of DTS audio from the standard 48 kHz up to 96 kHz. An encoder using this extension will start by performing a standard DTS 48 kHz encode of the core data, then decode that and subtract it from the original to determine which residuals were lost in the process. The X96 extension data is then generated by encoding these residuals. As with XCH, many legacy decoders and A/V receivers support this extension under the name "DTS-96/24™" ("24" refers to the 24-bit audio data that is output).

XXCH

Similar to the XCH Channel Extension, XXCH is a newer version that supports a greater number of additional discrete channels. For HD DVD and Blu-ray, however, the number of additional channels is limited to two, extending the core 5.1 channel configuration to 7.1 channels. Only new decoders and A/V receivers that support DTS-HD directly will recognize this extension.

XBR

Like XXCH, the High Bit-Rate Extension, XBR, is a new extension for DTS that provides greater audio quality by increasing the available encoded data rate via the extension substream. In short, an encoder implementing the XBR extension would do so in a similar manner to X96, above. First, the data is compressed using the core encoder to form the core substream. It is then immediately decoded and subtracted from the original audio data to calculate the residuals that were "missed" by the core encoder. These residuals are then encoded into the XBR extension in the extension substream, resulting in an overall increase in data rate. When the core and XBR extension are later decoded and recombined, the result is a more accurate reproduction of the original audio data. Like XXCH, only newer decoders and A/V receivers that directly support DTS-HD will benefit from the XBR extension.

NOTE   New A/V receivers that support the new DTS-HD extensions XXCH, XBR and XLL will only benefit from them if they are able to receive the extension substream from the player. For instance, a player connected to one of these new A/V receivers via a legacy S/PDIF digital connection will only receive and decode the core substream because the extension substream requires more bandwidth than an S/PDIF link can handle.

XLL

The Lossless Extension, XLL, is perhaps the most exciting new extension to the DTS codec. With the ability to perform lossless compression of up to eight channels of 24-bit audio at sampling frequencies of up to 192 kHz, DTS-HD Master Audio promises to deliver all of the quality of a Linear PCM stream at a fraction of the bandwidth and disc space. However, support for the XLL extension in both HD DVD and Blu-ray players is only optional, and it will require a new A/V receiver that supports DTS-HD Master Audio to decode it, and only if it is able to receive the extension substream from the player. Like the other variations, DTS-HD Master Audio includes a DTS core 5.1 channel component that can be decoded even by legacy devices.

Like Dolby TrueHD, DTS-HD Master Audio is a *lossless* compression scheme, which means that every bit that is encoded from the original Linear PCM source audio will be recreated at the decoder with exact precision. However, it should also be noted that both compression methods have variable data rates that are content dependent. In other words, some content, such as 24-bit audio with a lot of noise in the least significant bits, can require significantly more bandwidth and disc space than other content. In extreme cases, this can cause a DTS-HD Master Audio or Dolby TrueHD stream to take up nearly as much room as an uncompressed Linear PCM stream, though this is quite rare. One recommended technique in such cases is to consider "bit shaving," in which the least significant one or two bits of the audio samples are removed (set to zero) prior to encoding. This often results in a stream that is far more efficient to encode, while having little or no impact on the audio quality of the stream. In any case, this is a technique that the mastering engineering ought to perform, as is seen fit. It is not something the encoder itself would do, since it would no longer be considered a lossless encode if this took place inside the encoder.

WARNING

One of the key production challenges when working with either Dolby TrueHD or DTS-HD Master Audio is that one cannot know the size or peak data rate of the audio stream until after it has been encoded. This may be an issue because data rate and size is usually needed to properly budget for other types of content on the disc. This can be particularly problematic when audio is not delivered until late in the production process. It is even possible that some audio content cannot be encoded losslessly for a disc because peaks in the data rate of the audio would coincide with peaks in the video and other content, causing the combined streams to exceed the maximum available rate for the disc format. (E.g., an HD DVD disc with 20 Mbps primary video and a losslessly compressed audio stream will have problems if the audio stream ever reaches its specified maximum data rate of 18 Mbps since 20 Mbps + 18 Mbps = 38 Mbps, which is approximately 8 Mbps over the limit for HD DVD.)

Choosing a Secondary Audio Codec

Both HD DVD and Blu-ray support the notion of *secondary audio*, which refers to a second audio stream that may be simultaneously decoded and mixed with the primary audio output. This has significant implications for the design of the player, since the player must include a second audio decoder that can run in parallel to the main audio decoder, as well as have a panner/mixer that can map the audio channels from the secondary stream and mix them with those of the primary stream.

Secondary audio is intended for use in situations in which additional audio content is available that needs to be heard in parallel to the primary audio and has to remain synchronized with it. For example, an audio commentary by an actor or director of a film is usually just one or two audio channels mixed with the feature audio. In DVD, this often required creating an entirely new audio stream that had both the feature audio (usually at a reduction in quality and channels) and commentary audio mixed together. In the new formats, the commentary can be provided as a separate one or two channel secondary audio stream that can be mixed with the multi-channel feature audio of the primary audio stream. The result is better quality and smaller file space requirements.

Besides being included on the disc itself, secondary audio may also come from other sources that the player supports. Both HD DVD and Blu-ray mandate the use of local storage, such as internal hard disks or Flash drives. In addition, the HD DVD-Video specifications and Profile 2 of the BDMV specifications mandate the presence of a network adapter in the player, which would allow content to be downloaded from a remote network server. Secondary audio can be played from either of these locations. For instance, a director's commentary that was not prepared until after a given disc was released could later be made available for that disc via network streaming (assuming the proper disc programming was already in place).

The most common uses for secondary audio will tend to be for access via network or local storage. For network storage, data rate tends to be the chief concern as it is difficult to guarantee high data throughput across the Internet, or even on some local area networks.

Similarly, for audio stored on local data, file size tends to be a primary concern since players in each format are required to provide only minimal amounts of local storage.[2]

In general, when selecting a secondary audio codec, you want to apply many of the same criteria as was used for primary audio. However, there tends to be much more restriction for secondary audio. For example, HD DVD limits the number of channels to two, while Blu-ray supports up to a 5.1 channel configuration. In addition, both formats place significant restrictions on data rate. For HD DVD, data rate limitations are dependent on the codec used, with the maximum rate at 512 kbps for DTS. Blu-ray, on the other hand, restricts secondary audio to no more than 256 kbps, half that of HD DVD, which is ironic since Blu-ray offers more than twice the channel count. This should not come as a surprise, however, since secondary audio is optional in the initial Blu-ray players. One other difference between the two disc formats is that HD DVD-Video offers a much larger selection of codecs in comparison to BDMV, which offers only two.

Table 4.3 HD DVD-Video Codec Support for Secondary/Sub Audio

Codec	Mandatory	Sampling Frequency	Bits per Sample	Max. Data Rate	Max. Channels
Dolby Digital Plus	Yes	48 kHz	compressed	504 kbps	2
DTS-HD[a]	Yes	12/24/48 kHz	compressed	512 kbps	2
mp3	No	24/48 kHz	compressed	320 kbps	2
MPEG-4 HE AAC v2	No	48 kHz	compressed	288 kbps	2
WMA Pro	No	24/48 kHz	16 bits	440 kbps	2

[a]DTS-HD uses a special LBR encoding scheme that consists of only an extension substream for data rates below 192 kbps.

Table 4.4 BDMV Codec Support for Secondary Audio

Codec	Mandatory	Sampling Frequency	Bits per Sample	Max. Data Rate	Max. Channels
Dolby Digital Plus[a,b]	No	48 kHz	compressed	256 kbps	5.1
DTS-HD LBR[a,b]	No	48 kHz	compressed	256 kbps	5.1

[a]Secondary audio streams must include additional mixing metadata.

[b]Support for secondary audio will become mandatory in all players after June 1 2007.

---

[2]HD DVD-Video specifies a minimum persistent storage size of 128 MB, while BDMV specifies a minimum of 256 MB.

4

Dolby Digital Plus (DD+)

Dolby Digital Plus is ideal for network and persistent storage applications of secondary audio. It is able to offer excellent audio reproduction at low data rates, in some cases matching or exceeding DVD quality at the rates which are now allowed by the two disc formats. In addition, playability is guaranteed for any player that supports secondary audio.[3]

> **NOTE**
>
> As a subset of Dolby Digital Plus, you might expect that Dolby Digital would also be supported for secondary audio. However, due to its inability to carry mixing metadata, it is only allowed to be used when substituting main audio in HD DVD-Video, and is prohibited from use as secondary audio in BDMV.

DTS-HD LBR

DTS-HD LBR is a special configuration of DTS-HD in which there is no encoded audio data in the core substream and only Low Bit-Rate Extensions (LBR) in the extension substream. This provides DTS-HD with a means of storing highly compressed audio in a manner that is able to reproduce audio effectively at lower data rates than a standard DTS core would normally be able to provide. Like DD+, DTS-HD LBR will play on any player that supports secondary audio, making it another strong choice for secondary audio applications.

Optional Codecs for Secondary Audio

The HD DVD-Video specifications define several optional audio codecs for secondary audio besides the DD+ and DTS-HD LBR mandates. These codecs were primarily included due to their great popularity and ubiquity in the consumer market, and the effectiveness with which they deliver high quality audio at extremely low data rates. When selecting from among these codecs, it is important to keep in mind that they are optional, which means that there is no guarantee that a given player will be able to play the stream.

Only after more players are released onto the market will we truly have a sense for the fate of these audio codecs in HD DVD-Video. Any one or all of them could become very popular in players, as happened with DTS support in DVD players. Or, they could all but disappear, as did Sony's SDDS® format and the Philips-backed MPEG Audio formats in DVD. In the meantime, note that the BDMV specifications do not define any optional audio codecs for secondary audio.

MPEG-1 Audio Layer-3 (mp3)

Perhaps most well known of all the audio compression codecs, mp3 has been widely used by consumers for recording music and playing it back either via computer, personal media player, or even in their cars. Although somewhat antiquated at this point compared to the other codecs discussed here, mp3 has such a strong following and widespread tool base that it was a natural addition to the HD DVD-Video specifications. Many players will already

---

[3]Secondary audio is optional in profile 1 BD players manufactured prior to June 2007. Subsequent to that date, all profile 1 and profile 2 players will be required to support the feature.

have mp3 audio decoding capabilities for playing back content from CDs and DVDs, so it is likely that a large percentage of players will support mp3 for secondary audio.

MPEG-4 HE AAC v2 (aacPlus v2)

Marketed under the name *aacPlus v2*, MPEG-4 High Efficiency Advanced Audio Coding version 2 is perhaps one of the most advanced of the optional audio codecs. To get a better sense of this claim one must understand its lineage. The codec is based originally on AAC, which was developed as a superior successor to the mp3 codec, offering high quality at lower data rates. AAC was made popular by its use as the default codec in Apple® Computer's iTunes® software and iPod® media players. The *aacPlus v1* codec superceded AAC by adding Spectral Band Replication (SBR), another encoding mechanism that offers significant increases in coding efficiency, allowing quality to be maintained at even lower data rates. Finally, aacPlus v2 was released and standardized as MPEG-4 HE AAC v2, which further adds Parametric Stereo (PS) to provide additional coding efficiency improvements, that are especially relevant at very low data rates. The result is a codec that consistently scores above almost every other available codec in its class at very low data rates such as 32 or 48 kbps. It is even considered capable of effectively reproducing audio at rates as low as 18 kbps though, of course, not without noticeable distortion.

Because of the extreme efficiency of the aacPlus v2 codec, it is likely to gain popularity, especially in wireless applications such as cell phones and other low data rate environments. In time, it could even begin to compete significantly with AAC and mp3 codecs, though its use has not yet been widely established at the time of this writing.

Windows Media Audio Professional (WMAPro)

Like AAC, Windows Media Audio (WMA) was developed by Microsoft Corporation to be an improved successor to the popular mp3 codec. In this case, however, the primary emphasis was licensing revenue, as Microsoft had to pay royalties to include support for mp3 in its Windows® operating system. So, Microsoft set out to design an alternative codec that would be able to compete with mp3 on quality and efficiency for audio applications and launched WMA under the name *Windows Media Audio 7*. After several major updates, *Windows Media Audio 10 Professional* (WMAPro) was released, utilizing a completely different compression technology, which gave WMAPro significantly improved performance over the original WMA technology. Nonetheless, many still hold the opinion that WMAPro is comparable only to the standard AAC coding, which is perhaps just slightly better than mp3. However, given that all recent versions of the Microsoft Windows Media Player and Windows operating systems are likely to support this codec, it is very possible that its popularity will reach a level that allows it to be widely supported in HD DVD players, too.

Video Coding Technologies

Just as with audio, the HD DVD and Blu-ray Disc formats offer several options for video compression. Fortunately, both disc formats support essentially the same options, and their characteristics make it relatively simple to choose from. The three primary codecs from which to choose are MPEG-2, SMPTE VC-1 (based on Microsoft's *Windows Media Video 9*

4

codec), and MPEG-4 AVC. Each has its own strengths and weaknesses such as encoder price, efficiency, quality, and decoder performance, but otherwise all three deliver exactly the same thing — compressed high-definition video.

Differences Between HD DVD and Blu-ray

Although both disc formats support the same three high definition codecs, there are some subtle differences in *how* they are supported. For example, a VC-1 or AVC stream that was created for HD DVD will not work for Blu-ray because each format specifies a different order for some of the data structures in the video streams. In other words, the VC-1 and AVC specifications may not require that the data structures appear in one order or another, but the disc format specifications do. So, a stream that is valid as AVC in general may not be valid for HD DVD or for Blu-ray (or both!).

The two disc formats also vary with regard to the codec features that they support. For example, the BDMV specifications allow film content to be encoded and stored at its original 24 progressive frames per second (fps) rate. Yet, the HD DVD-Video specification allows the compressed video data to be stored as 24 progressive frames per second, but requires additional indicators in the video stream that tell it how to playback on a 60 Hz video system (e.g., at 30 fps), which involves inserting *2:3 pulldown*. In a Blu-ray player, the player figures out where to insert the 2:3 pulldown, to playback on 60 Hz systems. It's a subtle difference — the video stream maintains the same data rate and results in the same size file on disc — but it can prevent you from using a stream that you may have spent a lot of time preparing.

One additional important difference between the video compression support of HD DVD and Blu-ray is that HD DVD mandates players to support *Film Grain Technology* (FGT) when decoding MPEG-4 AVC video. This topic is discussed in more detail at the end of this chapter, but suffice it to say that FGT provides a way in which HD DVD titles can more efficiently encode material that contains film grain, producing a higher quality reproduction at lower data rates.

Choosing a Video Codec

The HD DVD and Blu-ray specifications each support the same three high-definition video codecs: MPEG-2, SMPTE VC-1 and MPEG-4 AVC.[4] When attempting to choose from among these codecs, availability alone will play a significant role since VC-1 and AVC are newer and have fewer tools available, whereas MPEG-2 is well established and has a wide range of hardware and software based solutions. Although the newer codecs are significantly more powerful, allowing them to achieve the same or better picture quality at much lower data rates. Not only does this leave room for higher quality audio streams (e.g., Dolby TrueHD), but also allows for more content to be included on the disc.

One issue that has not been clearly addressed at the launch of these new disc formats is playability on computer systems and settop devices. Generally speaking, most settop devices

---

[4]In fact, HD DVD also supports the MPEG-1 video codec, but only for video at lower resolutions, and only for use in the Interoperable VTS, a special construct for HD DVD-VR (Video Recording).

are specifically designed to handle the maximum data rates and processing requirements defined by the format specifications. However, as home entertainment systems begin to incorporate more personal computer (PC) components, they also begin to take on some of the architectural issues that most PCs face. For instance, the VC-1 and AVC video codecs are significantly more complicated to decode than the older, simpler MPEG-2. They require more memory and more processing power to work. On a traditional settop device with dedicated components that serve specific purposes, this is not a significant concern. However, on PC systems that tend to rely on a single processor (or perhaps two) and a single memory pool, high complexity video decoding has to compete with other processes such as audio decoding, graphics animation, and navigation. As a result, the higher the complexity of the video codec, the greater the chance that lower-powered PCs and PC-like systems will have trouble playing the disc. As such, if the target platform for your content is likely to be a PC, you may find yourself favoring the lower impact MPEG-2 codec, or perhaps VC-1, which has more highly optimized PC implementations available compared to AVC.

MPEG-2

MPEG-2 was created as a higher quality successor to MPEG-1, with which it shares many fundamental concepts. Originally published in 1996 as ISO/IEC 13818-2, the MPEG-2 specifications were later revised and republished as a second edition in the year 2000. Like VC-1 and AVC, MPEG-2 is a *blocked-based* video compression codec, which means that video frames, or *pictures*, are divided into rectangular areas for encoding. For example, MPEG-2 uses 16×16 pixel rectangular areas that are known as a *macroblocks*. Each frame or picture of an MPEG-2 video sequence can be one of three types, an *I*-picture, *P*-picture or *B*-picture. The first, also referred to as an *intra*-coded picture, uses a purely spatial compression method through the use of a *discrete cosine transform* (DCT) and subsequent *quantization* of the resulting DCT coefficients. Suffice it to say, an *intra*-coded picture has all of the data necessary to recreate the entire picture that it represents. The *P*- and *B*-type pictures are referred to as *inter*-coded pictures, which means that they utilize not only spatial compression, but also temporal compression. In other words, *inter*-coded pictures only include the portions of the image that have changed over time, and use *motion compensation* to track portions of the image that have moved or remained stationary. Once the picture data is encoded, a final step referred to as *entropy coding* is applied, which further reduces the size of the data. Figure 4.4, presented at the end of this chapter, shows a generalized block diagram of an MPEG-2 encoder.

Among the chief benefits of MPEG-2 is the fact that it is a very well understood technology with a wide range of tools available that support it. MPEG-2 encoders have been available for nearly a decade and, in that time, the algorithms used to generate the encoded streams have improved dramatically. Making use of the latest information related to the human visual system, MPEG-2 encoders have been able to greatly refine their application of the codec, resulting in higher quality at lower data rates and in smaller file sizes. In further support of this, MPEG-2 used for high-definition tends to appear even better than standard definition partly because the size of the macroblocks relative to screen size is smaller. For instance, a macroblock corresponds to approximately 2.22% of the 720-pixel width of a standard definition DVD picture, but only 0.83% of the 1,920-pixel width of a full high-definition

4

picture. Therefore, coding artifacts (e.g., blockiness) in MPEG-2 become less noticeable in a high definition video stream.

Besides the large number of low-cost, high quality MPEG-2 encoders available, the MPEG-2 codec requires much less processing power for both encoding and decoding than the more advanced VC-1 and AVC codecs. This translates to faster encoding times (often real-time or faster), and less strain on the player during decoding, which is particularly important for PC systems that may already have difficulty supporting the two disc formats.

NOTE   It is interesting to note that today's HD DVD and Blu-ray players have approximately the same or greater processing power as that found in the real-time MPEG-2 video encoders that were commonly used for DVD titles less than five years ago.

Despite all of the benefits, MPEG-2 also comes with several detractions. First, although MPEG-2 can deliver good quality, it is not nearly as sophisticated as the more advanced codecs. As a result, MPEG-2 requires significantly higher data rates, as much as 25% to 50% more, to achieve the same level of quality as today's VC-1 and AVC encoders. Likewise, MPEG-2 is considered to be at its limit with regard to new improvements in compression efficiency. Many believe that there will be no more significant breakthroughs in the improvement of MPEG-2 compression efficiency since all of the available opportunities have already been seized. The two advanced codecs are relatively new and still hold a great deal of promise for new approaches to improving coding efficiency and quality. For example, some believe that over the course of its lifetime, AVC will deliver equal or better picture quality at data rates that are as low as 25% of a comparable MPEG-2 stream.

In addition to having less efficiency for encoding video data, the failings of MPEG-2 tend to be much more apparent than the other two codecs. Both VC-1 and AVC implement a sort of *de-blocking filter*, which smooths the edges of the encoded blocks, therefore obscuring the most common coding artifact, blockiness. Other characteristic problems of the MPEG-2 codec include *intra*-picture pulsing (at a rate of typically once or twice per second), *mosquito noise* (stray pixels that break up high contrast edges, as in text), and *color banding* (contouring of colors caused by aggressive quantization of the color components of a picture).

SMPTE VC-1

SMPTE VC-1 was standardized in 2006 as SMPTE 421M, but was originally created by Microsoft as *Windows Media Video 9* to be a higher quality successor to MPEG-2. One can only imagine that the impetus was to eliminate licensing royalties associated with MPEG-2 that would be required to include support in the Windows operating system. However, after becoming an open standard, over a dozen companies claimed ownership of essential patents used by the codec. This has, unfortunately, resulted in a somewhat clouded licensing situation for VC-1.

On the surface, the structure of a VC-1 encoder does not vary greatly from that of an MPEG-2 encoder, as shown in Figure 4.5 (also presented at the end of this chapter). It still uses a block-based motion compensation and spatial transform scheme like MPEG-2, but adds several innovations that greatly increase the efficiency of the codec, including adaptive block size transform, 16 bit transforms, improved motion compensation, and a simple deblocking filter, among others. The end result is a codec that offers encoding efficiencies similar to MPEG-4 AVC, while also allowing for optimized PC implementations that require only a small amount more processing power than MPEG-2.

When compared to MPEG-2, it is clear that VC-1 has many benefits with regard to offering improved quality at lower data rates. In addition, like MPEG-2, high definition VC-1 also benefits from the smaller block size relative to screen size. However, VC-1 is able to take this benefit even further by supporting adaptive block sizes, which allow more complex areas of an image to be represented by successively smaller blocks, such as, 8×8 or 4×4 pixel blocks.

Although the SMPTE VC-1 standard was released just recently, the core codec has existed for several years as *Windows Media Video 9*, which in turn means that developers have had some time to develop hardware and software tools for the codec. Unfortunately, at the time of this writing, the availability of HD DVD and Blu-ray compliant video encoders was still quite low. On the contrary, with a major powerhouse like Microsoft backing the codec, one can expect that quite a bit of support is being given to help facilitate the development of such tools. Already we have seen highly optimized PC implementations of the decoder, and work has been done on a high quality encoder. Development of AVC encoders, in contrast, has been relatively slow as companies have had to rely more on their own resources. This is one of the reasons why virtually all of the initial launch titles in HD DVD have been created with VC-1 rather than AVC.

Unfortunately, the added sophistication of the codec also adds to the processing time required to encode material with VC-1. Whereas, an MPEG-2 encoder could encode an entire feature film in roughly the time it takes to play it, VC-1 requires dramatically more processing power, and even then may take several times the duration of the video just to encode it. Alternatively, many professional users are creating *render farms* in which multiple computers connected via network are able to share the task of encoding, thus reducing the total amount of time needed to process a sequence.

One very important limitation of VC-1 as it relates to HD production is the fact that the codec does not support embedded source timecode. Both MPEG-2 and AVC offer methods for storing the original source timecode of the content within the stream itself, but VC-1 does not. Although not often used by players, this embedded timecode information can be critical to maintaining proper synchronization between audio, video and subtitle data during the authoring process. Without it, anyone handling the video, whether they be the compressionist, asset technician, or the author, must be careful to keep track of the timecode information, and will always run the risk of a mistake that results in a loss of synchronization. The lack of such a core, professional feature, however, is easily explained given that VC-1 originated as a purely consumer product where timecode is rarely used. MPEG-2 and MPEG-4 AVC have been largely developed for professionals who use timecode on a daily basis.

Characteristic problems in material compressed with the VC-1 codec include heavy noise or overly intense film grain, disintegration of moving foreground elements, occasional macroblocking or *patchiness* (e.g., when the simple de-blocking filter is insufficient), as well as various types of motion artifacts at lower data rates.

4

MPEG-4 AVC

Of the three video codecs supported by HD DVD and Blu-ray, MPEG-4 AVC (also known as *H.264*) is the newest and most sophisticated. Although it follows the same block-based structure as MPEG-2 and VC-1, AVC incorporates a wider range of advanced features, including flexible configuration of blocks and *slices* (contiguous groups of multiple blocks), integer-based DCT algorithms that eliminate reconstruction errors on decode, a strong de-blocking filter, flexible multi-picture inter-picture prediction, greater precision for motion compensation, and *Context-Adaptive Binary Arithmetic Coding* (CABAC) for more efficient entropy coding of syntax elements, among others. Please see Figure 4-6 at the end of this chapter, for a generalized block diagram of an AVC encoder.

Created in concert between the ISO/IEC Moving Pictures Expert Group (MPEG) and the ITU-T Video Coding Experts Group (VCEG), the combined groups were referred to as the Joint Video Team (JVT), and the resulting specification was published both as ITU-T H.264 and ISO/IEC MPEG-4 Part 10 (ISO/IEC 14496-10). It was later extended with the publication of the *Fidelity Range Extensions* (FRExt) amendment, which added new integer transforms for better reproduction of fine textures, as well as support for film grain coding, described in the following section.

Of course, the greatest benefit of AVC is its ability to provide exceptional picture quality at surprisingly low data rates. In most cases, a good AVC encoder can achieve comparable quality at half the typical rate of MPEG-2. However, for some content, such as animated features, it is not unusual to see full HD resolution (1920×1080 pixels) content encoded with excellent quality at rates as low as 4 to 6 Mbps (nearly 25% of the comparable MPEG-2 encoding data rate). Although rare, these cases provide a glimpse at what many believe may be possible for a majority of content over the next several years.

Unfortunately, the extra power and coding efficiency of AVC comes at a steep price in complexity of both the encoder and decoder. Of the three codecs, a good MPEG-4 AVC takes far more time to encode a given video stream than either VC-1 or MPEG-2. This will improve over time but, at present, high-end facilities are experiencing encoding times upwards of 8 times the running length of the video being encoded, and that is on render farms with as many as 64 or more processors. That is not to say that there aren't faster AVC encoders available, or encoders that run on a single machine in a reasonable amount of time, but most of them achieve their speed by taking shortcuts and ignoring many of the features that make AVC such a powerful codec.

Not only does the encoder pay the price of complexity — so does the decoder. For instance, most current PC implementations of AVC decoders require so much processing power to decode high-definition video in realtime that even the latest, fastest multi-core PCs have a difficult time keeping up, much less leaving enough processing power for advanced interactive graphics and the other features of the new disc formats. Fortunately, affordable computer systems continue to increase in speed and power at an astounding rate, such that by the time this book is published, a new generation of PC computer systems will already be available. In addition, decoder manufacturers continue to find new, innovative ways to increase the performance of their software, much as Microsoft has done for VC-1.

Characteristic problems with AVC encoded material tends to include monochromatic patchiness (or blockiness if the de-blocking filter is not active), occasional partial disappearance or fading of foreground elements (such as during fast action), and odd motion artifacts.

Film Grain Technology

Most of our readers have likely experienced film grain either through still photography, such as in dark scenes, or motion pictures. (Remember the last Tarantino flick you saw?) What many do not realize is the importance that film grain plays in the visual narrative of film. Grain has become an important tool in a director's toolbox for helping to engage the audience. It can be used to establish a time period, such as using characteristics of an old film stock from the 1960s or 1970s. Or, it can help build suspense by obscuring one's view and creating a sense of suppressed atmosphere, as used in virtually every horror film to date. The lack of film grain, on the other hand, can be used to create an artificial, plastic-looking world.

Film Grain Technology (FGT) was developed by a group of engineers at Thomson S.A. as a means of achieving higher AVC compression efficiency for content that contains film grain. Film grain itself is a type of noise caused by the chemical emulsion on the film negative. While different film stocks and exposure settings will have consistent film grain characteristics (size, shape, intensity, color, et cetera), the specific layout of the grain "particles" is completely random from one frame to the next, which makes it relatively easy for the human eye to avoid being distracted by it except under the most extreme circumstances. As such, it provides a mechanism for reaching the viewer's subconscious in a way that many filmmakers find to be quite effective. Unfortunately, this randomness is also extremely difficult to encode using predictive coding technologies like AVC, often resulting in either a total loss of grain (causing flat, plastic-looking images) or, even worse, "clumps" of colored pixels that appear to stick to the screen for brief periods of time. Film Grain Technology uses the consistent characteristics of film grain to provide an alternative approach to coding the information, maintaining the look of the original film source while preserving the randomness and characteristics of the grain itself. Thus helping to reproduce the original presentation intended by the filmmakers.

FGT works by analyzing the characteristics of the grain in the source material, identifying the size, shape, intensity and color components for each group of frames in the video sequence. After optionally applying film grain filters and then encoding the picture data, it compares the source to the subsequently decoded pictures to determine what amount of film grain was lost in the process. Finally, a data structure is created that represents the film grain characteristics that must be reapplied to the decoded output, to make it resemble the original source. In MPEG-4 AVC encoding, this data structure is stored as a *Supplemental Enhancement Information* (SEI) message, one per frame. During decode, an HD DVD player reads the Film Grain SEI message and generates an additional layer of simulated film grain over the top of the decoded picture data, the result of which is an image that more accurately reproduces the look of the original content.

Of course, Film Grain Technology is generally applied only to content in cases where the process is able to improve overall picture quality. The first way in which this is achieved is through the use of FGT combined with film grain filtering. If FGT is used to wholly recreate

4

the film grain appearance in the decoded image, then the video encoder can be fed "clean" images with little or no grain in them. This results in far more efficient video compression, allowing the encoder to focus on the details of the pictures rather than struggling to recreate the randomness of grain. FGT has the added benefit of helping to hide common coding artifacts, such as blockiness, because the frame by frame randomness of the grain helps to break up the visual appearance of edges on blocks, resulting in a much more natural image. Finally, because film grain is usually very fine, and Film Grain Technology is able to fully reproduce that level of fine granularity, the experience of high definition can be further enhanced because, even though the content being filmed may not have any fine details, the grain itself does, and the HD DVD player is able to fully reproduce it.

Unfortunately, at the the time of this writing, no released HD DVD movie titles have employed the use of AVC video compression with Film Grain Technology, so viewers have yet to see the benefits in mainstream content. However, as AVC encoders become more widely available and the tools and procedures for adding FGT improve, customers of the format will be able to experience those benefits.

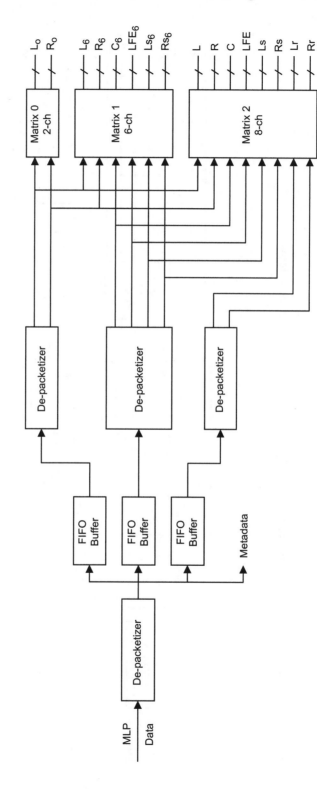

Figure 4.1 Block Diagram of a Dolby TrueHD Decoder

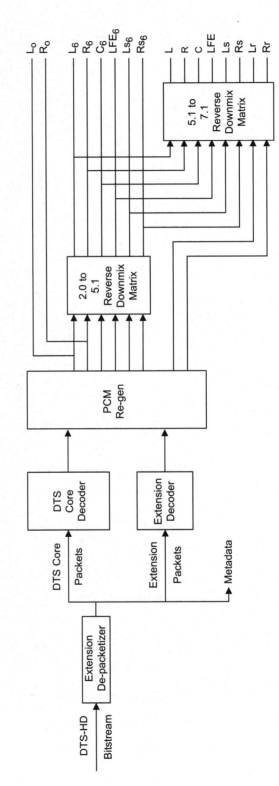

Figure 4.2 Block Diagram of a DTS-HD Lossless Decoder

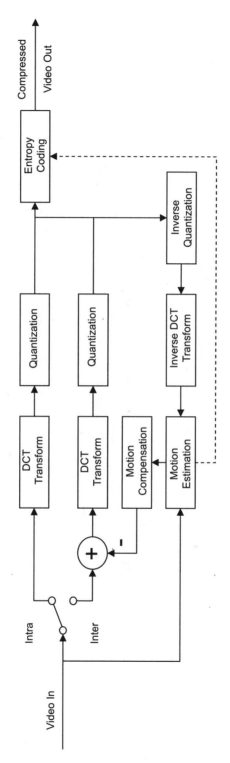

Figure 4.3 Generalized Block Diagram of an MPEG-2 Encoder

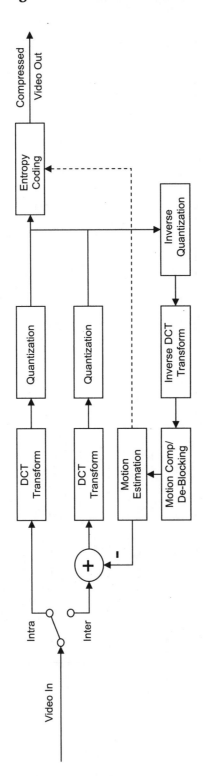

Figure 4.4 Generalized Block Diagram of a VC-1 Encoder

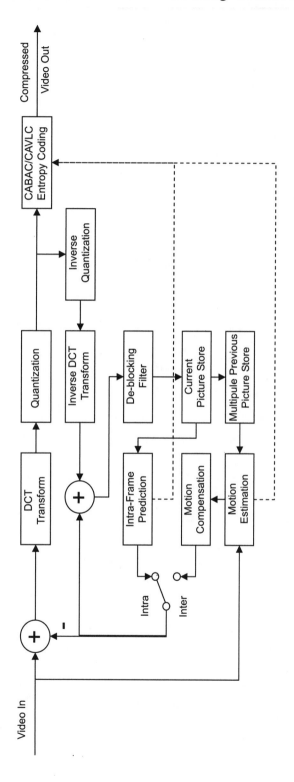

Figure 4.5 Generalized Block Diagram of an AVC Encoder

Chapter 5
Persistent Storage

A great new capability of next generation dics players is persistent storage, enabling many new applications even at the minimum of 64 KB for Blu-ray or 128 MB for HD DVD. Persistent storage is principally controlled through advanced applications — Advanced Content for HD DVD and BD-J for Blu-ray Disc™. Standard Content titles are not generally allowed to access the persistent storage, although in Blu-ray's BDMV, you can take advantage of persistent storage via a virtual file system that has been updated with an advanced application written in BD-J.

The closest DVD came to persistent storage was in the use of non-volatile random access memory or NV-RAM. Only available on a very few select players, such as the Philips Professional DVD Player 170 and the Pioneer DVD-V7200 and V7400, it was as large as 512 Kbytes (256 registers of 16 bits each).

Persistent storage allows for web-like capabilities. Not only can the disc experience be programmed in real programming languages, but information can be stored on the player, enhancing the user experience, which can change and evolve even after the disc has shipped. This chapter focuses on considerations for implementing persistent storage on the next generation disc players:

- Technical description: size, speed, format fit and relation to disc.
- Basic capabilities and potential uses.
- Design and planning to optimize capability
- Building the experience: implementation checklists
    - Data security using Globally Unique Identifiers (GUIDs) and unique IDs
    - Adding AACS
    - Testing

Technically Speaking

Also referred to as non-volatile storage or local storage, persistent storage requires the use of the advanced application modes of HD DVD and Blu-ray. Also, discs using persistent storage must use AACS to prevent malicious applications, unauthorized changes or player damage. AACS is further described in Chapter 17. Table 5.1 details the specifics of the amount of available memory for each format and some of the performance characteristics to expect.

Table 5.1 Comparison of Blu-ray and HD DVD Persistent Storage Capabilities

Format	Access Method	Minimum Size	Sustained Transfer Rates	Data Cache	Maximum number of concurrent open files	Player Version/ Profile
Blu-ray	BUDA/ VFS and URI[a]	256 Mbytes[b]	Level 1: >= 16 Mbps sustained read performance: required Level 2: >= 32 Mbps SRP: optional Level 3: >= 48 Mbps : optional	42 Mbytes	128	Profile 1 player (pre-June 2007)
Blu-ray	BUDA/ VFS and URI[a]	1 Gbytes[b]	Level 1: >= 16 Mbps SRP: required Level 2: >= 32 Mbps SRP: optional Level 3: >= 48 Mbps SRP: optional	64.5 Mbytes	128	Profile 2 player (after June 1, 2007)
HD DVD	URI	128 Mbytes[c]	33Mbps sustained	64 Mbytes At 7.5 Mbps	Not specified	All players

[a]BUDA/VFS is the Binding Unit Data Area which is where the Virtual File System is staged.

[b]Can be either built-in or on a removable device (that may not be plugged in). So, effectively, a player may have only 64 Kbytes of persistent storage if the user has removed or disconnected the persistent storage device and the manufacturer chose not to implement a fixed built-in persistent storage. Also, the user can remove the removable device at any time.

[c]This is the minimum fixed storage capacity. It is mandatory for all HD DVD players. Players therefore will have a minimum of 128 Mbytes of storage and can optionally have more.

Access methods for Blu-ray include both a file reference approach, known as URI (Uniform Resource Identifier), and a Virtual File System (VFS), explained later in this chapter. HD DVD uses only the more flexible URI approach, requiring file-use planning for both disc and persistent storage devices.

In addition to the minimum of 64 KB of fixed memory, Blu-ray players must have 256 MB (standard player) or 1 GB (network player) of persistent storage. The odd thing about the requirement though is that the additional persistent storage can be on removable devices instead of built in, leaving only 64 KB available inside a Blu-ray player.

Both formats allow persistent storage on removable USB devices such as HDDs or memory cards. For HD DVD, network attached storage (NAS) can also be used. With a terabyte or more of storage available in persistent storage, the next generation players could become home entertainment centers serving user preferred media clips.

The data transfer rates and the size of the data cache become important considerations when setting design guidelines for a title. Because the memory is finite in size and speed, the title design must recognize how much data will be loaded into memory and how long that operation will take. Failure to pay attention to these limitations results in titles that take a very long time to load. It is very possible to have 60-90 second load times. Even a 15 second load time gives the user the perception that something is broken.

Potential Uses

Possible applications include storing user preferences, playlists, game scores, bookmarks, online transactions and configuration information, as well as, downloaded material such as trailers, bonus materials or new material and updates to the application logic (e.g., bug fixes). The user may save the disc state, returning to the same point at the next viewing or interaction. With removable persistent storage devices such as a USB Flash drive, information can be moved from player to player. Watching a movie could be finished at home after starting to watch it at a friend's house. Students could track their progress through a course on a Flash drive and then submit the Flash drive for grading or with network access, transmit their course work on-line.

Additional uses of persistent storage include —

- additional purchased content
- status of content and copyright notice viewing
- user preferences between discs for language, subtitles, etc.

Persistent Storage Implementations

Implementation of persistent storage basically handles how the files are written and read and how those files are isolated from the files of other discs or content providers. Of course, HD DVD uses a different model for accessing persistent storage than Blu-ray.

The speed at which data can be written and read depends on the performance characteristics of the persistent storage media, whether it is removable or an optical disc. This difference can be used to create a more responsive application. While the disc transfer rate is 30 to 48 Mbps with an average latency[1] of about 1/8 second (125 ms), persistent storage has a minimum transfer rate of 16 to 48 Mbps with latency of less than 1 ms (see table 5.1). Persistent storage devices such as a typical USB flash drive can have transfer rates of 24 to 80 Mbps with a latency of less than 1 ms. HDDs via USB have sustained transfer rates of at least 160 Mbps with latencies often less than 10 ms. Reading from and writing to the persistent storage device can be two to four times faster than optical disc with latency times that are ten to 125 times less. An application can be designed to use persistent storage as intermediate storage so that there are not long pauses in the disc title interactivity.

HD DVD

HD DVD allows for the retrieval of audio/video stream files from local storage, a network, and of course, the disc. HD DVD also has a data cache. This data cache is volatile and only accessible with advanced navigation. Files can be moved between volatile memory, the data cache, and the non-volatile memory of persistent storage.

---

[1]Latency is the time it takes for a device to process a command and begin to read or write data. For a hard disk drive this includes the time to process the command, position the head over the read/write sector, and then wait for the data to be positioned for access. Hard disks typically show 5 to 13 ms latency, while optical disks such as CD, DVD, BD and HD DVD average 125 to 250 ms latency.

5

The speed that data can be transferred between memory devices will vary from one player implementation to another. The data cache has a minimum transfer rate (33 Mb/sec) that closely resembles the rate at which data can be read from the disc (36 Mb/sec).

> **NOTE** **Why transfer rates are important**
>
> The user's actions should always result in a visible or audible response. If data is streamed from a network server, the instantaneous response can be simulated by flashing a "Loading" message while data is transferred to a cache. Some of the data could even be pre-loaded so that the experience has a more instantaneous feel.

The player is responsible for providing a storage management mechanism. This mechanism is usually implemented in players using a Persistent Storage Manager function available in the player setup menus.

HD DVD comes with two types of persistent storage — fixed (required, minimum of 128 MB) and additional or removable (optional, variable size), such as USB Memory or USB hard disk drive or a Memory Card. Network attached storage may also be used as additional persistent storage.

Persistent storage devices are referenced via URI. The data in persistent storage are referenced by provider and content id or GUID (Globally Unique IDentifiers). A GUID is created by a GUID generation program. Several programs can be found by using the keywords "GUID generation" on internet search engines.

> **TIP** Specification RFC 4122 defines a Uniform Resource Name namespace for UUIDs (Universally Unique IDentifier), also known as GUIDs (Globally Unique IDentifier). A UUID is 128 bits long, and can guarantee uniqueness across space and time.

The URI explicitly defines a fully qualified filename and path. Addresses for files are based on the data source where the data source parameter can be either 'Disc', 'p-Storage', 'Network' or 'Filecache'. The File Cache manager controls where data sources are loaded and released via the Resource Management API. Data source identifiers are used at the following levels —

- **Device ID** Each persistent storage device is assigned a Device ID by the player when a device is read from the Device Information file. This is the value in the file for the Device_id key. The value is a GUID.

- **Provider ID** Another GUID ensures that each provider is held unique, and keeps the information stored on the persistent storage devices separate. This means that one content provider cannot access the information of another content provider.

- **Content ID** Each disc can have only one Content ID. An advanced content application can access content from the same Provider. In other words, an advanced application, if it knows other Content IDs, can access information from those directories.

The values for these data sources are defined in the file DISCID.DAT which is in the root of the ADV_OBJ directory. The presence of this file lets the player know that there is an advanced content structure on the disc.

Table 5.2 DISCID.DAT Configuration File

Order	Field	Length (bytes)	Contents	Description
1	CONFIG_ID	12	Identifier of this configuration file	Describes "HDDVD-V_CONF" of disc using any ISO-8859-1 characters
2	DISC_ID	16	Disc ID	to read 1b unless network access used
3	PROVIDER_ID	16	Directory of the Provider ID	GUID of the content provider, or to read 1b if persistent storage is not used.
4	CONTENT_ID	16	Directory of the Content ID	GUID of content id for this disc or '1b' if persistent storage is not used
5	SEARCH_FLG	1	Search Flag for disc start up	'0b' to indicate search for Playlist file on disc and in persistent storage or '1b' if persistent storage is not used.
6	Reserved	67	Reserved	Not currently used

Any of the advanced content files can be stored in persistent storage. Advanced navigation can copy files between data sources such as file cache, persistent storage and network. The Secondary Video Player can read Secondary Video Sets from persistent storage.

Network server is an optional data source for advanced content playback. Obviously, the player needs to have access to the network server to play this content.

An example of a directory structure for a persistent storage device is shown in Figure 5.1. In the root of the device is a directory named 'HD DVD'. If the device were loaded on a computer, this is the directory that is seen at the root. There is a common directory that can be used to store common files that are accessible by all discs. Each provider has a directory named by a unique ID (i.e., a GUID) that they have chosen. For an advanced application, only the directory level down from the provider ID is exposed. Each disc has its own GUID, the Content ID. An advanced application can access the files in the Provider ID directory and the Content ID directories. This means that an application can access files from another disc's persistent storage but the reference must be to the GUID name of that other disc. The Information Files named "info.txt" are used to store information that are used by the Persistent Storage Management Menu of the player. Advanced Applications can read and write these files. The keys in the file define the image icon file and explanation text for each language and the device_id GUID.

Figure 5.1 An example of Directory Structure of Persistent Storage Device

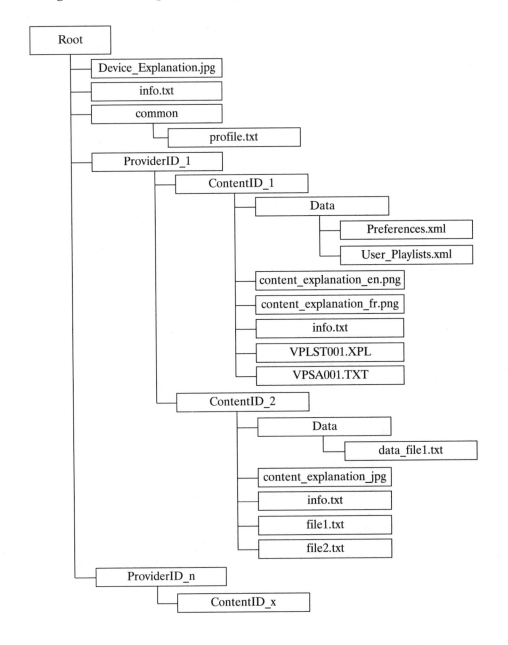

<table>
</table>

WARNING	AACS is required for accessing persistent storage. One of the gotcha's is that whenever accessing files on persistent storage, remember to use the same Title Key and Content_ID that was used to initially create and access them. Otherwise, a fatal player error can result and the player will stop playing the disc. If you've done this wrong, it only shows up when you are testing the replicated disc with AACS.

Figure 5.2 shows the data access model for HD DVD. Persistent storage can be network access, fixed storage or removable storage. The Data Access Manager reads and passes on secondary video streams (the S-EVOB files). For HD DVD, primary video cannot be retrieved from persistent storage.

Figure 5.2 HD DVD Data Access Model

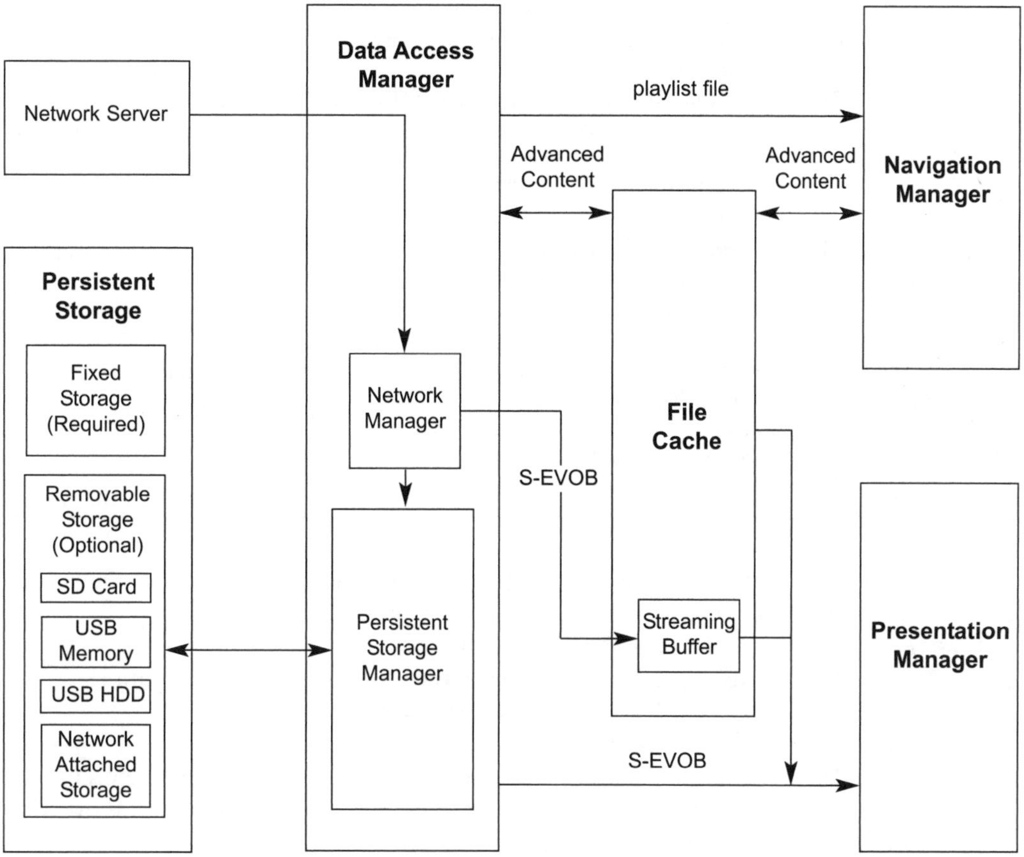

5

Advanced applications can find all removable storage devices connected to the player and are able to reference those devices by a base path, such as usb1. Table 5.3 provides examples of the URI references to persistent storage.

Table 5.3 HD DVD Local Data Source Areas

Device	Examples	Description
HD DVD disc	`file:///dvddisc/`	Disc
File cache	`file:///filecache/'file Name'`	Temporary (volatile) storage area managed by the API

Required Persistent Storage	`file:///required/'file Name'`
	`file:///required/'Content_ID'/'filename'`
	`file:///required/'Content_ID'/'direct oryname'/.../'file name'`
		Content provider area in fixed Storage

Additional Persistent Storage	`file:///additional/'Base Path'/'file name'`
	`file:///additional/'Base Path'/'Content ID/'file name'`
	`file:///additional/'Base Path'/'Content ID'/'directory name'/. . ./'file name'`
		Content provider area in removable storage BasePath includes the device (e.g. 'usb1') and the Device_ID.

Common Area on Required Persistent Storage	`file:///common/required/'file name'` `file:///common/required/'directoryname'/.../'filename'`	Common area in fixed storage
Common Area on Additional Persistent Storage	`file:///common/additional/'Base Path'/'file name'` `file:///common/additional/'Base Path'/'directory name'/. . ./'file name'`	Common area in removable storage

The player translates the file names to the full path. For example, for a file named datafile.txt on a USB device, the path that would be expressed as —

```
file:///additional/'Base Path'/datafile.txt
```

where 'Base Path' is [device_name] "Device ID GUID". An example of a full path is —

```
file:///additional/usb1/3c05d6a0-5d3f-11db-b0de-
     0800200c9a66/datafile.txt
```

The advanced content program as defined by the HDi XML and JS files will reference the persistent storage API using the above conventions. The actual code is beyond this overview of the capabilities of the HD DVD persistent storage model.

Blu-ray: Binding Unit Data Area (BUDA) and the Virtual File System (VFS)

The Blu-ray Disc Association's specification allows player manufacturers to include as little as 64 KB of persistent storage, unbelievably small compared with the size required by high definition media assets and sophisticated programming, and with the low cost of memory in today's market.

As players, content providers and authoring facilities implement this feature experimentally, the specification remains in flux. Currently, Blu-ray's persistent storage mechanism is only described in the specification and implemented in limited ways for test discs.

Blu-ray speaks of local storage, defining two types, the Application Data Area (ADA) and the Binding Unit Data Area (BUDA). The Application Data Area is located within the required persistent storage area of the player, but the BUDA, where media content is stored, may be located in optional local storage devices. The BUDA may include audio-visual files, subtitles, Java applications and databases. Local storage files are accessed using the Java IO package in BD-J mode.

While Blu-ray and HD DVD both use an explicit file path to access content from persistent storage, Blu-ray also uses a second approach called the Virtual File System (VFS), allowing a combination of assets from local storage, both persistent and disc, to be presented as one file system. The VFS combination, located in persistent storage, is used instead of the original.

Figure 5.4 illustrates how the merged file system is presented to the player, re-mapped using a manifest file. The manifest declares which files are to be used from the persistent storage and which are to be retrieved from the disc. The updated content is packaged in the BUDA. Once the content and/or code is staged in the BUDA, updating the VFS maps the new content/program over existing files, presenting new trailers, subtitles, content and program functionality. Easy to implement by specifying assets, the updated VFS can also be available to the HDMV mode.

Figure 5.3 Placement of Binding Unit-Root Certificate

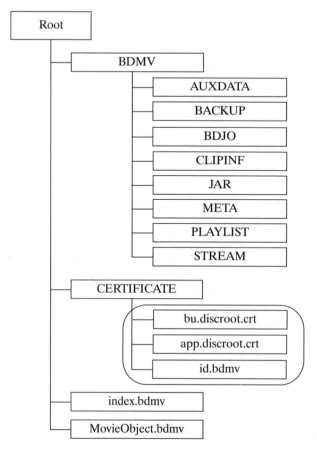

The BD-J application can request VFS updates, revising binding units and binding targets on the local storage (see Figure 5.4). The binding unit manifest file (.bumf) and signature file are used to map the VFS. Updating the Virtual Package is done by moving the new .bumf to the organization_producer/content_ID directory and renaming it to .bumf. Old versions of the .bumf, if they exist, are replaced with this new one.

Players manufactured before June 1, 2007 are not required to have any BUDA. After June 1, 2007, Profile 1 players have a minimum requirement of 256×2^{20} or 268,435,456 bytes and Profile 2 players have a minimum requirement of 2^{30} or 1,073,741,824 bytes. The manufacturer may choose whether to build the storage into the player or support removable storage with the above minimum requirement. Each BD player will have an unpredictable amount of storage beyond the 64 KB minimum.

It remains to be seen how player manufacturers will respond to disc author demand for 256 MB - 1 GB of storage, instead of requiring the user to provide it. New disc packaging will need to indicate minimum system requirements, and code testing this condition will become common practice, warning the user.

Figure 5.4 Example of Virtual Package

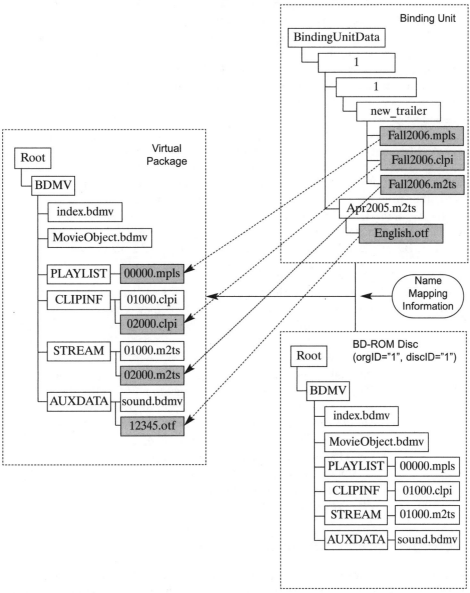

Practically, the required minimum persistent storage is only big enough for settings, preferences, state, high scores, etc. So the author must plan accordingly. BD-J also includes access to local and system storage using the Java™ IO package. The path to files is defined by the GEM (Globally Executable MHP[2]) specification.

For security reasons, the JAR (Java Archive) files must be signed and authenticated.

---

[2]MHP is the acronym for Multimedia Home Platform. Therefore, the full definition of GEM is Globally Executable Multimedia Home Platform.

5

Access to the local and system storage is shared by all discs that the player may play. Other discs may leave data in local storage so the amount available to any single disc is uncertain. In other words, there could be zero bytes available.

BD-J has a sandbox mode that limits application access to storage. To go outside the sandbox requires that JAR files be authenticated via a signing process. Some of the authoring tools handle the signing as part of adding the JAR files to the HDMV mode application. Authentication is required for reading and writing to local and system storage, or using the network connection to connect only to defined servers.

Read and write access can be defined per file using a disc root certificate associated with each studio, and leaf certificates for each producer. By default, data from one studio or producer cannot be seen from another's application, as each uses a different root certificate or Organization ID — each producer's ID is unique among a set of certificates created by the same studio. Each producer signs his own applications, and can use credentials to grant access to other producers such as for collaboration or peer review.

Disc ID (Content_ID) and Organization ID are stored in the file id.bdmv in the CERTIFICATE directory. The Disc ID is 128 bits and may be created by a GUID. The Organization ID, assigned by the Blu-ray Disc Association, is 32 bits (4 bytes).

TIP

Now you see it, now you don't —

- ▪ The Binding Unit Data Area may not be available as it is located on the local storage devices. Users may remove the device at any time. Application authors should include a mechanism for handling such an event cleanly, and for dealing with read or write exceptions.

- ▪ Minimize per disc data storage in the Application Data Area. The size of the ADA in local storage on a BD-ROM player may be limited. Per disc, the most important persistent data files should have a high priority, long lifetime and small size (4 KB or less in an area that may be 64 KB or less).

- ▪ Avoid additional directories in the Application Data Area, The number of files and directories supported may be limited. Avoid creating directories in the ADA, beyond those automatically created by BD-ROM players.

Conclusion and Comments

As of this writing, these persistent storage capabilities have been fully implemented for HD DVD players and most discs on the market use persistent storage for their bookmark feature. For Blu-ray players, the Application Data Area has been implemented on most players, but the Binding Unit Data Area and therefore the VFS feature has not been implemented. As early adopters and other developers experiment with implementation, you can expect to see better support in the authoring tools and more program guidelines and tips. Information, cooperation and implementation resources will be available from the Blu-ray Association and the DVD Forum as well as from the authoring tool vendors. In time, implementation details will stabilize and become less visible.

Chapter 6
Network/Streaming
Technologies

Next generation DVD is distinguished from standard definition DVD by network and streaming features. On a settop box in the living room, with network connectivity, it is no longer necessary to have a dedicated computer in order to create a highly interactive, Internet-connected experience. Adding the ubiquitous presence of high bandwidth network access means that the consumer can now have a seamless integration of content from both a network and a disc.

> **NOTE** Network access can only be programmed using the advanced application modes of Blu-ray and HD DVD. The player must be connected to a network for this feature to work.

As of the beginning of 2007, no discs have shipped in either format using network capability. The first networkable releases are expected during Q1/Q2 of 2007 and will just scratch the surface of the feature's possibilities.

Challenges include —

■ Disc format specifications need to clarify details regarding the implementation of this capability.

■ Players currently released and shipped are not consistently equipped for this capability. The network port is equipped only to download firmware updates in many cases. Blu-ray player manufacturers are not required to support networking until Profile 2 players slated for the end of Q2 2007 and later.

■ None of the current authoring tools support this capability.

■ AACS copy protection complicates any implementation of networking.

■ The typical user setup (wireless, DHCP, firewalls) and connection speed are unknown.

The good news for endusers is that the potential for network connectivity is so great that everyone in the industry is motivated to work together to get this feature to work. The studios, the player manufacturers, the compression and authoring houses are all working closely together to make these first titles. And, there is a strong likelihood that the first of these networkable titles will cause the specifications to be refined and clarified, that the hardware players will have their firmware updated and that new authoring techniques and tools will be developed to create a robust and pleasantly delightful user experience.

As an author and title developer, with the network/streaming feature, now you can do all those things that you wanted to do with DVD but had to either explain could not be done or

6

you had to resort to trying to do them with Adobe's Director or InterActual® Player or Sonic's eDVD® or Apple's dvd@ccess™. Now there is a platform that is consistent without all of the same hassles as before. Applications of this network capability can include —

- Trailers — streamed and downloaded
- Purchased content
- Live events
- Messaging
- User registration
- Promotions and special offers
- Shared playlists
- Rate discs
- Get recommendations
- Online storage and retrieval of preferences
- Online quizzes
- Unlocking hidden content on disc
- Downloads to persistent storage of audio, video, subtitles and data
- Updating disc functionality (content, updates)
- Multi-player games

By being connected, the user experience may be kept fresh and alive. New content can be added such as subtitles in another language or a new commentary track. The disc can be updated so that new features can be added, such as new interactivity to search the content for scenes with a particular actor or a new game. The code for a disc can be updated to fix or enhance playback. Or updates may be made to enhance the code that allow for new features on newer players introduced after the disc was originally produced. The potential uses are limited by your imagination and, of course, budgets of time and money.

Though, here's another reality check on the capability. Not all players are network capable and not all users will be connected to the network and there is no way to know how fast the network access will be for a particular user. Developers will need to plan for these situations in their design and to test for them in the disc implementation. *"Server not found"* or *"Error 404: page not found"* are not living room friendly experiences. A *"Loading"* message that plays for more than a few seconds will not win awards either.

Figures 6.1 and 6.2 show the basic flow of data for a network connected experience. There are differences in the terminology and the cache and buffer implementations for each format. The basic architecture of each format accommodates common functionality for the user experience. In essence, you have the options of using downloaded content and data or streaming or files that can be accessed (opened for read/write access) from a server.

Figure 6.1 HD DVD Data Flow

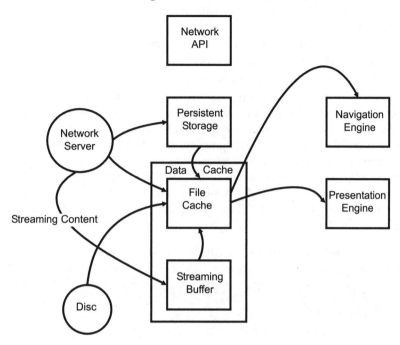

Figure 6.2 Blu-ray Disc Data Flow

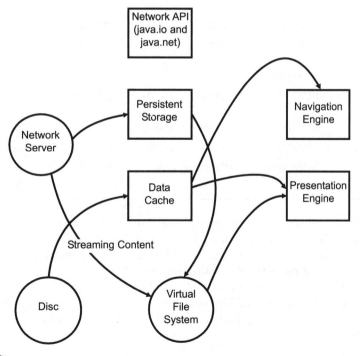

Looking past the hype of what can be done with a connected next-generation disc, some practical limitations include —

- How much can be stored locally in persistent storage?
- How long does it take to download an asset?
- What are the types of media that can be used?

Table 6.1 lists some typical file sizes for types of media assets and the length that will fit in the various sizes of persistent storage. For example, at 30 Mbps (megabits per second), a high-definition video and audio stream of 128 MB can be stored in persistent storage in 34 seconds. 8 Mbps (Megabits per second) is equivalent to 1 MBps (Megabytes per second). Considering that trailers are typically 90 to 150 seconds, this means that only one very short teaser will fit in 128 MB of persistent storage. If the player had more storage, for example 1 GB, then there would still only be room for 4 minutes and 27 seconds. Perhaps two trailers could be stored, but that still does not fulfill the vision that is implied by the promise of trailers and new scenes or alternate endings.

Table 6.1 Buffer Capacity for Various Assets

Type	Buffer Size (bytes) Frame size (bits/second)	64,000	64,000,000	128,000,000	256,000,000	1,000,000,000
Audio (mp3 quality)	64,000	0:00:08	2:13:20	4:26:40	8:53:20	34:43:20
Audio (DVD AC3 Audio)	192,000	0:00:03	0:44:27	1:28:53	2:57:47	11:34:27
Audio (PCM)	1,536,000	0:00:00	0:05:33	0:11:07	0:22:13	1:26:48
VHS Quality 720x480	3,000,000	0:00:00	0:02:51	0:05:41	0:11:23	0:44:27
DVD Video	4,500,000	0:00:00	0:01:54	0:03:48	0:07:35	0:29:38
Cable/Satellite HD Video	10,000,000	0:00:00	0:00:51	0:01:42	0:03:25	0:13:20
ATSC HD Video	18,000,000	0:00:00	0:00:28	0:00:57	0:01:54	0:07:24
Feature HD Video	20,000,000	0:00:00	0:00:26	0:00:51	0:01:42	0:06:40
High Quality HD Video	30,000,000	0:00:00	0:00:17	0:00:34	0:01:08	0:04:27

Looking at how long it takes to download assets, the reality of downloads is even more humbling, as reflected in Table 6.2. For a 34 second trailer, at typical DSL or cable modem speeds, the download time is over 17 minutes when using full network bandwidth, if available. Practically, it could take 50-100% more time, which means that it could take about 30 minutes to download the 34 second trailer. A straightforward implementation would be to download the asset and then play it. This could mean, though, a delay of almost 30 minutes to get the 34 second experience. Although this is the common approach for a web page, this is not a very DVD-like experience.

If you can count on a medium DSL or cable speed of about three Mbits/second or faster then you could use streaming video and audio, not for a high definition video stream, but for a VHS-quality stream of video. This streaming mode could allow for the experience of many trailers with minimal waiting time for the user. Currently, in the US, 60% of the homes now have high speed internet access but only at one Mbit/second. To be practical for even VHS-quality streaming the user would need to have a minimum of three Mbps. Until network connection speed increases, even VHS-quality video streaming is impractical.

Table 6.2 Time to Fill Buffer at Various Connection Speeds

Connection Type/Speed in bits/second	Buffer Size (bytes)	64,000	64,000,000	128,000,000	256,000,000	1,000,000,000
High Speed Modem	56,000	0:00:09	2:32:23	5:04:46	10:09:31	39:40:57
Upload ADSL	128,000	0:00:04	1:06:40	2:13:20	4:26:40	17:21:40
Upload Cable	256,000	0:00:02	0:33:20	1:06:40	2:13:20	8:40:50
DSL/Cable	1,000,000	0:00:01	0:08:32	0:17:04	0:34:08	2:13:20
Medium Speed DSL/Cable	3,000,000	0:00:00	0:02:51	0:05:41	0:11:23	0:44:27
High Speed DSL/Cable	5,000,000	0:00:00	0:01:42	0:03:25	0:06:50	0:26:40

Creativity Requirement

The practical limitations of the network connection — how big the assets can be and the time to download these assets — mean that there is room for lots of creativity. If you were to take the one Mbits/second bandwidth and use it for a multichannel audio stream with a slide show or for a low frame rate animation, the user experience could still be spectacular even when using a lower bandwidth connection. Another option would be to use a smaller frame size for the video, say 360 × 240 pixels, with flashing lights or text and a rich audio track using, for example, 448 kbps 5.1 AC3. This could create a very compelling audio and video experience yet fit within the bandwidth constraint.

Another creative example takes advantage of the fact that even though a download might take one to two hours that is the time it might take to watch the main feature. New, extra content could be downloaded in the background without making the user watch a *"Please Wait"* screen. The user may also have a large drive connected as persistent storage. It is not unthinkable to foresee a user having more than a terabyte connected to their next generation player as extended persistent storage. In this scenario, the limitation for extra downloaded content would be time not space. During the playing time of the feature, you could download another episode in full HD. Being creative with the limitations will result in many very exciting and intriguing user opportunities.

6

Design and Planning

Your design and planning can be best managed by keeping the following parameters in mind —

- Data access method(s)
  - Downloading
  - Streaming
  - File server access
- Security method and extent
  - Signed applications, certificates
  - AACS
  - No encapsulation or signing
- Performance of connection and player
  - Minimum system requirements
  - Player capability
  - Network access and connection
  - Network connection speed
  - Usability for streaming

Implementation

Network access is an extension of the persistent storage capabilities of Blu-ray and HD DVD. As such, for both formats, you can use the network IO calls to access network files — java.net for BD and network API for HD DVD. This is useful for accessing status files, databases, and multimedia assets. The Virtual File System available to Blu-ray is a very powerful way of integrating online media with disc media in a seamless way. Updating playlists that reference network content is a way to implement this same functionality for HD DVD.

As shown in Figures 6.1 and 6.2, data in the form of either multimedia assets or database records can be retrieved either from the network or from a data cache to persistent storage. The workflow for creating and adding the necessary file signing and/or AACS encoding is diagrammed in Figure 6.3.

Basically, the files are prepared as they would be for inclusion on the disc. The advanced application is written to access the files from the network and then download them to local storage. The files can be media or data files. File access can be read/write when accessing files such as a database or preferences file, et cetera. Both HD DVD and Blu-ray support streaming, but in different ways. With HD DVD, you declare a streaming file by using the playlist XML (see the example below). Blu-ray handles streaming files using either Java code via the java.net package or by defining "progressive" files in the binding unit manifest file (i.e.'<disc_id>'.bumf) and initiating access when the Virtual File System is updated through the binding unit data area. Java.io can also let you access continuous streams that aren't files at all. This is pretty common - sometimes you'll continuously monitor something that streams data to you, like a live feed, or a sensor.

HD DVD can initiate network access either through the playlist or via APIs in the script. The playlist plays video using a Secondary Video Set. This Secondary Video Set can contain data from disc, file cache, or from persistent storage. The Secondary Video Set can be played from the Streaming Buffer when downloading data from a server.

Figure 6.3 Workflow for Network Access

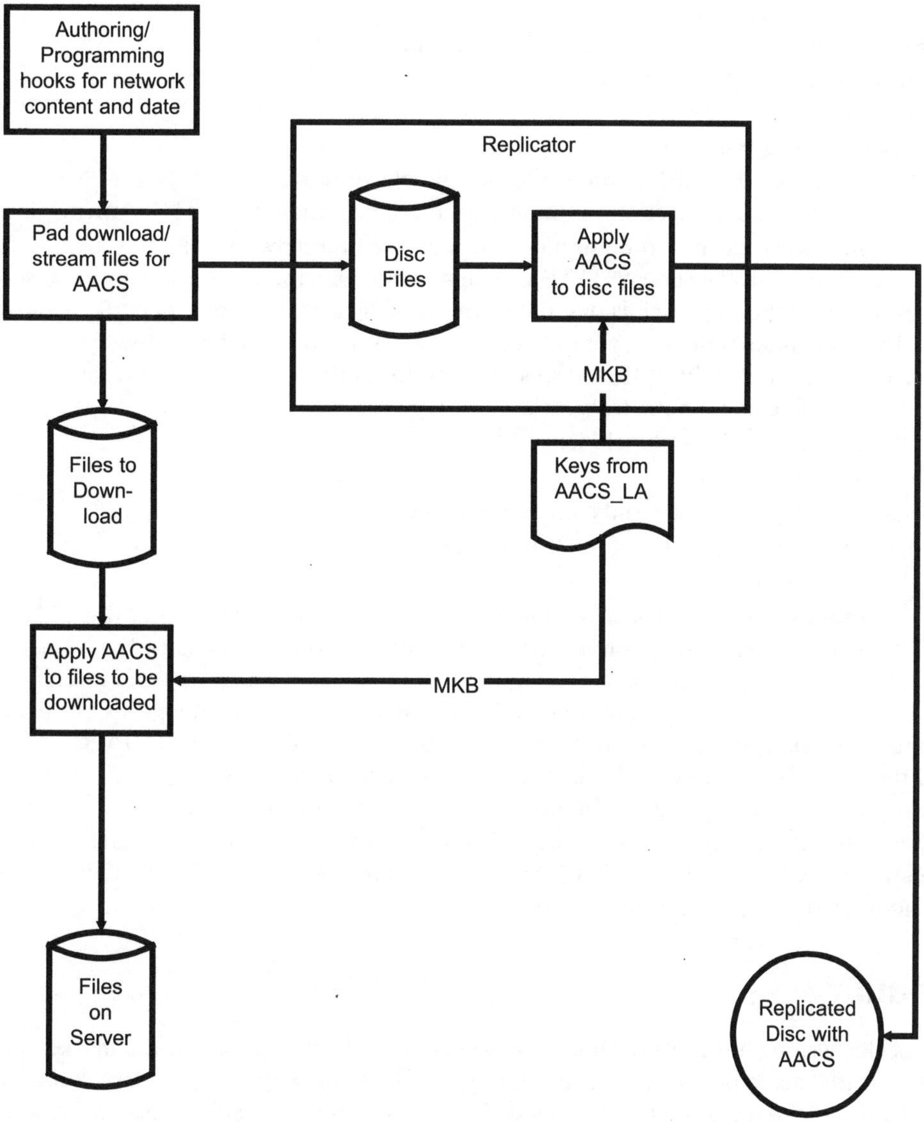

```
┌─────────────────┐
│  Authoring/      │
│  Programming     │
│  hooks for       │                    ┌────────────────────────────────────┐
│  network         │                    │            Replicator                │
│  content and     │                    │                                      │
│  date            │                    │                  ┌──────────┐        │
└────────┬─────────┘                    │   ┌──────┐       │  Apply   │        │
         │                              │   │ Disc │       │  AACS    │        │
         ▼                              │   │Files │       │ to disc  │        │
┌─────────────────┐                    │   └──────┘       │  files   │        │
│  Pad download/   │───────────────────┼──────────────────│          │        │
│  stream files    │                    │                  └────┬─────┘        │
│  for AACS        │                    │                       │              │
└────────┬─────────┘                    │                      MKB             │
         │                              └────────────────────────────────────┘
         ▼                                                       ▲
┌─────────────────┐                              ┌──────────────┴──┐
│  Files to        │                              │  Keys from      │
│  Down-           │                              │  AACS_LA        │
│  load            │                              └─────────────────┘
└────────┬─────────┘
         │
         ▼
┌─────────────────┐
│  Apply AACS      │◄───────── MKB ──────────
│  to files to be  │
│  downloaded      │
└────────┬─────────┘
         │
         ▼
┌─────────────────┐                              ┌─────────────────┐
│  Files           │                              │  Replicated     │
│  on              │                              │  Disc with      │
│  Server          │                              │  AACS           │
└─────────────────┘                              └─────────────────┘
```

Here are examples of the playlist code that can initiate a network transfer —

```
<TitleResource src=http://www.example.com/disctitle/picture.png
titleTimeBegin="00:01:00:00" titleTimeEnd="00:01:30:00"
loadingBegin="00:00:40:00" />
```
Or,
```
<SubstituteAudioVideoClip dataSource="Network"
Src=http://www.example.com/disctitle/Video_clip1.map
 preload=00:14:40:00"
titleTimeBegin="00:15:00:00" titleTimeEnd= "00:30:00:00"
clipTimeBegin="00:00:00:00" >
<Video track="1" />
</SubstituteVideoClip>
```

One advantage of the XML approach is that the player handles all of the interaction with the server, such as the connection to the server and the creation of the HTTP messaging.

A network transfer can also be initiated by script with a network API call. In this case, when connecting with the server, HTTP message handling is the responsibility of the script. The advantage of this approach is that the error handling can be more explicit than with the XML. The good news is that you get to handle the messages but you have to process them to complete the transfers. RFC2616, the specification for HTTP, details the handshakes necessary to set up and complete the transfers.

Challenges to be addressed include:
- Server authentication
- Buffer flow, pause mode and connection timeout
- Nonexistent resource (Error 404) resolution, such as default page/logic
- Server no longer exists

Blu-ray uses the java.net package to download files from a server to local storage. The files are first copied to local storage and then either accessed with the virtual file system or by leaving them in local storage and accessing the file(s) via the java.io package. Files may also be moved to the 64 KB Application Data Area. Progressive or streaming files are used by opening the stream with java.net (openStream), moving that file to the BUDA, passing the .bumf file to the Virtual File Package to initiate a new VFS and, voila, the progressive file is now available to your application. In other words, the trailer can now play.

When using the java.net package, as with the network API method for HD DVD, the application is responsible for the HTTP messaging and flow control. The same issues need to be addressed in the code as listed above.

Summary

As of this writing, no Blu-ray Discs™ or players have been released in the market which implement this network connectivity capability. For HD DVD, the players have some capability but the firmware is still being refined. The specifications are still in flux and modifications are required as users experiment with [development and] implementation. As such, expect to see changes, updates, programming guidelines and tips on the Blu-ray Disc Association and the DVD Forum websites as well as from software vendors of authoring tools. Microsoft's HDi website also has helpful examples and programming references.

In the long run, implementation details will become increasingly transparent to DVD authors as standard modules are developed to handle the complexities of network connectivity. Until then, early adopters will need to work closely with player manufacturers, and be members of the Blu-ray Association or the DVD Forum in order to have the information, cooperation and resources necessary to implement network and streaming technologies. Good luck and have fun!

Chapter 7
The HD DVD Format

When work began in the DVD Forum to create a disc format for high definition video, it was considered to be just an incremental change to the existing DVD-Video specifications. The new format, called *HD DVD9*, was intended to provide a way of including high definition content on existing dual-layer "DVD9" discs — the dual-layer type of disc used for standard DVD-Video. Although the new format added support for high resolution audio and video, plus improved graphics, and removed many of the more annoying limitations to disc navigation, some were concerned that these improvements would not be enough to compete with the independently developing Blu-ray Disc™ format, which at that time already included an option for highly programmable, interactive applications.

In parallel with the development of HD DVD9, the DVD Forum was also working on an extension to the DVD-Video format called the *Enhanced DVD Specifications*, or *ENAV* ("Enhanced Navigation"). ENAV had been under development for over two years, and was intended to provide an upgrade for the existing DVD format, adding new browser-like capabilities, such as the ability to display HTML pages with interactive graphics, animation and sound, as well as an ability to render text and run JavaScript™ programs. An interface was defined that would allow JavaScript code in HTML pages to control the playback functions of a DVD player. The intention was to create a new set of advanced interactive features that DVD players could support, to help distinguish medium and high-end players from the bargain basement devices that were beginning to flood the market. Over time, it became clear that this new interactive format could provide important insights for the new high definition formats being contemplated.

Roll forward a few years and you have the HD DVD-Video format. The name HD DVD9 is no longer used, but you can still put HD DVD-Video content on a DVD9. Likewise, the Enhanced DVD Specifications never saw the light of day. By the time ENAV was ready for final approval, many Steering Committee[1] members were afraid it was too late for such a format since HD was just around the corner. They did not want to risk undue market confusion and potentially spoil the HD market in the process. If you look closely you can see some striking similarities between the now-existing HD DVD-Video specifications and what ENAV was going to bring.

After a long and winding path, what ultimately came out of the years of work in the DVD Forum was a format that defines two distinct types of content, *Standard Content* and *Advanced Content*. More like different modes of operation, both Standard Content and Advanced Content provide a means of delivering high-definition video and audio, but they differ significantly in how that is done and the level of functionality within each. Standard Content is the direct result of the HD DVD9 specification work, and represents an incremental improvement to the existing DVD-Video specifications. Advanced Content provides enhanced interactive capabilities through a variation of HTML and JavaScript, making it possible to implement everything that Standard Content supports and significantly more.

---

[1]In the DVD Forum, the Steering Committee is the top-level group that sets the overall direction for the organization. This group also has final say as to which specifications get published and which do not.

7

Standard Content

As with DVD, HD DVD Standard Content defines two primary layers, a *presentation data* layer and a *navigation* layer. The presentation data layer contains the data that is presented from the disc, including all audio, video, still images, and subtitles. The navigation layer defines how that data should be organized and is presented to the viewer. Although smaller in size, the navigation data is critical to the proper operation of the disc. For this reason, the specifications call for a backup copy to reside in a different location on the disc in case the primary location becomes unreadable, due to a scratch or other physical imperfection.

The presentation data on an HD DVD-Video disc is stored in the form of MPEG-2 *program streams* (sometimes abbreviated as PS). To create a program stream, the presentation data is first *packetized*, meaning that it is broken up into small, equal-sized amounts of data that are then wrapped into a logical package with information in the pack header, such as the type of packet the data belongs to and its size. Once the data is broken up into packets, it is *multiplexed* together into the program stream, which means that the packs are sorted together into a single stream such that each appears at approximately the point needed to be delivered on time to its corresponding decoder, as shown in Figure 7.1. In a typical HD DVD-Video title, video has the highest continuous data rate, so video packs appear most frequently. In contrast, subtitles appear only periodically, so their corresponding packs tend to spaced far apart within the program stream. In addition to audio, video and subtitle packs, HD DVD-Video program streams include specialized navigation packs (NV_PCK) that hold information for fast-forward and reverse play, as well as *highlight information* (HLI) packs that define button highlight information for menus.

Figure 7.1 Multiplexing Structure

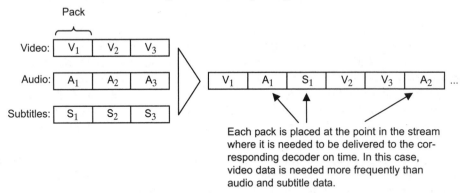

Each pack is placed at the point in the stream where it is needed to be delivered to the corresponding decoder on time. In this case, video data is needed more frequently than audio and subtitle data.

Taken as a whole, the presentation data in an HD DVD-Video title is referred to as an *enhanced video object set* (EVOBS). Each EVOB within the set can be further subdivided into one or more *cells*, which contain one or more logical access units called *enhanced video object units* (EVOBU). Each EVOBU starts with a navigation pack and is followed by the content that is presented within that logical unit. These EVOBUs represent the entry points into the presentation data that may be accessed by the playback engine during normal playback and *trick play modes*, such as fast-forward and reverse. Whereas, cells represent the entry points into the presentation data that may be accessed by the navigation layer, which is discussed later.

The HD DVD-Video navigation layer defines collections of relatively small data structures that provide information about the presentation data. For example, the navigation layer includes the language settings for each of the audio and subtitle streams in a title. It defines what user operations, such as Skip Forward and Skip Backward, are allowed at different times during playback. In addition, the navigation structures define the sequence in which the presentation data is played, as well as the navigation commands that may be executed before, during, or after the playback of a sequence. Without the navigation layer, the content on a DVD would play in one linear sequence, similar to a CD-Audio disc or a videotape.

Similarities to DVD

Although the names have changed a bit, the structure of presentation data defined in the HD DVD-Video specifications for Standard Content is very similar to that defined by the DVD-Video specifications. Although DVD-Video referred to its program stream data as *video object sets* (VOBS) rather than *enhanced video object sets* (EVOBS), the structure and contents are nearly identical. HD DVD-Video supports additional audio and video codecs, as described in Chapter 5, but the way the data is stored in the stream is still the same. Only the button highlight information (HLI) data has changed significantly in that it is now stored in its own pack type, whereas in DVD-Video the HLI data is stored inside the navigation packs.

Likewise, the navigation structure between HD DVD Standard Content and DVD-Video is nearly the same. Both formats include four primary *domains* or modes that the player may occupy: the *first play* (FP) domain, the *video manager* (VMG) domain, the *video title set menu* (VTSM) domain and the *title* (TT) domain. Combined, the first three domains create what is referred to as *system space*, whereas the last domain may be referred to as *title space*. The purpose of defining these domains is to identify specialized behavior that occurs either within a specific domain or during the transition from one to another. For instance, the first play domain represents the starting point for an HD DVD Standard Content title. A player will seek out the first play domain and begin playback there. Another example of this type of specialized behavior is found when playback transitions from the title domain to any of the domains in the system space. Unless specifically blocked by a disc's authoring, the player will store information that can be used to later *resume* playback in the title domain at the point from which it left. Resume is an important topic that we will discuss later.

A common navigation structure that is found in all four domains is the *program chain* (PGC). A program chain is a linear collection of zero or more *programs* (PG), which in turn may contain one or more *cells*. Cells are, of course, the common interface between the navigation layer and the presentation data and represent the logical entry points into the presentation data that may be accessed by the navigation layer. In addition to the logical organization of presentation data, program chains may also include zero or more navigation command sequences — collections of individual commands that may be used to redirect playback to another location, read or modify player registers, or select specific streams in the presentation data for playback, such as audio and subtitle streams or video angles.

7

Key Differences from DVD

About three years ago, a small group was formed among DVD authors from several major studios to look at how the DVD-Video specifications might be improved. Work was just beginning on the HD DVD-Video specifications, then referred to as *HD DVD9*, with the intention of adding support for high definition video to the DVD specifications.[2] This was seen as an opportunity to fix many of the problems and little annoyances that had been found with the DVD-Video specifications as they were applied to real world situations. These issues ranged from minor annoyances, such as having to always navigate through the video manager when jumping from one video title set to another, to much more critical, such as the manner in which DVD-Video supported multiple menu languages, a condition that had been rendered virtually useless due to the limited way in which some player instructions had been implemented. Many of the differences between the DVD and the HD DVD formats are summarized in Table 7.1.

Table 7.1 Comparison of DVD to HD DVD Standard Content

Feature	DVD	HD DVD
Number of SPRMs (16-bit each)	24	32
Number of GPRMs (16-bit each)	16	64
Max. video angles in a menu[a]	1	9
Max. audio streams in a menu	1	8
Max. subtitle streams in a menu	1	32
Max. buttons in a menu	36	48
Max. number of VTSs	99	511
Max. number of titles	99	511
Max. number of PTTs[b] in a sequential title	99	511
Max. number of PTTs[b] in a VTS	999	1023
Max. number of programs in a program chain	99	511
Max. number of cells in a program chain	255	511
Direct VTS to VTS navigation	No	Yes
Jump by GPRM value	No	Yes
Resume commands	No	Yes

[a]Including the "main" angle.

[b]Commonly referred to as "chapters."

---

[2]When work began on the HD DVD-Video specifications, it was intended to be an incremental change to the DVD-Video specifications and offer a means of including high definition video on existing DVD media. The name HD DVD9 was given since dual-layer "DVD9" discs were expected to be used for the format.

High Definition Everything (Audio, Video and Subtitles)

One clear difference between HD DVD-Video Standard Content and DVD-Video is the fact that HD DVD-Video provides support for high definition video and audio content, as well as higher resolution subtitles. This core audio and visual content is referred to as *presentation data*. HD DVD allows presentation data to be composed of both high definition and standard definition video. Further, because the HD DVD-Video specifications started as an incremental improvement of the DVD-Video specifications, all of the modes that DVD supports are also included in HD DVD.

Included among the new types of presentation data that HD DVD supports are the Dolby® Digital Plus, Dolby TrueHD, and DTS-HD™ audio codecs. Each of these new codecs brings a collection of new features that were not previously possible with DVD-Video. Higher data rate support, statistically lossless encoding, lossless down-mixing, and support for up to eight audio channels are just some of the new capabilities that HD DVD-Video provides. HD DVD also adds the new SMPTE VC-1 and MPEG-4 AVC high efficiency video codecs, as well as the *main profile at high level* (MP@ML) mode of MPEG-2. Support for VC-1 and AVC are particularly important because they are capable of encoding video to nearly half the size of a comparable MPEG-2 video stream, and have additional features that help make the playback look even better. This is true whether working with high definition material or standard definition. To put this into context, a single-layer DVD disc could typically hold just over two hours of simple, standard definition audio visual content. Now, a single-layer HD DVD disc using the new audio and video codecs can hold over twelve hours of the same type of standard definition content, and a dual-layer HD DVD disc can hold twice that. Imagine having an entire 24-hour day's worth of video on a single disc![3]

Another benefit that comes with HD DVD is improved subtitle support. In addition to offering HD resolutions, the HD DVD-Video subtitle format supports both 4-color (2-bit) and 256-color (8-bit) subtitles. The reasoning behind this, however, does not necessarily make much sense. During the DVD Forum meetings when the HD DVD-Video format was being created, content producers (television and motion picture studios) claimed that an increased number of colors beyond DVD's original 4-color subtitle support was necessary for HD viewing modes due to the complex structure of Eastern language character sets, such as Japanese and Chinese. However, with almost six times the resolution of standard definition, individual pixels are actually much smaller (even on very large television sets) in an HD picture. As a result, the jagged edges and coarse shapes that were common on 4-color standard definition subtitles actually look much better at high definition resolutions because of the smaller pixel size. In other words, if anything, they really needed 8-bit color for standard definition where the pixels are larger and benefit much more from anti-aliasing and other techniques that can be used when more colors are available.

Nonetheless, 256-color subtitles are a welcome new feature. In fact, you can do quite a bit with 256 colors. At the most basic level, subtitle text can be made much more readable through the use of anti-aliasing and other font smoothing techniques. Subtitle text can be made more colorful, for example, using color to distinguish one speaker from another or to

---

[3]This gives new meaning to the popular FOX television series "24" in which an entire season of 24 one-hour episodes covers a continuous 24-hour period in the show. An entire season could be put on a single disc and watched from start to finish over the course of one complete day.

separate captions for background sounds from those for dialogue. Going even further, one can begin to imagine supportive imagery such as graphics and diagrams that are provided via the subtitle stream. Imagine having a detailed and colorful on-screen map that shows set locations, or popup diagrams that accompany training videos. Perhaps the most extreme use of the increased color depth of HD DVD subtitles would be for natural images, such as photographs. A great deal of work was done back in the 1980's to develop and improve image reproduction on computer systems that only supported 256-color displays at the time. This was further improved over the course of the 1990's when the World Wide Web was becoming popular and 256-color graphics were widely supported. As a result, one can now take any natural image and have a good chance of being able to create a compelling 256-color version of it that can be encoded into an HD DVD subtitle stream. This could, for example, provide an interesting means of adding a synchronized slide show to a video presentation.

When using 256-color subtitles for these more enhanced applications, there are certain limitations that must be kept in mind. For example, the HD DVD-Video specification defines restrictions on data size for these types of images. Subpicture images use a *run-length encoding* (RLE) technique to compress the image data, in which several pixels in a row of the same color (called a *run*) are coded simply as the number of pixels and the corresponding color value. A run of sixteen pixels that are all the same color would be reduced from sixteen individual color entries down to a single color entry plus a count. This technique provides excellent compression for images like subtitle text where adjacent pixels often have the same color. However, in natural images or computer-generated images that use dithering and other image enhancement techniques, it can be very rare to find runs of more than one or two pixels. As a result, the images do not compress well at all. For example, a typical full HD resolution subtitle image might compress to a size of 40 to 80 KB using an RLE compression scheme. A comparable natural image that employs dithering might be as large as 1,800 KB or more — almost as large as an uncompressed version of the image, and actually larger than what the HD DVD-Video specifications would allow for a single subpicture image.

File Structure

Another difference between HD DVD Standard Content and DVD-Video can be seen in the file structure of HD DVD discs. The HD DVD-Video specifications define an HVDVD_TS directory, rather than the VIDEO_TS directory used for DVD, to contain all of the discs audio visual content and navigation data. In addition, all of the files found in this directory have been renamed from their original DVD-Video counterparts accommodating larger numbers of items and file sizes. Yet the basic structure is still quite similar. For example, both HD DVD-Video and DVD-Video place all navigation structures in files ending with ".IFO" and backup copies in corresponding ".BUP" files. Likewise, where DVD-Video used to have the multiplexed audio visual program streams stored in files whose names ended with ".VOB" (standing for *video object*), HD DVD-Video uses the ending ".EVO" (short for EVOB or *enhanced video object*). However, like DVD, the HD DVD specifications allow additional user-defined content to be recorded on the disc in directories of its own, as long as it is not placed within the HVDVD_TS directory (see Figure 7.2). When Advanced Content is included on the disc, an additional directory structure is required, as described later in this chapter.

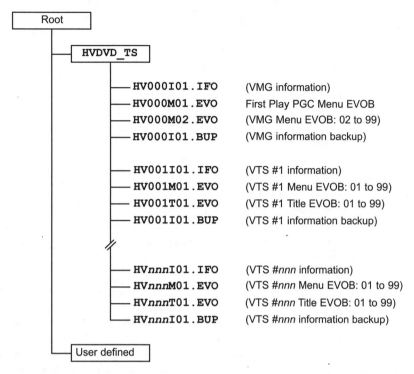

Figure 7.2 Directory Structure of HD DVD Standard Content Disc

```
Root
    HVDVD_TS
            HV000I01.IFO     (VMG information)
            HV000M01.EVO     First Play PGC Menu EVOB
            HV000M02.EVO     (VMG Menu EVOB: 02 to 99)
            HV000I01.BUP     (VMG information backup)

            HV001I01.IFO     (VTS #1 information)
            HV001M01.EVO     (VTS #1 Menu EVOB: 01 to 99)
            HV001T01.EVO     (VTS #1 Title EVOB: 01 to 99)
            HV001I01.BUP     (VTS #1 information backup)

            HVnnnI01.IFO     (VTS #nnn information)
            HVnnnM01.EVO     (VTS #nnn Menu EVOB: 01 to 99)
            HVnnnT01.EVO     (VTS #nnn Title EVOB: 01 to 99)
            HVnnnI01.BUP     (VTS #nnn information backup)
    User defined
```

Navigation Structure

Overall, the structure of navigation data, which controls how content is presented on a disc, is not very different between DVD-Video and HD DVD-Video. Nearly all of the basic structures, like video title sets, titles, program chains, programs, cells, parts of title, and so on, are still present and serve the same purposes as before. The most significant change, however, is the number of each of these structures that is allowed on a disc. Table 7.1, earlier in this chapter, reflects the dramatic increase in the limits for each of these items. For instance, HD DVD-Video supports over five times as many VTSs, titles, and programs, helping to relieve the pressure on the types of discs that used to push up against DVD's lower limits.

Navigation Commands

In order to accommodate the increased number of navigation structures allowed on a disc, most of the navigation commands defined by HD DVD-Video had to be revamped, as well. At the most basic level, the size of the command codes had to be enlarged to accommodate the larger numbers of navigation structures that one might jump or link to. In addition, navigation commands increased in numbers, as well. For example, the number of *general parameters* (GPRMs) available to navigation commands was increased from sixteen to 64. Each GPRM still represents a single 16-bit integer value, but this represents a four times increase in the amount of memory that is now available to Standard Content applications. In addition, more system parameters (SPRMs) have been added to track some of the changes in menu structure and other features.

In general, all of the same commands are present and they perform essentially the same operations as before. And a few extra conveniences have been added, such as the ability to link directly or jump straight from a GPRM value. Previously, the DVD-Video specifications defined link and jump commands such that one was required to provide an *immediate value* (a literal number) to the command specifying the destination. For more sophisticated discs, this was quite unpleasant as it would often require long sequences of "If...Then" statements to check the value in a GPRM and jump to a corresponding destination, (e.g., If GPRM0 = 1 Then JumpToTitle 1, Else If GPRM0 = 2 Then JumpToTitle 2...). The new format would allow something as simple as "JumpToTitle GPRM0," which is much easier to program and far less error prone to write.

Another convenient change that has been made to the navigation commands is the ability to jump directly from one location to another on the disc without always passing through the video manager (VMG) area. In DVD-Video, any time you wanted playback to go from one VTS to another, it was always necessary to jump first to a command sequence in the VMG that would then redirect the playback to the desired destination. Only the video manager had direct access to any individual VTS. In HD DVD-Video, that is no longer the case. Now, navigation commands make it possible to jump from within one VTS to a title inside another VTS, bypassing the VMG in the process.

Of all the improvements that have been made to the navigation commands, perhaps the most important have to do with cell and button commands. In both cases, the DVD-Video specifications allowed for a single navigation command to be executed either at the end of a cell or in response to the activation of a button. In HD DVD-Video that number has been increased to allow up to eight commands to be executed at the end of a cell or in response to a button. This allows much more to take place during these command sequences and allows the command sequences to be far more useful in general. In addition, the HD DVD-Video specifications have indicated that execution of these commands can take place seamlessly. While button commands were already seamless, meaning that video playback was not interrupted while executing the command, cell commands were always non-seamless in DVD. The addition of seamless cell commands allows for far more interesting navigation models to be implemented that are far less distracting to the user. For instance, imagine a game in which the user must choose to turn left or right at key points during the video presentation. Using seamless cell commands, checks could be made at key points in the video to determine if the user had gone the right way. If so, the video sequence would continue to play seamlessly. If not, the cell commands could cause playback to jump to a crash scene. Frankly, this has been a sorely missed feature in DVD interactive challenges for quite some time!

Menu Languages

When the DVD-Video specifications were created, a special navigation structure called a *language unit* was defined as a means of supporting menu sets for multiple languages on a disc. The concept was that the viewer could specify his or her menu language preference in the player's setup menu, and the player would automatically look for a corresponding language unit in the menu space of each disc that was inserted. If the disc included menus in that language, they would be played automatically. Otherwise, the disc's default menu language would be used. All an author had to do was make sure that the menus existed in the appropriate language unit and the player would take care of the rest.

Unfortunately, in practice, this approach has two fatal flaws: 1) DVD players are not required to provide a menu language setup option, and 2) no navigation command is provided to allow a disc's author to specify the menu language, such as in response to a menu of his or her own creation. As of this writing, numerous players have been released onto the market that either support only a subset of the available menu languages, or offer no menu language preference setting at all. So, an owner of one of these players who wishes to see menus in Japanese may not have the option of indicating this to the player. As a result, even if the disc inserted contains a Japanese menu set in the appropriate Japanese language unit, the menus may not appear at all.

To avoid such an unpleasant user experience, authors around the world have had to invent their own methods for supporting multiple menu languages on a disc. The most common is to place all of the different language menu sets in the default language unit of the disc, and then provide a menu listing of the available languages at disc startup, often referred to as an *option card*. The logic authored on the disc must then always choose the menu set to display based on the initial selection made by the viewer. Although this is an effective solution, it requires a significant amount of additional programming and leaves a key feature of the DVD-Video specifications unused.

The HD DVD-Video specifications address this problem in two ways. The *first play PGC* was modified to allow it to contain interactive content. In DVD-Video, the first play PGC can only contain navigation commands, simply directing the player to an author-defined location to begin playback. With the changes in the HD DVD-Video specifications, it may also be used as the location for an option card, allowing the viewer to select a preferred language or other options. The specifications also include an additional navigation command executed from within the first play PGC that sets the menu language to be used. In this way, the author of the disc can build the menu sets in separate language units as originally intended, but can control through navigation commands which menu set is displayed by the player. When the two features are combined, the result is an improved user experience for selecting menu language and a simpler method of authoring multiple language menu sets.

Menu Features

In addition to improved handling for menu languages, the HD DVD-Video specifications reduce the number of restrictions on content in the menu domain. In DVD-Video, the menu domain restricts content to accessing no more than a single audio stream and a single subpicture stream, whereas the title domain allows up to eight audio streams and 32 subtitle streams to be accessed. However, the structure of the presentation data in each of these domains is essentially identical. In other words, these restrictions were artificially imposed when the DVD-Video specifications were written.

Although the restrictions on audio and subpicture stream access were not derived from a technical basis, there was, in fact, some reason behind them. Specifically, no one creating the specifications thought you would need more than one audio or one subpicture stream in a menu. The reason for this has its roots in the *localization* of content — preparing it for application in other regions of the world and other languages. Unlike a feature video, where alternate languages are generally handled by changing to different audio or subtitle streams, menus tend to be localized in a different manner. In most cases, the text of a menu is what

needs to be changed from one language to another. However, that text is very often a fixed part of the background image, so to change the language would mean to change the background image. But, the audio and subpicture streams in a menu usually hold non-localized content like music and generic button highlights. These elements do not often need to change to accommodate a new language. For this reason, the creators of the DVD-Video specifications decided to limit access to just one audio stream and one subpicture stream, and then provide the multiple language unit areas described previously to allow the text in the background image to be changed from one language to another.

Unfortunately, what the original writers had not taken into consideration is that this artificial limitation on the number of audio and subpicture streams one could access in a menu only served to frustrate and limit the creative imagination of those designing DVDs. In fact, over the years, numerous authors and designers have envisioned a wide range of applications that would greatly benefit from having wider access to different types of streams within menus. For example, a feature found on several Hollywood titles includes a "call to action" audio clip in the menus. This is an audio stream in which something draws the attention of the user after a certain period of time to encourage them to interact with the disc. The famous Walt Disney picture *Snow White and the Seven Dwarves* was one of the first major titles to utilize this feature, such that the voice of the magic mirror would call out to the user to instruct them on what to do. The disc included several calls to action, and these were localized to nearly a dozen different languages across the various regional releases of the title. The downside of this is that because of the restriction on audio streams, a separate copy of the menu had to be included for each call to action and for each language.

Although some workarounds were possible for these artificial limits in menu space, such as putting menu content in title space instead, they often resulted in a loss of key functionality that was unique to the menu domain. To rectify this situation, the HD DVD-Video specifications were modified to allow all of the same audio, subpicture, and even video angle streams in the menu domain that are available in the title domain. This makes it possible for menus to maintain all of the characteristics that are important for menus, such as language unit structures and the ability to handle resume functionality (discussed later), but without limiting creative techniques that disc authors and designers may employ.

Resume Functionality

One unique feature that was introduced with DVD-Video was the concept of *resume* — the ability to transition from the middle of a video to a menu and subsequently return to the video at exactly the same location from which you left. Unfortunately, the way the DVD-Video specifications were written, there is no way to program the resume functionality directly other than to call the resume function from within the menu domain. For example, if your disc includes a feature video and several bonus video clips, but you want to make sure that resume only goes back to the feature video and never to the bonus clips, there is no way to define this in a straightforward way. You would have to use the strategy of motion picture studios, playing a special blank title every time anything other than the main feature finished playing. This extra title would automatically become the resume target, because it was the last item that had been played. Extra code would then be added to that title that could detect if the title were being played intentionally or if it had started playing as a result of the resume function. The programming could then route playback to the appropriate location. Workarounds like this are complicated, error prone, and negatively affect the responsiveness of the user experience when playing the disc. In the end, these techniques are really just hacks to get the DVD player to do something that it was not built to do, and it shows.

Fortunately, the creators of the HD DVD-Video specifications took mercy on all of the poor disc authors out there and made a few simple, but important modifications to how resume works. First, they added a member called the *RSM_permission* flag to program chains in the title domain that indicates whether the content in that PGC can be resumed to. In this way, program chains that exist in the title domain can be restricted so that they have no effect on where the resume function returns when activated. So, in the example above, the RSM_*permission* flag would be set to "permitted" for the feature video and "prohibited" on the additional bonus videos, so only the feature could be resumed to.

Another problem with how resume functionality was implemented in the original DVD-Video specifications is that there was no method for programmatically detecting that resume had been triggered and executing any navigation commands prior to resuming playback. As an author, sometimes one finds that it is necessary to execute a command sequence to initialize variables, resume a timer, or initiate some other process prior to resuming playback of a title. However, because the DVD-Video specifications do not provide a means of doing this, playback logic on the disc can sometimes become confused.

To address this problem, the HD DVD-Video specifications introduced an additional navigation command sequence on program chains that is specifically executed when resume is about to take place to that PGC. Containing up to 1,024 navigation commands, this command sequence can be used to initiate any number of different procedures prior to resuming playback of the PGC. For instance, audio, angle or subpicture streams can be selected. General parameters can be checked or modified, and playback can even be redirected elsewhere if necessary, interrupting the resume process, for example, if certain conditions have not been met. The end result is a powerful refinement to an already beneficial mechanism.

When is HD DVD Standard Content Appropriate?

Assuming that you have already settled on HD DVD as the disc format that you would like to use, the question remains as to whether you should employ Standard Content or Advanced Content. To help make this decision, it helps to consider the kinds of features that are unique to Advanced Content while also weighing the benefits of Standard Content. Standard Content tends to be significantly easier to create when DVD-style functionality is desired. High quality authoring tools already exist to help create Standard Content in a straightforward, well understood manner, leaving much less room for error or unwanted behavior than is currently associated with Advanced Content. From a production standpoint, this can give Standard Content a strong lead, especially for discs that focus on audio and video alone, such as film dailies and demo reels. However, if you need to utilize popup menus, interactive animations, picture in picture videos, or other advanced features, then Standard Content may be out of the question because it does not offer those capabilities.

Advanced Content

In HD DVD-Video, Advanced Content offers an entirely different approach to authoring content. Unlike Standard Content, Advanced Content does not rely solely upon a declarative model. Advanced Content is far more open-ended and allows for much greater flexibility while it also requires a completely different set of skills and expertise to implement well.

Advanced Content is created through a set of *extensible markup language* (XML) documents. Creating these documents is, in many ways, similar to website development since it is based on a variation of the *hypertext markup language* (HTML) which is used to create web pages. In general, XML provides a standardized way of storing information in a somewhat human-readable manner that is also relatively simple for computers to process. On its own, XML is just a file format — a method of organizing information and associating structure and attributes with that information. As such, XML is perhaps just as appropriate for keeping a list of grocery items as it is for defining HD DVD content. What is important about XML is its ability to organize and convey information in a manner that can be easily understood by people as well as machines.

In the case of Advanced Content, a specific form of XML is used that is related to the *extensible hypertext markup language* (XHTML), a stricter formalized version of HTML. In this form of XML, a document contains a series of elements that have specific meaning. An XHTML document would typically be divided into head and body elements, in which the head contains elements that describe the document itself, while the body contains the elements of the document, such as buttons and text, which in turn have their own attributes. This hierarchical organization of data in the form of elements and attributes leads to a model for conceptualizing documents, whether they are printed pages, web pages, timelines, or the contents of an entire disc.

Although XML is generally not considered a programming language, it can be created in much the same way a person types in the various elements and attributes necessary to create the document. Because XML has a standardized structure and syntax, computers can generate properly formed documents given a certain set of input data. At this point in time, however, there is very little computer software that knows how to author an HD DVD title, so it generally falls on the human to create the documents.

This is quite different from the typical model that has been previously used with DVD. In that model, one uses an authoring tool because the final output has to be represented in a binary format that is far from being human-readable and certainly could not be easily typed. Instead, the authoring software gathers information from the user based on what he or she intends to do, and then processes that information, translating it into the machine code format that a DVD player needs to see. Advanced Content uses a more human-readable format, allowing anyone who understands elements, attributes and how they are formed, to type the information directly.

This, of course, requires that the player can translate the human-readable format into a set of internal behaviors, actions and structures that result in items showing up on the screen and audio playing out of the speakers. As one might imagine, this means that the player must have a certain amount of additional processing power available to make this translation. The open-ended structure of XML allows for far more flexibility than traditional DVD navigation structures and, as such, the player must have the processing power necessary to handle this extra burden. Fortunately, Moore's Law indicates that in the ten years since the first DVD players were introduced, comparably priced processors have increased in power by a factor of 64, and will at least double again by the time HD DVD becomes a mainstream format.[4]

It is important to understand that this XML approach is actually still *declarative*, like DVD-Video and HD DVD standard content. This means that the author must still spell out everything that must happen and all of the different conditions that allow that to happen in a very deterministic manner. It is not terribly dynamic, and the author must predict all of the different ways in which the content might be used to create that behavior. To provide even greater power and flexibility, Advanced Content enables a *procedural* approach, as well, through the use of an additional scripting language called *ECMAScript*, named after the organization where it was standardized. In a procedural language, one can program algorithms that serve as a sort of intelligence that can determine what the condition is at any given moment and then determine what the behavior should be based on that condition. Algorithms allow an author to instruct the machine with a relatively simple sequence of commands organized into loops and conditional statements rather than having to explicitly map out every possible path through the content.

Structure of Advanced Content

Advanced Content is composed of three major items: a *playlist*, an *advanced video title set* (Advanced VTS), and *advanced applications* (see Figure 7.3).

Figure 7.3 Generalized Structure of HD DVD Advanced Content

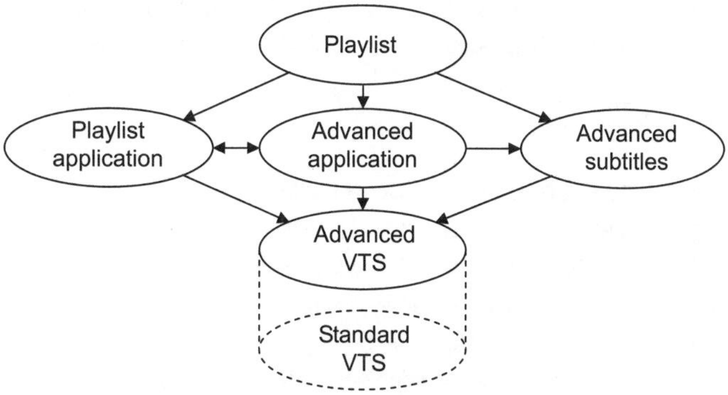

```
                    ┌──────────┐
                    │ Playlist │
                    └──────────┘
         ┌──────────────┐  ┌──────────────┐  ┌──────────────┐
         │   Playlist   │←→│   Advanced   │→ │   Advanced   │
         │ application  │  │ application  │  │  subtitles   │
         └──────────────┘  └──────────────┘  └──────────────┘
                         ┌──────────────┐
                         │   Advanced   │
                         │     VTS      │
                         └──────────────┘
                         ┌ ─ ─ ─ ─ ─ ─ ─
                           Standard
                             VTS
                         └ ─ ─ ─ ─ ─ ─ ─
```

The playlist defines the general structure of titles on the disc as well as the relationship between the presentation data in the Advanced VTS and the advanced applications. The Advanced VTS contains the Advanced Content presentation data, which includes the audio, video and subtitle data that has been multiplexed together for real-time presentation with great efficiency in the player. The advanced applications, on the other hand, include not only

[4]Moore's Law, attributed to Intel co-founder Gordon E. Moore, postulates that the transistor density of integrated circuits approximately doubles every 18 to 24 months, relative to minimum component cost. This translates to a doubling of processor power over any two year period.

all of the navigation logic for how and when to play back that content, but also include additional content of their own in the form of interactive graphics and sound effects that are mixed with the presentation data during playback, but controlled independently.

File Structure

The files that contain the elements of Advanced Content on an HD DVD are divided across two directories, the ADV_OBJ directory and the HVDVD_TS directory — the same directory used by Standard Content, as shown in Figure 7-4.

Figure 7.4 Directory Structure of an HD DVD Advanced Content Disc

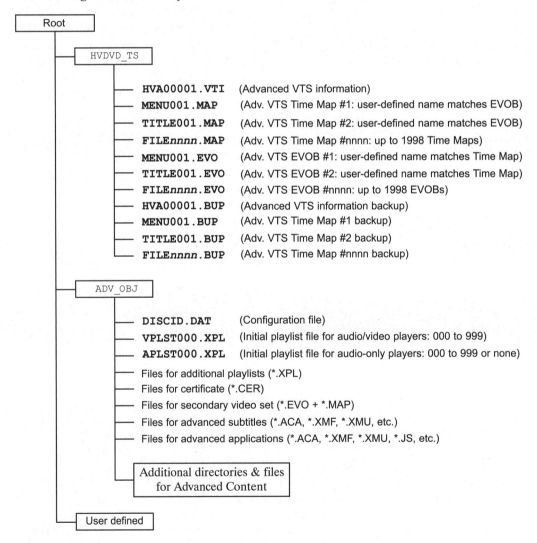

Unlike Standard Content, however, the actual names of the files are up to the author to define for everything except the Advanced VTS information file (HVA00001.VTI), the disc configuration file (DISCID.DAT), and the video and audio-only playlist files (VPLST*nnn*.XPL and APLST*nnn*.XPL, respectively). The author does have to be sure that the time map (.MAP) files, enhanced video object (.EVO) files that they refer to, and the time map backup (.BUP) files all have matching file names, as demonstrated in the figure.

Playlist

The general structure of a playlist is broken down into three major sections: *configuration*, *media attributes list*, and *title set*. The configuration section defines a few initial player settings, such as the aperture size for defining the screen resolution to use, the streaming buffer size for networked content, and the main video background color for areas of the screen that are not covered by the primary video.

The media attribute list provides information for the player regarding the coding parameters of each of the presentation data items, including codecs, data rates, number of channels, resolution, and so on. This information may be used by the player to help configure itself for optimal playback of the content.

The third and most significant section of the playlist is the *title set*. This is where the primary navigation structure and timelines are defined for the content on the disc. The title set may contain one *first play title*, which indicates where the player should start playback, a *playlist application*, which spans the duration of the entire playlist, and anywhere from one to 1,998 titles.

A title defines a timeline that is composed of a series of segments. Those segments may include *primary audio/video clips* that contain the presentation data, *title resource* segments that identify file resources that are loaded and unloaded at certain times, and *application segments* that can also be loaded and unloaded at specific times. In addition, a title includes specific navigation information including a *chapter list* and *schedule control list* whose items can specify frames to pause on or times to trigger events that the application programming can respond to. These are all examples of items that fall within the title set section of a playlist.

Advanced VTS

The advanced video title set represents the collection of presentation data that is used by the advanced content on the disc. However, unlike the standard VTSs found in HD DVD Standard Content, it is somewhat difficult to relate to a particular data structure or group of files. In fact, the advanced VTS includes an advanced VTS information (.VTI) file and a disparate set of time map (.MAP) files and their backups, as well as a collection of primary EVOB (.EVO) files. Of these, only the advanced VTS information has a fixed name. All of the others can be named more or less as the author sees fit, which is why it is sometimes difficult to see a tangible set of files that correspond to the advanced VTS. Despite the wide array of names that may be possible, the HD DVD-Video specifications indicate that these files must be physically located within a common area of the disc that immediately follows the standard VTSs and precedes the rest of the Advanced Content and user-defined data.

7

The key importance of the advanced VTS is that it contains all of the multiplexed audio, video and subpicture data that can be used by the Advanced Content. The presentation data of an advanced VTS differs from that of Standard Content, however, in that it can also contain sub-video and sub-audio streams multiplexed in with the normal audio, video and subpicture data. These additional streams provide the basis for picture in picture and secondary audio applications in which the Advanced Content can independently control and mix this secondary content with main audio and video while maintaining full synchronization.

Advanced Applications

An advanced application includes navigation logic, its own graphics plane, a secondary video plane, and an additional audio decoder that allows images, animations and sound effects to be composited with the video from the advanced VTS during playback. Unlike Standard Content, which is limited to just playing the content in the primary presentation data, advanced applications can independently control not only the primary presentation data, but also a secondary set of audio and video presentation data, as well as a third set of interactive graphics, animations and sound effects. These additional resources may be composited together or individually enabled or disabled as they are presented. An advanced application can, for example, display a slide show or popup menu over the primary video as it plays from the EVOBs while mixing in a director's commentary and subtitles that were downloaded from the network.

Advanced Content also includes a font rendering capability that allows text to be dynamically generated and displayed without having to operate from predetermined sets of images as was needed for DVD. That text can either be pre-defined or generated on-the-fly by the advanced applications and then rendered to the screen in whatever location, size or style that the font supports. For example, in Standard Content, if one wanted to show scores in a game, that would require that the author determine every possible value for that score and have a graphic stored on the disc for each value. In Advanced Content, one can simply include an OpenType font on the disc and have the application generate the character codes (e.g. "10", "20", "340") for the score and render them to the screen. In this way, it is far more dynamic than anything previously possible with DVD.

Structure of an Advanced Application

Advanced applications are generally composed of four parts: a manifest, markup, script and resources. The *manifest* is an XML document that indicates which script, markup and resource files are needed by the application. In addition, it defines a few basic properties of the applications such as the portion of the screen that it covers. This is required, to let the player know which resources it needs to make available, allowing the application to execute properly. Those resources can include script and markup pages, as well as content resources such as image files, fonts, sound effects and animations (see Figure 7.5).

Figure 7.5 Generalized Structure of an Advanced Application

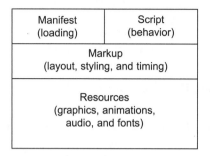

Manifest (loading)	Script (behavior)
Markup (layout, styling, and timing)	
Resources (graphics, animations, audio, and fonts)	

Markup files are another type of XML document that represent the declarative aspect of an advanced application. The markup files, similar to an HTML document, define the layout of objects that are used in the advanced application, the style or appearance attributes of those objects, and the timing elements that will be followed by the application during play-back. *Layout* defines which objects are present in the document, such as paragraphs, buttons, or divisions, as well as their relation to one another and default positioning. *Style* defines the specific look of the objects such as size, position, color and opacity. In all, the HD DVD-Video specifications define nearly 70 style and state attributes for markup. *Timing*, of course, refers to how the document changes over time. This generally consists of animating style attributes as time progresses, but it can also relate to changes in state, such as buttons become selected or activated, or events being triggered for handling by the script.

The timing block of an application consists of three separate clocks: the title clock, the application clock, and the page clock. The *title clock* is a timer that increments as the current title's timeline plays back. When the title is paused, such as from pressing the pause key on the remote control, the title clock pauses as well. When scanning backwards, the title clock decrements, as well. The *application clock* begins incrementing the moment the application starts running and continues to increment for as long as the application is present and still active. Similarly, the *page clock* begins ticking as soon as the current page is loaded and continues incrementing until that page is unloaded and a new one is loaded, at which point the clock resets. In some cases, a page can cause itself to be reloaded, which also resets the page clock.

In some ways, script is even more interesting than markup. *Script* in an advanced application provides a mechanism for procedural programming of the content on the disc. This means that you can create algorithms that have a sort of intelligence of their own, allowing them to be far more dynamic. The script in advanced applications is a set of programming code based on the *ECMAScript* programming language[5]. ECMAScript is a simple script language that was designed to allow programmatic extensions of existing applications. In this

---

[5]ECMAScript, named after the ECMA organization where it was standardized, refers to ECMA-262 3rd edition Standard, which was originally based on Netscape's JavaScript scripting language popularized by its use in the Netscape Navigator web browser.

case, ECMAScript can be used to add interesting dynamic behaviors, such as maintaining score for a game or controlling playback and stream selection of the primary presentation data. In fact, the script provides the primary mechanism by which Advanced Content may control playback, such as pausing, or jumping to another title. All of the default player behaviors that respond to the remote control keys, such as play, pause, and fast forward can be overridden by an advanced application's script.

Special Types of Advanced Applications

In addition to the normal advanced application structure, there are two additional types of advanced applications: the Playlist Application and Advanced Subtitles. Each of these are similar to a normal advanced application in that they have a manifest, markup, and may utilize additional script or resources. However, as shown in Table 7.2, they each have additional restrictions because they are intended to be used in very specific ways.

Table 7.2 Restrictions on Each Type of Advanced Application

Feature	Playlist Application	Advanced Application	Advanced Subtitle
Can be loaded from disc	Yes	Yes	Yes
Can be loaded from persistent storage	Yes	Yes	Yes
Can be loaded from network	No	Yes	Yes
Persists across titles	Yes	No	No
Can use Title Clock in markup timing	No	Yes	Yes
Can receive application focus	Yes	Yes	No
Supports inline styles	Yes	Yes	No
Supports script	Yes	Yes	No
Supports <Event> element in markup	Yes	Yes	No
Can be Hard Sync or Soft Sync[a]	Hard Sync	Both	Both
Max. number that can be active at one time	1	> 1[b]	1
Graphics plane rendered to	Graphics	Graphics	Subpicture

[a]Hard Sync will pause the title timeline while an application loads, whereas Soft Sync will not.

[b]The HD DVD-Video Guideline for Player and Content suggests no more than 12, including the Playlist Application.

Playlist Application

The *playlist application* is an advanced application that is designed to persist throughout all of the titles in a playlist except the specialized First Play Title, which loads resources for the playlist application. It starts at the beginning of the first normal title that plays and remains active throughout the playback of the playlist, terminating when the playlist finally concludes. However, because the playlist application spans titles, it does not have access to a title clock. Only the application and page clocks function in the playlist application.

Advanced Subtitles

Advanced subtitles differ from a normal advanced application in that its output is rendered to the subtitle graphics plane rather than the normal interactive graphics plane. As its name implies, advanced subtitles are intended to provide a method of using advanced application features to output subtitles. As such, there are numerous restrictions on how styling must be provided, what objects are allowed in the layout, as well as how objects may behave. In general, the restrictions for advanced subtitles are designed to keep them non-interactive. Events and scripting are prohibited, as are button and input objects since they would allow interaction.

When is HD DVD Advanced Content Appropriate?

In general, Advanced Content can be used to do essentially anything you could do with Standard Content. Advanced Content is capable of all of the same functionality — playing back audio and video, handling navigation, providing interactive menus, changing audio and subtitle streams, showing still images and slideshows. However, in some cases, Standard Content may actually be much easier to create than the equivalent Advanced Content. This is because Advanced Content can often require a significant amount of programming even to create what would be relatively simple authoring in Standard Content. Until some future date when tools become available that do some of this programming for you, which no doubt they will, the author has to program it himself. That can be a daunting task as it is often much more difficult to type in all of the commands, navigation structures and values than it is to work with a Standard Content authoring tool. Working with Advanced Content tends to be far more difficult and far more error prone, and is more likely to lead to unwanted or unexpected behaviors until one has developed a significant amount of expertise and experience working with it.

There are however, a number of features supported by Advanced Content that Standard Content cannot do. For example, picture in picture video — having a secondary video stream that is overlaid on top of the primary video stream playing back at the same time while maintaining independent, dynamic control of that secondary video. In addition, Advanced Content provides methods for mixing audio on-the-fly, such as button sound effects or secondary audio streams that can be combined with the feature during playback. Other features include network support, drawing, screen capture, and persistent storage access to name a few. All of these are features that Advanced Content supports but Standard Content does not.

Perhaps the most important distinguishing characteristic of Advanced Content is its ability to control interactive graphics, secondary audio and video, subtitles, and primary video independently, mixing and compositing them during playback. Whereas Standard Content is tied exclusively to the primary presentation data, Advanced Content maintains independence between the different types of presentation data. This affords far greater flexibility and programmability, although it does come at the cost of significantly higher complexity.

Chapter 8
The Blu-ray Disc Format

8

The Blu-ray Disc™ format, like HD DVD, offers two different modes for accessing content on the disc, *HD Movie Mode* (HDMV) and *BD-Java*™ (BD-J). The first mode, HDMV, is a declarative navigation structure that defines titles of video content that may be further subdivided into chapters or scenes, which themselves may contain multiple audio and subtitle streams, similar to HD DVD's Standard Content. The second mode, BD-J, builds on top of HD Movie Mode, providing a far more sophisticated level of programmability using the Java™ programming language. As such, BD-J is open-ended and allows for a much wider range of different applications and specific features, as shown in Table 8.1.

Table 8.1 Comparison of HDMV and BD-J Features

Feature	HDMV	BD-J
Application types	Declarative	Procedural
Programming logic	Simple	Complex
General purpose memory size	16 KB[a]	9 MB
Font buffer size for text rendering	4 MB	4 MB
Maximum sound effects buffer size	2 MB	5 MB[b] / 6.5 MB[c]
Maximum graphics buffer size	16 MB	45.5 MB[b] / 61.5 MB[c]
Graphics bit depth	8 bpp (Indexed)[d]	32 bpp
Text-based subtitle rendering	Yes	Yes
Popup and multipage menus	Yes	Yes
Alpha-blended graphics	Yes	Yes
Overlapping alpha-blended graphics	No	Yes
Image scaling	No	Yes
Frame accurate animation	No	Yes
Resize primary video	No	Yes
Secondary video	Yes[e]	Yes[e]
Persistent storage access	Read-only	Read/Write
Network access	No	Yes[e]
Applications (e.g. menus) may persist across titles	No	Yes
Applications (e.g. menus) may persist across discs	No	Yes

[a]In the form of 4,096 32-bit general purpose registers (GPRs).

[b]Profile 1 players

[c]Profile 2 players

[d]HDMV graphics use an 8-bit (256-color) palette of Y'CbCr colors with 8-bit alpha.

[e]Only available on Profile 2 players and Profile 1 players released after June 2007.

8-18 - 1

8

Although the two modes offer similar basic features, the approach to creating content for each is completely different. HD Movie Mode, utilizing a declarative approach, requires the producer to use authoring software to create the complex hierarchy of logical structures that define the relationships of content on the disc. For instance, a clip may be created that contains audio and video material. That clip may be combined with others into a *playlist*. The items that refer to the clips from within the playlist are called play items, and they may contain all or a portion of each clip. In addition, playlist marks may be added to define chapter stop locations within the playlist. The playlist may then be included in a Movie Object that contains a few basic commands that define when the content should play and how. This set of relationships, combined with the list of simple commands defines the resulting playback behavior of the content. The creation of even moderately complex interactive content in this manner requires a tremendous amount of time and effort, and ultimately is faced with significant limitations.

BD-Java, on the other hand, uses a procedural approach in which programming code is written in Java that controls the player's behavior. Because the Java programming language is highly adaptable, the algorithms of the programmed software are able to operate in a far more sophisticated and dynamic manner. The producer must know how to program in Java, which can be quite complicated.

HD Movie Mode and BD-Java use the same lower level navigation structures, such as playlists and clips, the same transport streams containing audiovisual content, and the same high level navigation structures, titles. HDMV content resides, however, in *Movie Objects* (MOBJs) while BD-J content exists in *BD-J Objects* (BDJOs), discussed further in this chapter.

Format Comparisons

On the surface, Blu-ray Disc and HD DVD seem very similar in the fact that they both support two modes for playing back content. In fact, in both formats the modes can easily be classified as declarative and procedural, respectively, and offer many of the same features. Significant differences in format scope are illustrated in Figure 8.1.

Figure 8.1 Comparison of Interactive Functionality Between Formats

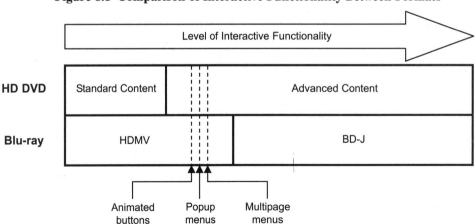

Comparing HDMV to HD DVD Standard Content

Both HDMV and HD DVD Standard Content are built around a declarative navigation structure, and both offer limited programmability through a list of simple commands. In addition, both approaches offer similar navigation structures for organizing and accessing content, including playing tracks, accessing chapter locations, and providing the ability to restrict user operations such as fast forward, rewind or pause.

HD Movie Mode offers a richer set of menu features, however. HDMV provides for click sounds and animated button states so that you can create menus with more engaging aural and graphical effects when menu buttons are selected or activated. And, HDMV supports popup menus over video. For example, during playback of feature video content, one can have a menu that appears over a portion of the screen when the user presses the menu key. This can be done without any interruption of the video as it plays, allowing for a much more enjoyable user experience than HD DVD Standard Content, which would generally require the user to stop playback and go to full screen static menus, instead. In addition, HDMV supports *multi-page menus* in which one can seamlessly change out portions of a menu to show additional options without interrupting the underlying video.

Another interesting feature that HDMV offers is access to both *secondary audio* and *secondary video* — alternate audio and audiovisual sequences that may be played in parallel with the primary content such as in a *picture-in-picture* window. Through the HDMV navigation structure and commands, it is possible to author content that controls when secondary audio and video is available, when it is viewed, and where on the screen secondary video is placed during playback. HD DVD also supports these features, but only through the use of Advanced Content programming. Standard Content provides no access to secondary audio or video.

Unfortunately, these and other additional features that HDMV supports come at the cost of complexity. For today's DVD author, it is far more difficult to author an HDMV title for Blu-ray Disc than it is to create an HD DVD Standard Content disc. In part, this is due to the fact that HD DVD Standard Content is very similar in structure to DVD, reducing the learning curve for an existing DVD author. Standard Content employs a simpler navigation structure, without all of those extra features. This, too, helps to make HD DVD Standard Content easier to learn than its Blu-ray counterpart, as well as reducing the time and number of operations required to create a given title.

Comparing BD-Java to HD DVD Advanced Content

When comparing BD-Java to HD DVD Advanced Content, one sees an array of similarities between the two formats. There are probably fewer differences between BD-J and Advanced Content than there were between HDMV and Standard Content. For example, both BD-J and HD DVD Advanced Content offer 32-bit graphics.[1] In addition, both

---

[1]Refers to images in which each pixel is represented by 32 bits of data: 8 bits each for the red, green and blue color components, plus an additional 8 bits for opacity, otherwise referred to as *alpha*. 32-bit graphics can differentiate over 16.7 million distinct colors with 256 levels of opacity for each, creating a palette of over four billion different combinations.

8

approaches allow graphics to be layered and blended together during playback, making it possible for the player to create complex graphical compositions on the fly. Both are capable of complex animations and interactivity, and both offer dynamic text rendering using *OpenType*® fonts.[2] Of course, both formats support all of these features in applications that can be run either independently or synchronized with feature video playback.

Despite the tremendous similarities, there are several notable differences that can have a significant impact on the types of applications one can create, and the ease with which they can be developed. For example, BD-J operates at a much lower level and is therefore far more open-ended than HD DVD Advanced Content. This opens up a much wider range of possible applications, many of which have yet to be conceived. Additionally, BD-J is, by nature, less bound to the content itself, allowing for greater re-usability of applications and their source code. For example, if one were to try and separate an HD DVD Advanced Content application from its underlying content (audio, video, etc.) in order to re-use it on another application, one would quickly find the task to be quite challenging. Unlike BD-J code, which can be largely abstracted from any specific implementation, Advanced Content is more dependent on the context defined by the content, thus making it difficult to re-use elsewhere without significant changes to the code itself. In other words, as the content changes, so too must the Advanced Content code that is associated with it. With BD-J it is much easier to create code that is largely independent, with a much smaller percentage actually tied directly to the underlying content.

One of the great benefits of BD-J is that it is largely based on the Java programming language, for which there exist a multitude of tools, references, collections, and sample code that can be readily adapted for use with Blu-ray Disc. For example, one might seek out an existing MHP application written in Java to use as a reference for creating something similar for BD-J. Because both are based on the same programming language with many of the same core *application programming interfaces* (APIs) used, it can be a relatively simple process to convert from one to the other. This is actually quite a strong advantage for BD-Java, because there is this wide range of tools and know-how already in use with the Java programming language that can be immediately applied to creating content for Blu-ray disc players.

In addition, BD-Java generally offers a greater range of capabilities, which can be programmed in Java but that are specifically not supported by HD DVD's higher level APIs. For example, the HD DVD format provides no mechanism for direct pixel by pixel manipulation of graphics prior to displaying them. However, such a feature can be relatively easy to program in Java. You are really only limited by the performance of the player that your program will run on and your imagination as a developer. This is the difference between assembling a city from individual building blocks (BD-Java) and assembling it from pre-fab buildings (HD DVD Advanced Content).

---

[2]OpenType is a scalable font format developed initially by Microsoft and later supported by Adobe Systems Inc. and others. Intended to succeed Apple Computer's TrueType® font format, by 2005 there were approximately 10,000 OpenType fonts available, including Adobe's entire library.

Despite all of the benefits that BD-J offers, however, there are a number of detriments as well. The lower level nature of BD-Java often means that one must write a lot more code in order to achieve the same level of functionality as a comparable HD DVD application. In BD-Java the commands are much simpler, often requiring more to be assembled to form an application, whereas in HD DVD Advanced Content an individual command tends to represent a more complex operation. For example in BD-Java, if one were to draw an image on the screen, one would have to actually write the code that loads the image, decodes the image, and then copies that image to the appropriate location on the screen. One or more function calls must be made for each step in that process, resulting in several lines of code. In HD DVD Advanced Content one can achieve the same effect with just one or two lines of code.

There is a benefit to HD DVD Advanced Content for it operates at a higher level and there are more opportunities for a player manufacturer to optimize the performance of those commands. For example, if there is a single function call for displaying an image, a player manufacturer can refine the underlying mechanism in order to make it perform very quickly on the specific hardware that exists inside that manufacturer's player. Internal hardware decoders and accelerators might be utilized to improve the performance, as well as "look ahead" techniques that can intelligently prepare for future operations. The lower level components of BD-Java, however, tend to be more difficult to optimize to a point that they make a significant impact on performance simply because the mechanism they represent is already quite simple. This is not to say it is impossible, but the nature of the optimizations and their impact tend to be less significant or require more effort to make.

Another issue with BD-Java is that although it may be more open-ended, it is not clear that those extra capabilities will actually be in strong demand for Blu-ray Disc titles. The feature set defined in HD DVD Advanced Content was derived from direct discussions with several major motion picture studios. As a result, it largely covers the functionality that has been requested by a majority of the content producers. Therefore, it should be theoretically possible to satisfy a majority of content producers using the higher level (supposedly easier to program) HD DVD Advanced Content than it would be to use BD-Java. So, even though BD-Java may be able to do more, it's not clear that anybody necessarily will. That added flexibility may not really help anyone, but it certainly comes at a cost.

HD Movie Mode

HD Movie Mode (HDMV) is the name given to the declarative navigation structure that has been created for Blu-ray Disc. It is composed of a complex hierarchy of logical structures whose relationship to one another combined with a basic command set determines the playback behavior of Blu-ray content. HDMV is organized on three basic layers — file structure (the directories and files on the disc itself), presentation data (the audio/video content that is played), and navigation structure (the data structures and commands that define the behavior of the disc).

File Structure

The file structure of HDMV content is organized in a simple, logical manner. The files that make up a Blu-ray Movie (BDMV) disc are organized into a hierarchy of folders, each of which contain files that serve a particular function, as shown in Figure 8.2.

Figure 8.2 Directory Structure of a BDMV Disc

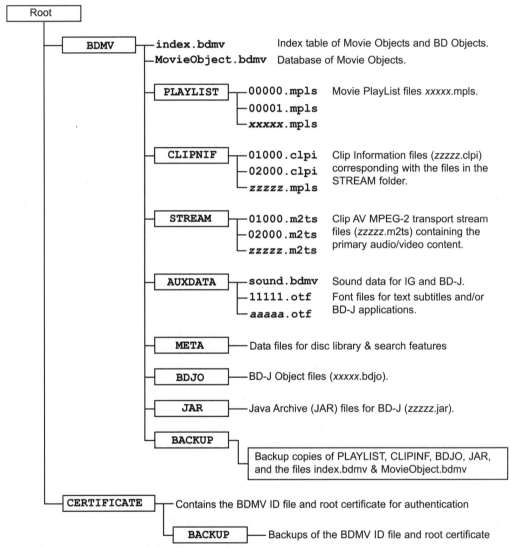

At the root level is the **BDMV** directory, which contains all of the files specifically associated with the Blu-ray content. Within this directory are found all of the files that contain the audio/video content and instructions that tell the player how to present it. Most notable among these are the **index.bdmv** and **MovieObject.bdmv** files, which define the basic organization of content and navigation on the disc. These files contain the *index table*, which identifies titles, and the list of *movie objects* for the disc, respectively.

The **BDMV** directory also contains several key sub-directories. First among these is the **PLAYLIST** directory, which contains, as one might guess, the playlist files that define the timeline on which the audio and video content is played. In addition, there is a **CLIPINF** directory, which contains the *clip information* files — indexed time maps into the *clip A/V stream* files. These stream files are the MPEG-2 transport streams that contain the multiplexed audio/video content and are themselves stored in the **STREAM** directory.

In addition, several auxiliary directories exist, such as the **AUXDATA** directory, which may hold the OpenType fonts and sound effects used on the disc. The **META** directory contains an optional set of files that can be used for the Disc Library and Search features that the BDMV specifications define. For BD-Java support, the **BDJO** and **JAR** directories hold the *BD-J objects* (BDJOs) and *Java archive* (JAR) files, respectively. Finally, a **BACKUP** directory exists that contains duplicate copies of critical files such as the playlists and the clip information files in case the originals are somehow damaged. For discs that employ the AACS content protection system, an additional directory structure is also required. This is addressed in a later chapter.

Presentation Data

The presentation data of a Blu-ray Disc is organized into MPEG-2 transport streams, also referred to as *clip A/V streams*, which are stored in the **STREAM** directory on the disc. MPEG-2 transport streams are similar in structure to the MPEG-2 program streams that are found on DVD and HD DVD discs, though there are several key differences. First, transport streams are subdivided into smaller *transport packets*, which are less than one tenth the size of a program stream packet and provide greater resiliency against transmission errors. Transport streams also provide increased flexibility for delivering multiple programs of content. For example, a transport stream is able to effectively handle multiple streams of program content even if they do not necessarily follow the same time line. Transport streams are often used for digital broadcast television and are the basis of the original BD-RE specifications, which defined a recording format for digital broadcast content. This is one of the reasons that we find transport streams in the BD-ROM format for prerecorded video content.

Although transport streams offer greater resiliency and flexibility over program streams, they also come with an additional overhead in bandwidth and processing required. While program streams typically have about 0.3% of overhead for headers and other information that is stored in the stream, a comparable transport stream will have closer to 2% of bandwidth overhead. While still a very small percentage of the overall bandwidth, on a 25 GB single-layer Blu-ray Disc that still amounts to approximately 500 MB of data.

Paths and Subpaths

The presentation data on a Blu-ray Disc is logically organized into *paths* and *subpaths*. A path represents the primary audio visual content on a disc. For example, the movie or other feature video content along with its corresponding audio and subtitles. A subpath, on the other hand, represents supplementary content, such as secondary audio or video, certain types of *interactive graphics* (IG) streams, and even audio for *browsable slide shows*, in which

slide show images can be accessed randomly by the viewer without interrupting the audio playback. Text subtitles, which are stored on the disc as plain text and rendered during playback, form another type of subpath defined in the BDMV specifications.

> **NOTE** Text subtitles require far less storage space than the *presentation graphics* (PG) streams normally used for subtitles, since they are stored as plain text rather than collections of bitmap images. Text subtitles support flexible rendering options such as user selected styles and dynamic positioning. However, PG streams have the advantage that they are composed of full bitmap images that can contain almost any sort of graphical information rather than being limited to only text.

Another type of subpath is the *out of mux synchronized elementary stream*. This generally means a file from either the network or the player's local storage that is played back, synchronized to the primary feature video content. For instance, a director's commentary that is streamed over the Internet and synchronized with the feature video playing off of the disc would be an example of this type of subpath. Likewise, *out of mux asynchronous picture in picture video* is another type of subpath, as is the *in-mux synchronized picture in picture video* — a secondary video that is actually multiplexed into the main transport stream with the feature video content and played back in synchronization with it.

Navigation Structures

The navigation structures for HD Movie Mode follow a strict logical hierarchy, as shown in Figure 8.3. This hierarchy defines the complex relationships between the different types of content, including various paths and subpaths, which in turn affect how the content is presented. To be able to author HDMV content well, it helps to understand this structure and the subtle implications that each tier in the hierarchy holds over the behavior of the player.

Figure 8.3 Structural Organization of BDMV Content

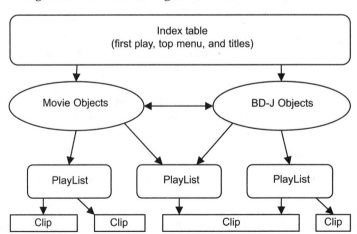

Index Table

At the top of the hierarchy is the *index table*, which defines the *first playback* movie object, the *top menu* movie object, and all of the other titles on the disc with an additional movie object associated with each one. Each entry in the index table is essentially a title, with the first playback and top menu being specialized types. The first playback entry identifies the title that is first played when the disc is inserted into a player. The top menu title, on the other hand, identifies the title that is played when the user presses the top menu key on the player's controller. Every title in the index table must in turn be associated with either a *movie object* or a *BD-J object*, which in turn are associated with the actual playlists and clips that contain content to be played.

Each title (including the first playback and top menu) can be one of two types, either a *movie title* or an *interactive title*. The primary difference between these designations is the set of user operations that are allowed during playback of each type. In addition, a movie title can contain only one playlist while an interactive title can include several. A movie title is intended to be presented as a simple linear playback of primary content such as a movie or music video. By default, during the playback of a movie title, one can use the standard play-back controls such as scan forward or reverse, pause, or skip to the next chapter. In addition, the current title, chapter and time offset are displayed by the player so the viewer can see where he or she is in the presentation.

An interactive title, on the other hand, is intended to be used with content that expects the viewer to interact via menus and other elements. Any video content contained within the interactive title is typically of a nonlinear nature, meaning that it may be accessed in differ-ent order, depending on the viewer's input. Likewise, from an interactive perspective there is a much wider range of user operations that the viewer will typically have access to, includ-ing the up, down, left, and right navigation keys for selecting menu buttons for navigating through the content. Because the playback of the content in an interactive title is typically not linear, showing a current title, chapter or time offset have no real value. In addition, cer-tain functions like pause, scan and skip to another chapter do not necessarily make sense. However, the Blu-ray player continues to allow all of these user operations except *chapter search* and *time search*, which would otherwise threaten the continuity and logic of the inter-active title. In terms of DVD, this would be similar to a sequential PGC title versus a non-sequential PGC title.

Movie Objects

A movie object provides basic setup information for the player, also known as *terminal information*, to allow HDMV content to be played back. In general, this is composed of a few flags to indicate how resume functionality is to be handled and whether or not the *title search* and *menu call* user operations may be applied. In addition, the movie object contains the command sequence that instructs the player on what to do when the movie object is accessed.

Similar to the DVD concept of pre-commands and post-commands in a program chain, the command sequence may be used to query or set values, branch to other locations on the disc, or select streams for playback such as audio and subtitle streams. Movie objects differ from program chains, however, in that there is no separation between pre- and post-com-mands. A movie object has a single command sequence, though within that sequence there

8

may be one or more calls to play a given playlist, at which point command execution is suspended while the content plays, later to return to the sequence to continue execution.

BD-J Objects

Titles may be associated with either movie objects or BD-J objects, which represent Java applications that are stored on the disc. BD-J objects contain a collection of settings including application lists and auto start information, as well as environmental information that the Java Application Manager requires in order to launch the applications. Included in this information are lists of events to which the applications may need access (e.g., specific controller key events), playlists that may be presented, and files that may be read or written.

Playlists

The next level down in the organizational hierarchy of a BD disc, playlists define the actual timeline that the content plays on. You can imagine a playlist as a series of parallel tracks that include individual items such as video streams, audio streams, presentation graphics streams (e.g., subtitle streams), and so forth. Each of these is organized in a linear fashion over a timeline and are all synchronized together within the playlist.

Specific segments of time in a playlist are referred to as *play items*. These define the specific in and out points within the clips that are being played, as well as other playback control information such as connection conditions (e.g., seamless playback), still mode settings, user operation masks and multi-angle settings.

Playlists also include information about the presentation subpaths that may be accessed, as well as their corresponding subpath play items. This allows otherwise independent subpaths to be associated with the playlist timeline.

Finally, each playlist includes a list of playlist marks, which define entry points into the playlist that can be accessed programmatically. For instance, the command sequence in a movie object, or that of a menu button, can include a command to jump to a specific playlist mark to begin playback. Each playlist mark can be one of two types, *entry point* or *link point*. The former is used to define the start of chapters and thus the entry points for chapter search and chapter skipping user operations. The latter is ignored by chapter operations and provides a purely programmatic entry point to which navigation commands may link.

Clips

Clips are defined in the clip information files and represent the transition between navigation structure (playlists and above) and presentation data (the clips themselves). Clips define the information necessary to successfully access and play content from within the clip A/V stream files — the actual multiplexed transport streams. Clips are not directly accessible themselves via user operation or navigation command, except insofar as they happen to line up with playlist marks within a playlist. Instead, they provide the player with the information it needs in order to properly play the underlying content.

Navigation Commands

As was described earlier, movie objects and menu buttons can each have command sequences associated with them. These commands allow the author to perform various arithmetic and bitwise operations, set and read player registers, define conditional operations, and direct playback to specific locations on the disc.

Player Status Registers

The navigation commands defined in the BDMV specifications are able to access an array of *player status registers* (PSRs), which as the name implies provide information about the player's current status and initial settings. Comparable to system parameters (SPRMs) in DVD and HD DVD, the player status registers can be queried by navigation commands to determine information such as the current title or chapter playing, language preferences for audio and subtitle streams, region codes, and so on. Each register is stored as a 32-bit unsigned integer, of which there are a total of 128, though 82 are reserved and 14 are used to indicate the text subtitle features supported by the player, leaving only 32 truly useful parameters. Of these, eight operate as back-up registers for storing resume information when a given movie object is suspended such as by one of the *Call* commands. The remaining 24 registers are divided between *playback status* values, which provide constantly updated information about the current status of the player, and *player setting* values, which specify the initial player configuration settings, such as preferred language or country code.

Similar to those used in DVD, PSRs #0 to #7 indicate playback status such as which streams are playing, as well as the current location in the HDMV navigation hierarchy. PSR #8 is particularly interesting in that it gives the current play item presentation time based on a 45 kHz clock value. Likewise, PSR #9 provides the current number of seconds remaining in the navigation timer's countdown before it triggers a jump. PSRs #10 and #11 provide state information regarding the currently active menu button and page, while PSR #12 provides information about the viewer's selection of text subtitle style preferences. PSR #13 indicates the viewer's parental level selection, while PSR #14 provides playback status and stream selection for secondary audio and video streams. Finally, PSRs #15 to #20 and #29 to #31 provide player configuration settings regarding player capabilities and user preferences.

General Purpose Registers

In addition to player status, the navigation commands in HDMV have access to 4,096 individual *general purpose registers* (GPRs). Like the PSRs, each general purpose register is stored as a 32-bit unsigned integer and may be used in a manner very similar to that of GPRMs in DVD. They serve as a collection of individual memory areas that the navigation commands can use to store variables and other information during execution. Compared to HD DVD's lowly 64 registers, Blu-ray offers over 100 times more storage. However, 4,096 registers may be a bit overkill since one is still limited to working with unsigned integer values, and the available command sets are also limited in function. Therefore, the true usefulness of having over 4,000 of them is questionable, particularly given the nature of the commands and how they may be accessed.

Command Categories

The HDMV specifications identify three categories of commands — *branch*, *compare*, and *set*. Branch commands include execution control, such as *no operation*, *go to*, and *break*. In addition, there are navigation transition commands such as *jump*, *call*, and *resume*. Jump

refers to transitions from one location to another on the disc in which playback of the current movie object is terminated prior to the jump. Call results in the suspension of the state of the current movie object prior to jumping to the new location. Finally, resume allows playback to return to the movie object that was last suspended, restarting playback where it had left off. Branch commands also include playback control commands such as *play* and *link*, which differ in scope, with the former being used to jump to alternate playlists, play items and play marks, while the latter is used to simply change location within the current timeline.

The second category of commands includes the typical comparison operators, including *equal*, *not equal*, *less than*, *greater than*, and *bitwise compare* for checking actual bit values. The comparison operators provide a means for defining conditional branches in your command sequences, a fundamental requirement for virtually any application.

The third and final category of commands, the set commands, is itself subdivided into three groups, including arithmetic operations, bitwise operations, and set system commands. The arithmetic operations include standard functions such as *move*, *swap*, *add*, *subtract*, *multiply*, *divide*, *modulus*, and the *random number* function. Likewise, the bitwise operations offer standard functions for manipulating values on a bit by bit basis, including *and*, *or*, *xor*, *bit set*, *bit clear*, *shift left* and *shift right*. Finally, the set system group of commands let you manipulate the playback system. For instance, set stream can be used to select audio, subtitle or angle streams. The set navigation timer command lets you define a time out period after which playback jumps to a new location. Likewise, the set button page, enable button, and disable button commands let you manipulate menu states.

In general, the navigation commands supported by HDMV are not significantly different than those found in DVD and HD DVD Standard Content, as they represent the basic set of commands necessary to drive a declarative navigation system and allow it to be reasonably useful.

Special Features

HD Movie Mode offers several special features that are rather unique. For instance, the *interactive graphics* (IG) streams defined in HDMV are quite sophisticated in the range of features that they support in order to provide a more dynamic user experience. In addition, HDMV supports a special mode of operation called the *virtual file system* (VFS), which defines a mechanism in which the contents of a disc may be seamlessly updated with new material.

Interactive Graphics

Most declarative navigation environments provide some sort of basic menu capability in order to allow the user to interact with the content and make selections. However, rarely do these environments implement the breadth of features that are found in HDMV's interactive graphics streams. For example, IG streams support multiple menu pages within a single menu. Each page defines a complete set of buttons, and switching between pages can be performed seamlessly, making it possible to create the effect of menus that grow or display state changes in response to user selections.

Although HDMV graphics are limited to only 256 colors, each menu page has its own

8

independent color palette. All of the graphics associated with that page must conform to this limited palette of colors, but other pages can utilize their own palettes. When used wisely, this can make it appear as though more colors are, in fact, available.

In addition, each menu page has multiple *button overlap groups* (BOGs). A button overlap group defines the rectangular area of an individual button on the screen. One BOG is not allowed to overlap another since this would require support of overlapping layers of graphics, however, each BOG may contain more than one button. Only one button within a BOG may be active and shown at any one time, a point that can cause confusion for new authors. The intention is that a BOG represents a single interactive hot spot on the screen. Within that overlap group, however, one may wish to define multiple variations of a button that convey different meanings or have different functions. For instance, in an audio setup menu, one may have a BOG for selecting the first audio stream on the disc. This BOG might include two separate buttons, one showing that the audio stream is already active, and the other showing that it is not active. As the user makes different selections, one or the other button may be displayed within the BOG.

Each button overlap group may contain up to 32 different buttons. A button includes images to represent its normal, selected and activated states, and these may be image sequences in order to provide a cell animation type of effect. In addition, the button can have audio sound effects associated with its selected and activated states in order to provide aural feedback in addition to visual. Finally, a button includes information such as the command sequence to execute when activated, a flag to indicate if the button should be automatically activated when selected, and the navigation linking to other buttons on the menu, corresponding to the up, down, left and right arrow keys on the controller.

Unfortunately, because the navigation linking is contained within individual buttons and refers to other specific buttons rather than the corresponding overlap group, the use of multiple buttons within a BOG is often more trouble than it's worth. Although it would be nice to use multiple buttons in a single BOG to indicate the "on" or "off" state of a feature, for example, the navigation links between the buttons on the menu are not automatically updated to incorporate any newly displayed buttons. For instance, if one button in an overlap group shows an "off" state and all the other buttons on the menu link to it, when the alternate "on" state button in the same overlap group is enabled, all of the other menu buttons will still be linked to the "off" version. As a result, attempting to navigate back to that button will become a problem.

TIP Some times it is easier to create an entirely new menu page in order to show changes in state rather than trying to enable and disable entire collections of buttons within a single menu page using the button overlap groups. In practice, there are really only a few cases in which enabling and disabling buttons within the BOGs is actually effective.

Another unique characteristic of IG streams are their support for effect sequences that may be displayed when transitioning in and out of a menu page[3]. Such effects sequences

---

[3]Technically IG effects and button sounds are optional for the player. However, they are so commonly used that it is unlikely any player will not support them.

8

include fade, scroll, wipe, move, and other types of animation that can be played. However, there are quite a few restrictions. For example, you can have an object move on the screen as part of a transition, but the viewer cannot interact with it during the transition. In addition, although two effect sequences can be performed at one time, no overlapping of the affected areas is allowed. In addition, although several effects are supported, image resizing is not one of them. In fact, HDMV provides no support for resizing images at all. If you wanted to have an image start small and grow quickly, you would not be able to do that using a transition effect. You would, in fact, have to implement the effect using a cell animation as part of a button selection or activation sequence. There are some interesting techniques that one can employ with various menu pages and different buttons so that you can have these kinds of animations automatically triggered and controlled by the button navigation commands, but they usually result in a large number of images having to be stored in the image buffer, which can quickly run out of room.

Virtual File System

Another special feature of HD Movie Mode is the *virtual file system* (VFS). Although not mandatory on all Blu-ray players, on those that do support it, the virtual file system allows you to update content on the disc with new material from the local storage system in the player. For example, new content that has been downloaded from the network to the internal flash memory or hard disc of the player can be seamlessly incorporated into the HD Movie Mode presentation.

The virtual file system essentially defines a means of binding files securely from the local storage to the file structure on the disc. In doing so, it creates an alternate "virtual" version of the disc file structure that is still completely valid. For example, if you have a pair of simple stream files that contain a series of static images, such as actor biography pages, you can use the virtual file system to update those pages periodically, replacing what is on the disc with new versions of the streams. In fact, the files on the disc are not physically replaced. However, when the disc is loaded, the player automatically checks its local storage to determine if a *virtual package* exists for the disc. If so, the player reads both the disc and the virtual package in order to determine how the new files in the package should map into the original disc structure. The player creates a copy in memory of the disc's file structure along with the changes incorporated from the virtual package, creating a new "virtual" file structure for the disc that contains the new material and is subsequently played. As a result, you can then access the disc's contents the same way that you did before, except that now there is new material presented. That new content could even include new menus, buttons, and capabilities that the original disc never included. Although the VFS is listed here as a special feature of HDMV, it is closely tied to advanced BD-Java functionality. In fact, the VFS feature is mandatory for BD players that support network connectivity via BD-Java. Likewise, BD-Java is required for downloading and storing the virtual package. Although HDMV can then access the updated content, it can also be accessed from within BD-Java applications.

BD-Java

BD-Java is based on the Java 2 Micro Edition (J2ME) platform with Personal Basis Profile, a subset of the standard Java platform created for embedded devices. Creating con-

tent using BD-Java is true software development in the sense that it requires the creation of software applications completely within a true programming language, just like writing software applications for a computer. At the time of this writing, there are few if any tools that abstract the programming part of BD-J development, though there are several integrated development environments that can be used to help facilitate the programming process. As a result, one must understand going in that, at this time, BD-J development requires a level of magnitude higher commitment compared to authoring content for other disc formats.

There are several benefits of BD-Java and the full programmability that it offers. Most notable among them is the flexibility that it provides. As a developer, you are only limited by your own imagination and the capability of the hardware that you are running on. Reusability is another great benefit because you can create abstracted code that can be more easily separated from the content itself, which makes it much easier to reuse. A third benefit of full programmability is the extensibility that it offers. Programming languages provide building blocks that can be used to create nearly anything. As the devices they run on become faster and more powerful, the features you can implement effectively can grow as well, without necessarily changing the underlying format specifications. For instance, it is probably not feasible to implement 2D graphic rendering in Java on today's Blu-ray players simply because the processor performance would be too slow to make it very useful. However, as the processors in these devices get faster, such a feature may actually become feasible.

Figure 8.4 Spectrum of Content Authoring and Programming

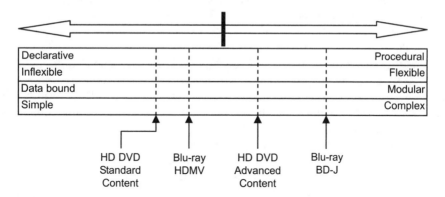

There are, however, certain detriments to using a full software development approach to creating content for discs. For player manufacturers, the challenge comes with flexibility. First, such an approach often results in fewer opportunities to optimize the hardware, making it more difficult to find ways to reduce the price of components in the players. When working with a programming language that is composed of low level building blocks rather than large logical structures, the individual components are already so simple that there is rarely much opportunity to optimize them in any significant manner. Second, programmatic approaches tend to require more general processing power and memory, which tends to add cost to a device. Third, "flexible" devices tend to be the most difficult to test simply because they are so open ended. Although this is nothing new to the PC world, it can be a major prob-

lem for consumer electronics manufacturers whose equipment is not expected to hang up, crash or reboot. The simple fact is that consumers have different expectations for electronic equipment than they do for computers. After all, you wouldn't be happy if your microwave oven suddenly required a reboot each morning, would you?

Another significant problem that a programmatic approach introduces for both player manufacturers and content producers is the question of consistency. Unfortunately, different hardware components excel in different ways. Each new player will have its own strengths and weaknesses, resulting in potentially dramatic inconsistencies between devices. For instance, one player may include hardware acceleration for decoding JPEG and PNG images, allowing it to very quickly load and display images. Another player, however, may have a purely software implementation for decoding images, resulting in the same process taking several seconds just to prepare the images for display. Depending on the application, this could cause the viewer to think something was actually broken in the player.

Short of requiring every player to use the same physical hardware, the Blu-ray Disc Association has taken steps to help alleviate this issue of consistency and performance. Their approach has been to work together to develop a set of guidelines and benchmarks for both player manufacturers and content developers. The guidelines recommend approaches to creating content and designing the players so that the content plays most effectively across all devices. Likewise, the benchmarks provide a set of performance tests intended to exercise the players in a manner that is similar to how real content applications would. The guidelines specify performance levels that all players should, at a minimum, be able to achieve while running the tests. Application developers can then design their software with the understanding that the software should function effectively even on players that perform only to this minimum level.

Unfortunately, the use of guidelines and benchmarks is not a complete solution. First, they can only serve as recommendations to player manufacturers and software developers. There is no strict enforcement, so it is still possible that players released will perform below recommended levels on certain functions. In addition, the benchmarks and guidelines make no attempt to place an upper limit on performance, so there will always be those players that perform much faster than the others on certain functions. In fact, this is generally encouraged, as well it should be, but if the differences between players become too severe, it will become far more difficult to create effective software that spans the gap. Perhaps the biggest problem is the fact that there is no requirement to maintain proportional performance across the functions of the player. For instance, one player may demonstrate strong gains in the performance of its graphics engine, but other general processing operations remain at the minimal acceptable level. Another player may make dramatic improvements to its disc seeking and playback performance, but makes no improvements to graphics rendering.

From the standpoint of disc production, this all results in more time and money. Simply put, it requires more skill to create the programming for a BD-Java disc versus an authoring approach for a declarative system. Writing source code will always be more challenging than using an authoring tool that takes you step by step through the process and only lets you do things that are valid. When programming, every keystroke offers an opportunity for catastrophe since a simple typo can be enough to fundamentally change the inner workings of the program. Finally, given the lack of performance consistency among players and the undiscovered problems that they are each likely to have, platform testing of the software on each make and model of player is likely to be necessary.

Fortunately, once the code is written and tested across a wide range of devices, it can become part of a code library that can be reused on later projects. In time, these libraries of existing software can begin to play an essential role in helping to reduce the time and cost associated with BD-Java development, while also addressing the problems of player compatibility and consistency. It is even possible that such libraries could become the basis of a new set of authoring tools that open up the features provided by BD-Java without requiring the user to write any Java code at all. Already we have seen such developments underway, though it may be years before these solutions become available to the general marketplace.

The Multimedia Home Platform

BD-Java is based largely on the *Multimedia Home Platform* (MHP), 'www.mhp.org', a specification for delivering and presenting interactive applications on broadcast television that was standardized by the European Telecommunications Standards Institute in 2000. The MHP specification was originally created by the Digital Video Broadcasting Project, a consortium of over 250 representatives from European cable, satellite and terrestrial television providers and consumer electronics manufacturers. Although MHP was originally based on European television signaling standards, it was later abstracted to form the *Globally Executable MHP* (GEM) standard, which defines a method for extending MHP to work with other signaling standards and platforms. The GEM standard was subsequently selected as the basis for Blu-ray Disc's BD-Java specification since it already defined much of the functionality that would be required for interactive video applications, and only required a small amount of work to adapt it from a broadcast television model to a disc-based platform.

The GEM specification essentially defines a collection of *application programming interfaces* (APIs) and functional equivalents that must be supported on a consumer device in order to be considered GEM compliant. These APIs are largely based on the Java programming language and are particularly well suited for interactive video applications and devices. In particular, the APIs were specifically selected for operation on embedded devices, such as settop cable boxes, which tend to have tight constraints on memory, processing power, local storage, and other resources that PCs often have in large supply. In addition, they address other challenges, such as the lack of a keyboard for entering user input, or window-based operating system with standardized user interface controls like text fields, buttons, check boxes, etc.

Extensions for BD-Java

Due to a few fundamental differences between broadcast television and packaged media (e.g., optical discs), several extensions had to be made from the original GEM specifications in order to effectively support disc-based content. For example, a broadcast environment has no concept of random access within the timeline of a transport stream. The closest it comes

8

is the ability for the viewer to change television channels at any time midstream, but the idea of jumping back to the beginning of a stream or providing fast forward and rewind functions just doesn't make sense in the broadcast world. Likewise, because of the ethereal nature of broadcast content (it's there for a moment and then gone) application data has to be delivered via a data carousel mechanism in which it is repeatedly transmitted with the hopes that the viewer will remain on the channel long enough to receive the complete package. A disc simplifies packaging, in terms of file accessibility.

In general, adaptations to the GEM specifications by BD-Java come in three forms —

- Specification of functional equivalents that, although different in mechanism, provide essentially the same capability (e.g., using OpenType fonts instead of another font type, such as TrueType)
- Removal of areas for which no functional equivalent is meaningful (e.g., elimination of the data carousel concept)
- Specification of additional extensions for functionality that is necessary, but for which there is no functional equivalent defined in the GEM specifications (e.g., adding player controls such as jumping to another title or playlist, or changing video angles)

Approximately one third of the BD-J specifications are devoted to defining functional equivalents between an MHP terminal and a Blu-ray Disc player. In the following sections we discuss changes in mapped features.

GEM Features Removed

As mentioned earlier, the concept of a data carousel that repeatedly broadcasts applications and their data to any terminals that may be listening has no direct relation to disc-based media. Although it is possible to create a mapping, for example, to data that is multiplexed into the transport streams on a disc, it is somewhat unnatural and generally unnecessary. As such, this is one feature that has not been propagated from the GEM specifications to BD-Java.

Although GEM itself does not reference it, MHP supports a multimedia format called DVB-HTML, which provides Web page style documents delivered over broadcast channels. This feature is specifically not supported in BD-Java.

Many other minor features have also been left behind in the GEM specification that simply did not make sense in the BD-J environment. For example, tuning and conditional access APIs have been removed, which have no meaning in a disc-based environment. Also, APIs regarding persistent storage access have been supplanted by the BD-J concept of the *application data area* (ADA), which essentially corresponds to secure access to local storage in the player.

Extensions Added

BD-J adds several APIs that are critically important to the disc-based model of Blu-ray. Of course, the most basic of these are the APIs and mechanisms defined for loading a BD-J application from the disc and starting its execution. In addition, a full suite of playback control APIs are provided to allow the BD-J applications to override default behaviors for common controller inputs such as scanning forward or reverse, skipping chapters, or pausing playback, among others. The playback control APIs also provide a mechanism for accessing playlists, content streams (e.g., audio, video and subtitles) and jumping to other titles on the disc, as well as tying into the virtual file system functionality of HDMV.

One unique feature that has been added is the concept of a multi-disc application life cycle. Using this feature, an application may be launched that continues to remain resident on the player even when the disc that it came has been ejected and a new disc is inserted. Of course, the new disc has to identify itself in a manner that allows the application to continue running in order to avoid security concerns. However, this feature makes it possible for an application such as a game or menu to remain operational as the user switches discs, such as to access a movie's sequel or a subsequent episode in a television series.

Another new feature added in BD-J is support for a quarter resolution (960x540 pixels) graphics mode. By reducing the number of pixels that must be drawn for each screen update, graphics performance can be greatly improved for graphically intense applications, while still providing a very high quality, high resolution experience. Theoretically, such a mode should make it possible for a player to deliver an increase in animation frame rates by a factor of four!

One of the most interesting extensions that BD-J adds is *frame accurate animation* (FAA), a mechanism for presenting high performance image sequence animations on the display. The system allows the BD-J application to pre-load a series of images into an image buffer, which the player subsequently displays one by one according to timing information provided by the application. These animation sequences can either be displayed independently, or they may be synchronized frame accurately with the feature video content that is playing at the same time. The intent is that frame accurate animation would offer animation rates that exactly match the frame rate of the content that is being played back, and can even be synchronized with specific video frames. However, there are limits to the amount of memory that players provide for the image buffers, thus limiting the duration and size of the animations.

BD-Java System Model

The BD-Java system is composed of four general components, the BD-J applications themselves, the BD-J application programming interfaces, the BD-J player module, and the player resources. As shown in Figure 8.5, these components have an almost linear relationship to one another, in which each provides the gateway for access to the next item in line. In other words, the BD-J applications only have direct access to the APIs, which in turn provide a gateway to the player's BD-J module. The BD-J module controls the application and the APIs, and marshals access to the player resources at the furthest end of the chain.

Figure 8.5 BD-J System Overview

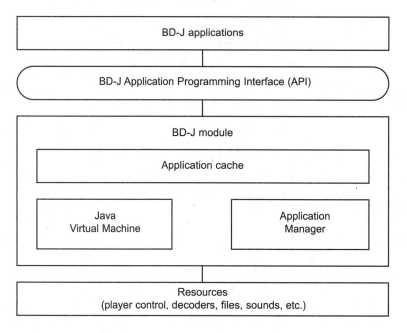

BD-J applications are referred to as *Xlets*, a term carried over from MHP. Similar to applets, which are commonly used on Web pages, Xlets provide a lightweight application model in which individual programs follow a simple life cycle state machine as they are loaded, paused, activated and destroyed.

The BD-J APIs are the collection of functions and objects that the applications use to perform any action of significance in the player. Without these APIs, an application could do little more than perhaps some basic math and assignment operations. For example, any attempt to manipulate text, load files, access the network, control playback or any number of other functions that would be interesting in an interactive media application must be performed via one or more API calls.

The APIs reside somewhere between the BD-J application and the underlying BD-J player module, which includes the application cache from which the Xlets are loaded, the Java Virtual Machine on which the applications run, and the Application Manager, which determines what applications may run and when. It is this underlying system that also controls access to player resources, translating the API functions to underlying system calls.

The player resources themselves represent the full range of player capabilities and functions. Included among these are playback controls, decoders, files stored locally and on the disc, network resources, memory buffers, sound effects, and so on. The BD-J applications do not have direct access to any of these resources. Instead, they must always access them through API calls, which in turn are marshalled to the system by the BD-J player module.

Relation of BD-J to HDMV

Unlike the relationship between HD DVD Standard Content and Advanced Content in which either may exist on a disc independently of the other, BD-Java is entirely encapsulated within the HD Movie Mode navigation structure. BD-J applications themselves provide an alternative control mechanism to the movie objects and interactive graphics streams of HDMV, but they can only be accessed through the HDMV index table and titles. In addition, BD-J applications can only play multiplexed audio/video content via HDMV playlists and clips. As Figure 8.6 illustrates, the BD-J objects (BDJOs) that encapsulate the BD-J applications on a disc are just one component within the larger HD Movie Mode hierarchy.

Figure 8.6 Relationship of BD-J to HDMV

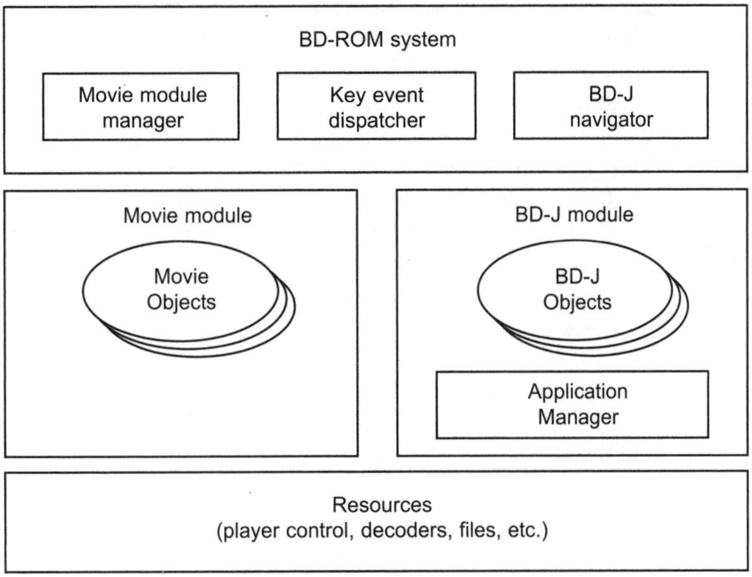

```
┌─────────────────────────────────────────────────────────────┐
│                        BD-ROM system                         │
│   ┌──────────────┐   ┌──────────────┐   ┌──────────────┐     │
│   │ Movie module │   │  Key event   │   │    BD-J      │     │
│   │   manager    │   │  dispatcher  │   │  navigator   │     │
│   └──────────────┘   └──────────────┘   └──────────────┘     │
└─────────────────────────────────────────────────────────────┘

┌───────────────────────────┐   ┌───────────────────────────┐
│       Movie module        │   │        BD-J module        │
│                           │   │                           │
│        ╭─────────╮        │   │        ╭─────────╮        │
│        │  Movie  │        │   │        │  BD-J   │        │
│        │ Objects │        │   │        │ Objects │        │
│        ╰─────────╯        │   │        ╰─────────╯        │
│                           │   │   ┌───────────────────┐   │
│                           │   │   │   Application     │   │
│                           │   │   │     Manager       │   │
│                           │   │   └───────────────────┘   │
└───────────────────────────┘   └───────────────────────────┘

┌─────────────────────────────────────────────────────────────┐
│                          Resources                           │
│            (player control, decoders, files, etc.)           │
└─────────────────────────────────────────────────────────────┘
```

HDMV is restricted to only allow control to be passed to a movie object or a BD-J object, at any given time. While a BD-J object (and the BD-J applications contained within) has control, no movie object may be executed, and no IG stream may be presented. This ensures that there is no conflict in control between the HDMV content and the BD-J applications.

BD-J objects are similar to movie objects in the sense that they provide the terminal information and metadata that is necessary to manage the applications that are to be run. The BD-J object includes references to the Java archive files (JARs) that contain the BD-J applications and their code. In addition, the BD-Java object defines information for the Application Manager, such as which events should be exposed to the BD-Java application. The BDJO also provides auto-start information to indicate which BD-Java applications should be started up immediately upon entry, as well as application cache information to help control the pre-loading of files that are needed from the disc. In addition, the BD-J

8

object also provides a list of the playlists that the application may need to access, as well as a flag indicating whether or not the first playlist should be started automatically.

When is BD-J Appropriate?

As mentioned earlier, there are numerous challenges in trying to work in a fully programmable environment like BD-Java. It tends to be much more complex and requires more time to program and test. So the question must be considered whether or not BD-Java should be used for any given project. Making the wrong choice can be costly. For example, if you fail to realize that your project requires a specific feature that is available only in BD-Java and started down the wrong path, it could mean building a large portion of the project all over again.

The fact is that HD Movie Mode, though having the benefit of supportive authoring tools, is still quite complex to work with and has many restrictive limitations that can render some applications impossible to implement. In addition, some operations, such as complex popup menus and animations that are difficult to implement in HDMV, may actually be easier to create using a programmatic approach in BD-J.

Another issue that should be taken into consideration is the fact that as you create BD-J applications, you can start to build up code libraries that you can reuse on future projects. Therefore, the increased time that might be required for the first project may result in a significant decrease in time on each subsequent job. This is quite different from HDMV authoring in which each new project has to be started pretty much from scratch. In the end, the choice will depend on the needs of the product being created, the resources available, and the skill set of those doing the work.

Chapter 9
Compression and
Authoring Tools

This chapter is a reference for selecting the best compression and authoring tools for your projects. Each section in the chapter lists the manufacturers who have released or announced products by the time of this writing. We have included list of features that you may use as a starting point in defining the specific needs of your operation.

Audio Compression

The tools that you need for audio compression depend on what you are doing in-house versus using third parties, the codecs that are needed for your projects and whether the authoring tool is handling audio compression as "transcoding". Also, the advanced software suites from Dolby® and DTS™ include the ability to modify compressed streams and to change some of the parameters such as dialog normalization.

Please remember that there are differences in how the streams are prepared for Blu-ray Disc™ and for HD DVD. A compression that was prepared for BD may not work or may need to be converted for use with HD DVD and vice versa.

Here are the tools that you should consider using for professional audio compression —

- Dolby Media Producer software suite comes with Encoder, Decoder, and Media Tools.
- DTS Master Audio™ Suite, http://www.dtsonline.com/pro-audio/mas_prodspec.-php. A version of the encoder is included in some versions of the Sonic® Authoring and Compression tools. Check with Sonic Solutions for the latest offerings.
  - DTS-HD™ Encoder
  - DTS-HD StreamPlayer™
  - DTS-HD StreamTools™
- SurCode™ DTS & Dolby Digital Encoders (http://www.minnetonkaaudio.com/products/products.html)
- PCM and the wide range of tools that support digital audio in 48 kHz/96k Hz 16- and 24-bit formats.
  - Basic tools include Windows® Sound Recorder (good for 16 bit, 48 kHz)
  - Most audio software, MacOS® or Windows based support output to PCM
  - Check that the audio tools and the authoring tool work together. For example, although many audio tools use Broadcast Wave Format

9

(BWF) because it includes the SMPTE timecode in the file, BWF is not supported currently by most of the authoring and multiplexing tools. You also may need to use a separate file for each channel of the audio instead of a multi-channel wave, again depending on the capability of your authoring and multiplexing tools.

Video Compression

You need to select a tool that meets all of your needs. Depending on the types of codecs needed for your projects, you may need more than one encoder. Just as we have seen in standard definition DVD with different MPEG-2 encoders, some encoders do a better job with some types of materials than other encoders. For a large operation, you may need all of the encoders listed below and then some.

TIP
In the early stages of the next generation disc format, you can not assume that an encoder is generating specification compliant streams. You may need to verify that, after encoding, your streams are specification compliant.

The selection of a video compression tool is governed by —

- Budget for hardware and software
- Speed — this is the raw encoding speed per pass. This should not be confused with the actual throughput of your workflow that you have defined for your operation.
- Workflow — how are two and three pass encodes processed? How long does it take to get a final encode, e.g., a full-length feature film can take from one to three weeks until you have a stream that meets the clients requirements and can be multiplexed into the final project.
- Quality and Flexibility — the more advanced compression tools yield higher quality results and offer more control over the compression parameters, resulting in better quality encodes for difficult scenes.
- Required Codec(s) — depending on the type of projects and project bit budgets, you may be able to use just one of the three video codecs specified.
- Video Source format — do not forget about specifying the types of tapes or files that you are expecting to use as input for the compression. Often, the video capture step is as capital intensive as the compression capability.

Most of the consumer-level authoring suites include support for built-in compression. They will take a Quicktime® or Windows Media® Format or HDV file and automatically compress. The control over the compression is often limited to "High", "Medium", and "Low" quality settings.

Here is a list of the professional video compression tools that were available as this book was going to press —

- Sony MPEG-2 HD Encoder — this is the only hardware-based encoder in this list. It supports both 50 Hz and 60 Hz MPEG-2 encodes for all frame rates and resolutions for Blu-ray Disc. The to-be-encoded source video can be HD-SDI or YUV file-based. The quality and speed of this encoder are very good.
- Microsoft VC-1 Parallel Encoder Project (PEP) — this encoder is available from Microsoft on a limited basis. It currently is not a product but is offered as a "technology". The software has been offered as "free", although the hardware requirements for the parallel encoding are extensive. Microsoft includes a utility that transcodes the streams from one disc format to another.
- Cinemacraft HD encoder (CC-HDe) — this AVC H.264 encoder workstation allows access to a full range of capabilities of the AVC codec. It also includes pre-filtering. The Cinemacraft MPEG-2 encoders for SD DVD have a reputation for being dependable, high quality encoders.
- Sonic Cinevision™ (MPEG-2, VC-1, AVC) — handles compression for HD DVD and for Blu-ray Disc. If you pick a stream with a bitrate that is compliant for both formats, it is possible to encode once.
- Apple Final Cut Studio®, Compressor 2 (MPEG-2 & AVC) — this encoder gives good results although it has historically not allowed control over many of the parameters of encoding that are often needed in professional quality encoding.
- MainConcept's H.264 Encoder v2 for Microsoft® Windows — is a new version of their encoding engine that encodes MPEG-1/2 and AVC. This engine has had a reputation for reasonable speed and good quality results. It exposes a number of the codec parameters to help get the best quality out of the codec.
- PixelTools' Expert-H264® — encodes to the AVC codec from most common video file formats. It allows for control of the basic parameters.

Authoring Tools

Choosing the right tools is based on your project requirements. You want to consider how productive your operation can be with the tool or toolset. Does it support multiple users, the sharing of projects, and/or breaking up the workflow for different steps (e.g., can programming be separate from multiplexing)? Consider which of the features that you want to use when selecting the right tool —

- Authoring mode (Standard Content or Advanced Content) — do the projects that you have only require Standard Content mode support or do you need the Advanced Content authoring mode, too? For Blu-ray Disc, how well does the tool support the integration of both modes for that format?
- Codec support — ensure that the codec that you will need for your projects is supported by the authoring software. Not all of the tools support all of the permutations of the codecs that are included in the specifications.
- The ability to select portions of audio, video and subtitle streams when creating playlists — you want to be able to adjust the streams with trimming (head and tail), appending and offsetting (forward and back). To reduce the duplication of

assets and to accommodate the actual delivered assets, as opposed to what you specified, you want to be able to adjust the assets within the authoring tool. You must be able to modify start and stop times so that you can sync assets, as necessary.

- Simulation ability — simulation is a must for quickly checking button function and logic accuracy. For advanced modes, details of the player status (e.g., register contents, current instruction being executed, etc.) are very helpful to be able to quickly troubleshoot problems with a project.
- Output media supported — single and dual layer versions of DVD+/-R/RW, BD-R/RE, and HD DVD-R. Also, you want to have the output of the image as the whole disc image as well as the L0 and L1 layers.
- CMF/DDP formats — to be used for replication master
- Output Format control — ability to set layer break and control layout of files on the disc
- AACS — this is a must for replication for Blu-ray Disc and is recommended for HD DVD when copy protection is needed. Look for ease of use and flexibility. Mistakes in this step are costly. You want a tool to make this as foolproof as possible.
- BD+ — for Blu-ray Disc, does the tool support integration with the Blu-ray Plus protection level
- Graphics — how easy is it to create, load and modify the graphic elements used for button states, backgrounds, and transitions/animations. What degree of re-authoring is required if/when the graphics are updated? For Blu-ray's BDMV mode, does the tool manage palette optimization to minimize the impact on the authoring?
- Subtitling — check that the tool supports the subtitle format that you or your subtitle vendor are able to deliver
- Advanced feature support including —
  - Multiangle
  - Multiple audio tracks
  - Seamless branching
  - Picture in picture
  - Slideshows with non-stop audio
  - Integration with advanced authoring (i.e., BD-J)
  - Persistent storage and networking features
- Verification and debugging tools — depending on the complexity of the discs and the target market (e.g., Hollywood titles), you will need tools that verify that your final disc image is compliant with all of the specifications (file format, file layout, correct fileset, layer break, bitrate/buffer usage, code validity, etc.). Debugging tools and emulators allow you to inspect the state of the player so that you can ensure the player is doing what you expected.

HD DVD Standard Content

Depending on the complexity and flexibility needed for your projects, there are a number of authoring software tools to consider. For simple titles, here is a list of vendors and products to begin your search —

- Corel's Ulead® HD DVD Authoring Suite allows creating HD DVD discs from VC-1 streams, Dolby® Digital and Dolby Digital Plus audio as well as ACA objects.
- CyberLink PowerProducer 4 allows you to easily author HD DVD with a choice of base templates to customize your disc creation. This is based on their VCD and DVD version which allows the assembly of videos, trimming video and setting chapter points. The suite also allows selecting standard VCDs, DVDs, and Blu-ray discs.
- Pinnacle, a division of Avid, has the HD DVD Authoring Pack that enables users to create high definition DVD movie discs, complete with HD motion menus, within the Pinnacle Studio Plus software using current generation, non-HD DVD burners.

For more advanced prosumer and professional projects, these are some of the tools available —

- Apple DVD Studio Pro®
- Sonic Scenarist® HD

HD DVD Advanced Content

For Advanced Content, you may need to assemble a set of tools in order to implement your authoring capability. Here are some of the tools to consider —

- Sonic Scenarist Advanced Content Authoring (SACA)
- Toshiba Advanced Content Multiplexer (limited availability)
- InterVideo®/Ulead Advanced Content Multiplexer (limited availability)
- Editors for XML and Javascript™:
    - <oXygen/>®
    - Microsoft Visual Studio®
    - Notepad
    - Eclipse
    - XMLSpy®
    - others...

BD HDMV & BD-J Authoring

The consumer level authoring tool selection should include a review of the following —

- CyberLink PowerProducer 4 currently authors only BDAV and not HDMV discs for simple titles with no menus. Future versions may include more capabilities.

Menu options are currently only available if you are creating a VCD, DVD or HD DVD.

- ArcSoft TotalMedia™ Extreme is a suite of applications that allows consumers to play video, create and edit video and music discs, and backup files using either Blu-ray Discs, DVDs, or CDs.

Prosumer and professional authoring tools include the following options —

- Roxio DVDit® Pro HD
- Sonic Scenarist BRE
- Sony Pictures Blu-Print™

Java Development Tools —

- Eclipse (integrated development environment)
- Java™ Developer Kit (JDK)
- NetBeans™ (JDK + integrated development environment)

Subtitling

The next generation disc formats support higher resolution subtitles and the authoring tools support formats that are different than those of standard DVD. At the time of this writing, the existing standard definition software tools are being enhanced to support HD compatible subtitles. Although the script file may work, the graphic file formats will need to be converted to HD supported PNG format. Some of the items you should look for in selecting the right tools are —

- Timecode needs to match the video frame rate which for HD may be 23.976 fps
- Frame size of 1920x1080 support
- Enhanced features available to Blu-ray Disc and HD DVD such as alpha, increased bit depth, and increased number of colors

Summary

The tools for creating the next generation discs are evolving as they support more and more of the powerful capabilities of the specifications and as they implement improvements in workflow and productivity. When selecting the right tools, look to the vendor track record for support and for a viable product roadmap. Beyond the purchase price, many of these tools also require ongoing support contracts or additional costs upgrades. Identify these expenses and plan on them.

Another aspect to consider is the training required for effective use and project production, on-time and high-quality, with the tools you have selected. You want to ensure that the tool set you have created will support your current needs and will also be able to grow with your operation as the disc formats mature.

Chapter 10
Setting Up for Production

Whether your operation is large, small, growing or redefining, you will need the same basic components as in standard definition DVD — project design and planning, graphic design, audio and video compression, authoring and programming, quality control testing, duplication or replication, and packaging. As you build a facility to produce high definition DVDs, three factors will determine your production capacity — people, process and tools. Setup involves implementing the right combination for optimal production capability (see Figure 10.1).

Figure 10.1 Production Capacity Pyramid

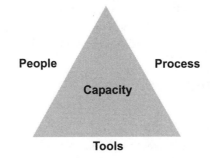

We thought it would be helpful to look at three categories of production operations —

❶ Large scale/Hollywood Production

- ▨ DVD production is essential to your organization's income.
- ▨ You produce multiple titles at the same time, using a large team of specialists.
- ▨ You compete by providing the most sophisticated and/or entertaining innovations in interactive user experience, compatible with even the least capable high definition DVD players.
- ▨ You work with a network of vendors with their own specialties, like 'Eastern European Subtitles' or 'Button Sounds'.
- ▨ Schedules include fixed shipping dates, with release dates publicized broadly, domestically and internationally.
- ▨ High quality is vital — small mistakes/compromises could anger thousands or even millions of customers.
- ▨ Large scale operations typically ship one or more titles per day. Facilities of this size may be working on twenty titles or more simultaneously.

❷ Small scale/Prosumer and Corporate Production

- ▨ You are getting paid for results that look good, function well, and

10

even sport a few cool features.

■ You have a small team of generalists working on more than one title at a time.

■ You must meet a definite deadline.

■ High definition DVD production is part of a set of activities — integrated with production or post-production.

■ Small scale operations finish four to ten discs per month. Discs may be moderately complex, each carrying 10 to 45 minutes of video and utilizing more than one menu layer.

❸ Single-user/Consumer Production

■ You produce one disc at a time, with a flexible schedule. You do it for fun or adventure, with results viewable by non-professionals, your Mom or spouse.

■ A personal or consumer producer may have a title with a basic chapter index, playing 15 to 30 minutes of video. The main purpose of this disc type is to share content, with minimal interactivity. This producer completes less than one disc per month.

Preparing for Production Capability

There are four steps to preparing for your production capability. The first step is to define your required capacity by characterizing the type, number and quantity of assets. These characteristics give you the input you need to define the process, tools and people required. Below are the questions you should be addressing at each step.

❶ Capacity Requirements - quantify:

■ Titles (per month)

■ Minutes (in minutes per title, week, month)

■ Menus

■ Buttons

■ Transitions

■ Audio and video assets

■ Subtitle requirements

❷ Process definition:

■ Identify the steps that establish your workflow.

■ What software and hardware tools are available to perform each step?

■ What skills will be required by by your team?

■ Start process definition at a high level, refining the definition as you identify ways to increase efficiency or productivity.

■ For enhanced team involvement, you may ask individual staff members to develop and maintain a procedure for their own roles in workflow.

❸ Tools:
- ▪ Identify the software, hardware and network configuration that meets your capacity and budget.

❹ People:
- ▪ What is the organizational structure you will use to support production, from sales to shipping? How does the structure support and compliment the workflow and toolset?
- ▪ Define requirements for each supporting group or role, and plan how these will be implemented within your organization. Take into account shifts per day/week, locality (networked or all in one location/floor) and use of contractors or suppliers.

Figure 10.2 shows the essential functions required for effective organization of a large-scale operation. You may need to address additional functions such as human resources or training. You can modify your organization to meet the unique needs, challenges and chemistry of your market, clients, and staff. Some specialized functions may be outsourced, such as graphic design or subtitles. In a large operation involving more staff members, highly specialized roles may support more efficient productivity, while in a smaller scale operation, individuals may be responsible for several functions.

Figure 10.2 Organization Chart for Large Scale/Hollywood Operation

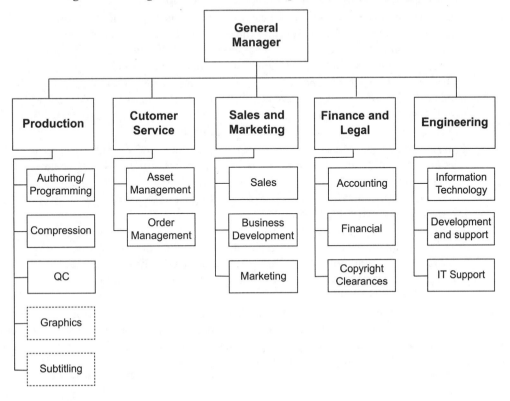

For a large scale operation, the functions may have the following roles and responsibilities. A small scale or consumer operation may have all of these functions, but one person wears many or all of the hats.

- **General Manager** (GM) Responsible for profit and loss, the GM sets the goals and direction for the organization while maintaining a smooth running operation that is responsive to its clients.

- **Production** The team that is directly involved in producing the title/s consists of one or more people in the roles of authoring/programming, compression, quality control, graphics and subtitling. Graphics and subtitling are often handled by subcontractors or vendors. Compression is responsible for encoding video and audio streams, and preparing/preprocessing these streams for the multiplexing step of authoring.

- **Quality Control** This group ensures that the final product meets all of the quality expectations of the client. Responsible for linear viewing of the all of the material on the disc as well as the navigational QC. This group ensures that the disc looks, sounds and behaves the way it is expected. Develops test plans for each title. Measures quality metrics for discs being produced and identifies areas in the process so that the same type of problems don't occur on future titles. If there is bad news to be learned about a disc, you want to hear it from this group first and not the client or worse yet the endusers.

- **Customer Service** This group takes order requests from clients, maintains capacity and delivery schedules, tracks the order for the clients and with vendors to ensure that all of the correct media assets arrive at the right time. This group may also log and stage assets for use by the production group.

- **Sales and Marketing** Responsible for understanding and helping define the needs of the market, this group works with clients to understand and develop product and service concepts that meet client needs and requests. Specific orders may be passed to the Customer Service group.

- **Finance and Legal** Measuring and reporting revenues and expenses for products and services. Finance monitors the return on investment (ROI) for hardware/software/personal expenses. Finance is also involved in the analysis of plans for new products and services, while the Legal team ensures that contracts are well formed, avoiding litigation.

- **Engineering** Maintaining the production infrastructure of the operation, Engineering manages information technology (IT) components of organization networks for both administrative and production needs. Your clever, resourceful Engineering team may also develop and support utilities that help the entire operation run more smoothly.

Figure 10.2 is an example of the core functions in an organization, and provides an opportunity to look at the elements that are involved in shipping a title. You may need to address other functions such as human resources or training or any one of a number of other items that may come up when running a business. Modify your organization to meet the unique needs, challenges, personalities and idiosyncrasies of your market, clients, and staff.

Single-User/Consumer Production

As a single-user/consumer operation, you may be a home movie enthusiast putting together a few discs to cover major holidays, perhaps as a video greeting card for family and friends. As an avid videographer or video artist, on the other hand, you may produce a number of titles for exhibition or limited distribution.

Single-user operation is the most tightly integrated. Look for ways to simplify the operation to keep it manageable. A very low capacity requirement can be misleading, because usually only a small part of each day is available for working on a title. Tools and processes must be easy to use with low error risk, easy to learn (or re-learn between titles!) and quick — long production time increases the risk of lost interest and aborted projects. Fortunately, most tool sets geared toward the individual consumer are designed to simplify processing and reduce production time.

If you are your own customer, creating products out of personal enjoyment rather than monetary gain, then you may minimize or eliminate sales and marketing, shipping, replication, special packaging, and other out of pocket investments.

Define Capacity

If you are an individual producer, defining capacity is a matter of setting time expectations for each step. Most of your time will be spent in the acquisition and preparation of the material. Once the creative process of preparing the sequence is complete, putting it onto a disc tends to be a predictable, technical process — sometimes referred to as "turning the crank."

Process Definition — Workflow for a Single-User

Operation workflow areas that are more precisely articulated in a larger operation, such as project planning and interactive design, are simplified or even automated by the individual producer's tools. As compression and authoring tools continue to develop, they become more efficient, while new tools also tend to come with new features that can slow down processing, resulting in no net time gain.

As a general rule[1], the production process or workflow (see Figure 10.3) can be divided into six steps, each step taking an hour or two, depending on the speed of your computer and the content being processed —

① Capturing the audio/video content
② Compressing the content
③ Authoring disc behavior
④ Multiplexing the disc contents
⑤ Recording the contents to disc
⑥ Testing the disc

---

[1]Some systems will allow you to author the behavior of the disc first, and combine the compression and multiplexing stages as the system can perform these operations unattended. Other systems allow you to test the playback of the disc contents before recording them to an optical disc, allowing you to catch and resolve errors without wasting recordable media.

Figure 10.3 Typical Workflow for a Single User Operation

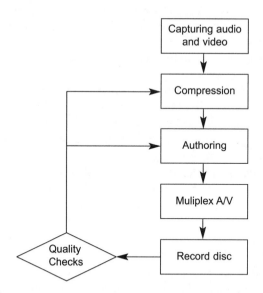

The disc contents of an individual production will often operate at such a straightforward level that very little planning is needed. In addition, since there is only one person involved in the operation, there is less need for ongoing documentation to be shared among the many people who may be involved in larger operations.

Most consumer-level compression and authoring tools handle graphics creation automatically through "themes," "styles," or some other standardized manner, and minimize or skip functionality required by large producers.

Tools

A large operation requires a diverse set of tools and systems and greater parallelization of the production process, to reduce turnaround times on multiple, simultaneous projects. While a consumer operation works primarily with audio/video files captured directly from a camera or output by a video editing application, larger operations' audio and video content is delivered separately and requires more complex handling, including synchronization. The single-user operation is performed as a linear process, often on a single computer system using a single integrated application.

Much of what can and cannot be accomplished is determined by the tools that are available. Tools designed for the solo producer tend to be relatively inexpensive, and built for ease of use. Like cameras, microscopes or other professional equipment, these tools are built for price, speed and convenience over professional quality. Even so, since most consumer projects are relatively short in duration, encoding data rates can be kept high, boosting disc quality.

So far, these tools also tend to be very specific in the features they offer — and they will do what you need them to do — but powered by structure rather than by versatility, with a significant impact on your production workflow. Software may require that the video content be delivered in a specific format, such as QuickTime® or Windows Media® format, or the software may only accept video directly from a camera.

Specialized hardware may be required to use the software effectively. Blu-ray Disc™ production requires a BD-R/RE drive, currently costing US $1,000 and up. In addition, the software necessary to create and/or playback HD DVD or Blu-ray Disc content may require a very powerful computer system, capable of fully supporting the processing load. To view high definition video, you will also want to use a graphics card and monitor that fully supports high definition resolution.

Recommendations for a Single-User Operation

At the launch of the HD DVD and Blu-ray formats, tools were being advertised with a surprisingly large range of features. It can be difficult to distinguish between the real capabilities of the products and the marketing hype that surrounds them. Many applications come with a load of features and formats, without clearly defining which are compatible and which are mutually exclusive. Our best recommendation is "buyer beware." Download trial versions or find someone who can demonstrate the specific features you want to use, before you commit to a system.

Small Scale Prosumer/Corporate Operation

As a corporate operation, you may have a small team of specialists working on more than one title concurrently. You may finish four to ten discs in a month's time, and discs may be moderately complex, each carrying 10 to 45 minutes of video and utilizing more than one menu layer. You will most likely use the Standard Content modes of Blu-ray Disc (HDMV) or HD DVD, and your authoring tool may or may not support subtitles, scripting, or AACS.

WARNING	BD and HD DVD Advanced Content modes require formal programming skills, at a level much higher than those required to author a DVD. We recommend that most users use the Standard Content modes of these formats until authoring tools better facilitate Javascript/XML and BD-Java™ development.

Define capacity

To help you answer the question, "How long will it take you to finish that disc?", Table 10.1 analyzes the work involved for a prosumer/corporate operation, and the complexity of titles typically created at this level. This table is also included in a file on the disc that accompanies this book.

You can define your capacity by characterizing your titles in terms of the number of minutes of video, number of chapters, number of buttons, and other terms modeled in Table 10.1. You can use the table to calculate the time it takes to complete a title.

25

Table 10.1 Sample Worksheet, Time Calculation Model for Small Scale Operations

	Small Title	Larger Title
Asset Characteristics		
		Minutes
Main Feature	15	30
Bonus/Extra	0	3
		Counts
Chapters/segments	2	5
Audio Tracks	1	1
Menus (Top and Pop-Up)	1	2
Buttons	3	8
Transitions	0	1
Subtitles	0	0
Language Sets	1	1
Game/Interactive Experience	0	0
Processes		
		Minutes
Compression – Video		
2 pass VC1/AVC encode	375[a]	825[b]
Re-encodes	0	825[c]
Compression – Audio	15[d]	33
Title Design/Architecture/Flowchart[e]	8	18
Graphics[f]		
Static	40	100
Animation	0	20
Authoring[g]	65	165
Programming	0	0
Multiplexing[h]	7.5	16.5
Record Test Disc[i]	7.5	16.5
QC - Viewing plus Nav[j]	40	90
Specification Verification	0	0
Premastering AACS	0	0
Premastering CMF	7.5	16.5
Total minutes	566	2126
Total hours	9	35

continues

Table 10.1 Sample Worksheet, Time Calculation Model for Small Scale Operations
(continued)

[a]For a single workstation encoding solution, expect that each hour of video will take 25 hours for two-pass encode. Adjust this benchmark according to your processor speed, drive features and other local factors.

[b]For a small cluster encoding solution, you may expect 10 times realtime encode speed.

[c]Even though you may only be re-encoding small sections of video, you should allow for the same amount of time as for the initial encode. Depending on the encoder, the source material and the client expectations, the time for re-encodes may be even greater.

[d]Audio compression often can be done in realtime. Validate this assumption in your actual implementation.

[e]Allow some time for the design of how a title will be built in the authoring tool. Detail the requirements for graphics. Here we have allowed one minute per button, chapter or transition. One minute per asset would be a viable number for larger projects.

[f]Graphics creation can vary considerably, but as a rule of thumb, start with 10 to 20 minutes per button and 20 to 45 minutes per animation.

[g]Authoring for standard content mode titles is very similar to DVD authoring. Figure about five to ten minutes per button or animation.

[h]Multiplexing takes approximately half as long as the total running time for all of the video assets.

[i]See table 11.4 for specifics on how to calculate the disc record time.

[j]QC takes about five to ten minutes per button, animation or chapter.

Notes:

1. Game design, implementation, and testing times are not included.

2. Other activities to consider include third party QC, check discs and usability testing.

3. These times assume that all tasks proceed linearly, with no concurrency or rework.

4. The time to multiplex subtitles into streams is included. Subtitle creation is not included.

In DVD, we estimate that authoring for a title takes about five minutes per button, chapter point or asset. This arbitrary estimate is tuned by our experience. You can start simply and make the model only as complicated as need be. In Table 10.1, the rule of thumb for the amount of authoring time for elements in an HD project is 15 minutes. For compression, an estimate of 15 to 30 times realtime is in the ballpark. Be sure to account for the time for re-encodes and re-deliveries.

At this point in high definition disc development, numbers shown in the Table 10.1 sample worksheet generate conservative estimates that are realistic. As tools improve and your people gain experience, your numbers may well improve.

The number of titles you need to complete each day will determine the number of workstations and staff used within your organization. Depending on the complexity of titles, one person with one workstation may be able to complete between four and twenty titles each month.

10 | Process Definition — Workflow for Small Scale Production

Setting up for production begins with a look at all of the steps needed to build a title. You can use each step in the process to prepare needed assets for the next step, minimizing error risk and minimizing asset rework as the project develops:

> **TIP**
> Workflow documentation reflects normal iterative procedures. Begin at a high level, and let the details evolve. Examine them to find opportunities to improve process efficiency. Capture video once, for compression, screen shots, audio conformance and subtitles.

The workflow (see Figure 10.4) process needs to address the following areas —

❶ Project Specification and Design
- Purpose of disc: audience, expected use
- Disc functions:
  - Menus
  - User interactivity
  - Sequence: first play, body sequence, end actions, other details
- Codec requirements: mandatory/options/recommendations
- Bit budget for all assets including a buffer for unexpected or yet-to-be-delivered assets

❷ Capture
- Format of the master video and audio
  - Tape (HDCAM, D5, HDV,...)
  - Files (YUV 8 bit, 10 bit,...)
- Pre-processing, such as format conversion to prepare for compression

❸ Compression
- Audio compression: which codecs do you need to support and what is the standard input required
- Video compression: which codecs do you need to support and what is the standard input required

❹ Creation of menu/interactive elements

❺ Multiplex all elementary streams

❻ Authoring: programming for interactivity, connecting elements

❼ Testing and QC
- Use software players, emulators and hardware
- Hardware testing using an emulator (Toshiba HD-A1) or recordable media (BD-R/RE, DVD+/-R, or HD DVD-R)
- Use known good, tested media, recorders, recording software such as

Nero or Roxio™, and playback software. Validate your combination during the early stages.

■ A change to any component should trigger re-checking and re-validating the workflow to identify any potential problems with the new process. If left until later, finding the problems necessitates a much larger investment of time to resolve.

■ Monitor the process: for example, using 2x media, it should take 45 minutes for a full image. If it takes substantially longer, then something is wrong in the project setup. Potential problem areas include network connection and traffic, hard disk speed/fragmentation, concurrently running processes or bad media. A process that has been set up and validated should deliver the same results and speed consistently.

⑧ Premastering
   ■ UDF
   ■ CMF
⑨ Check disc and QC
⑩ Replication

Figure 10.4 Typical Workflow for a Small Scale Operation

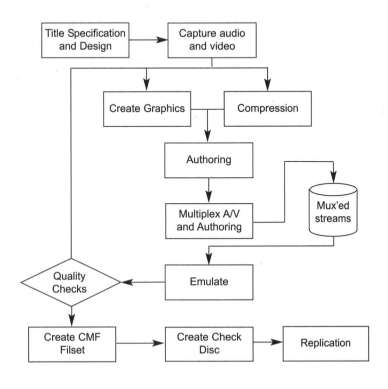

> **TIP** Using re-recordable media for testing and QC of authoring and content presentation saves time and money. DVD-R requires at least the equivalent of 1.5 Gbytes of data (3-7 minutes) to be written to the disc, but because BD-R/RE media uses a prerecorded lead-in and lead-out area, recording a 500 Mbyte disc takes less than two minutes to record to BD-R/RE, but more than five minutes at 4x to record to DVD-R. Not all players will play all formats of recordable media. Test beforehand to ensure that your player supports DVD+/-R, DVD+/-RW, HD DVD-R/RE or BD-R/RE media.

Next generation discs hold a lot of data. Recording a full HD disc can take a very long time when compared to DVD standards. Table 10.2 describes the times to record a full disc in the various shiny disc formats at 1x record speed. For CD and DVD, drives and media are available for faster recording times, but for high definition media, you currently should plan on the 1x times. Over time, the next generation recorders will be faster than the current 1.3-1.9x speed BD-R/RE recorders. Three hours (or more) is a typical time to record a dual layer, 50 Gbyte BD-R, or six hours if you include both a record and verify.

Table 10.2 Comparison of Media Record Times

	Disc Capacity (Bytes)		Minutes at 1x		
Disc Type	Single Layer	Dual Layer	Single Layer	Dual Layer	1x Speed bytes/second
CD	700,000,000	n/a	76	n/a	153,600
DVD	4,700,000,000	8,540,000,000	57	103	1,385,000
Blu-ray	25,000,000,000	50,000,000,000	93	185	4,495,625
HD DVD	15,000,000,000	30,000,000,000	55	109	4,568,750

Tools

The toolset for a prosumer/corporate production operation needs to cover the following task areas —

- **Graphics** Creating graphic elements for menus, transitions, motion buttons, etc. You will need to be able to optimize color palettes and to create and edit PNG (Portable Network Graphic) or a series of PNG files. Basic graphics tools may be included in your authoring tool. Graphics tool solutions include Adobe's creative suite (Photoshop, Adobe Illustrator, and Adobe After Effects) and Apple Final Cut Studio (compression, limited authoring, HD editing and motion generation). These applications are very useful and are some of the most popular graphic tools available.
- **Compression** Efficiently compressing the audio and video assets for a project is paramount for succesfully producing titles. A short list of useful compression products includes:
  - Sonic Solution's Cinevision™
  - Apple's AVC encoder

- Microsoft's VC1 encoder
- Main Concept AG's H.264/AVC encoder

■ **Authoring** Relatively few products are available for the prosumer/corporate producer for more advanced features that may not be supported by the individual/consumer tools:

- Blu-ray HDMV mode: Sonic Solutions Scenarist® BD or Sony Blu-Print™
- HD DVD Standard Content: Sonic's Scenarist SCA

When setting up to operate on the prosumer/corporate scale, the infrastructure of your network will utilize at least 1 Gbit Ethernet, and have a storage capacity of 1-4 terabytes. Storage capacity requirements are determined by whether you capture video assets from a tape or another input source, and whether you use a high definition disc format as output for an existing HD video production facility. If you already do HD video post production work, your operation may be configured to use existing storage capacity for delivery of input files for the compression process.

Note that even compressed video requires a large amount of space, and you will need room for multiple copies of assets for re-encodes, multiple multiplexed projects, or alternate versions of the full layout. The range for HD title storage requirements tends to be 1 to 1.5 terabytes per title for uncompressed video and audio assets, and 150 to 250 Gbytes per title for encoded files and working copies of the multiplexed files.

A formula for calculating storage space requirements is —

space required (Gbytes) = (minutes of video)×(Average or target bitrate for video encode) + (audio encode rate×number of audio streams)×(number of working copies of multiplexed files)

Setup costs for a small scale operation will vary. Table 10.3 identifies hardware and software required for a small scale operation, and gives a range of costs (available in 2006) by comparing an HDV videotape format operation with a DVC PRO-HD videotape format operation. This table is also included as a worksheet on the DVD that accompanies this book, and it can be used to estimate your setup costs for a small scale prosumer/corporate operation. Equipment selected for the DVC PRO-HD operation is more expensive for quality and/or capacity reasons.

TIP Optimize speed to avoid the risk of dropping frames while you are either capturing video or multiplexing: configure your hard drive array as RAID Level 0.

Table 10.3 Sample Worksheet, Setup Costs for Small Scale Operation

Master Tape format	HDV	DVCPRO-HD	Comments
Hardware subtotal	**$16,249**	**$61,295**	
HD VTR	3,300	18,500	The DVDPRO-HD is more expensive than a HDV deck.
HD Monitor	1,300	16,000	On the low end, you can use a LCD display. For more quality monitoring, you may want to have a CRT monitor.

continues

10

Table 10.3 Sample Worksheet, Setup Costs for Small Scale Operation *(continued)*

Master Tape format	HDV	DVCPRO-HD	Comments
Capture Card	249	995	Entry-level capture cards have HDMI or IEEE1394 inputs; higher end capture cards give HD-SDI capability. Some cameras are network capable or have hard disk emulation modes that can mean you don't need a capture card.
Workstation – Capture/compression	4,000	6,000	The difference here is in the graphics card to support quality, full speed playback.
RAID (1 – 4 Tbytes)	1,000	10,000	Entry-level firewire-connected disk arrays, vs. multi-user environments with greater capacity and speed.
Workstation – Graphics	2,000	3,000	Typical workstations are available with wide screen LCD monitors.
Workstation – Authoring	4,000	6,000	Consider multi-core processors with dual monitors.
"SneakerNet" disk drives (2-3 each)	300	300	
1G Switch	100	500	
Software Subtotal	**$57,200**	**$151,600**	
Production Suite (Apple or Adobe)	1,200	1,600	
Authoring	50,000	50,000	Currently, authoring seats are about $50k each.
Compression – video	5,000	75,000	Depending on whether you go for a powerful workstation or a small cluster license.
Total	**$73,449**	**$212,895**	

Toolset functionality outlined in the table will increase, and the price will decrease, just as it did for DVD authoring tools. An anecdote may put setup costs in perspective. In 1986, the Hewlett-Packard Company(HP) purchased their first CD recorder for about $149,000, then considered a bargain. The alternative was to send files out to create a check disc, at a cost of $10,000 and three weeks' time. CD-R media cost $70. As with projects today, it took several passes to achieve an acceptable master, so $149,000 was recovered within the first two projects. HP's second recorder cost them only $29,000 and recorded at 2x speed. The third generation of CD recorders cost only $5,000! HP has been making CD and DVD recorders for years, and is expected to be a major manufacturer in high definition DVD. Other manufacturers have shipped the first generation of HD recorders, selling them for less than $1,000. Recordable media is less than $20 ($60 for dual layer) and record speeds are already 1.3x to 1.9x. Within a few short years, the prices for recorders and the media will be where DVD is today, with recorders costing less than $100, media less than $1, and recording times less than 15 minutes for a full single layer disc. Ahhh, the price curve for new technology.

People

The team for creating the next generation disc titles for a prosumer/corporate operation will need to include the following skill sets —

- **Graphics** One or more designers fluent in the Adobe® Creative Suite and possibly the Apple® Production Suite. The complexity and number of titles in progress concurrently will determine the number of people needed.

- **Authoring** Your team needs someone who is good with details and has some experience in standard definition DVD. Depending on the authoring tool, this person may find basic proficiency with Adobe Photoshop® useful.

- **Compressionist** The best kind of compressionist knows how to see, hear and recognize encoding artifacts. The compressionist is also knowledgeable about various codecs, and can configure codecs for the best quality, within the given time and bit budget constraints.

- **QC** This person is responsible for evaluating the compression quality, and the integrity and robustness of the disc's navigation. This person is detail oriented and meticulous in executing test scripts and quality checklists.

Recommendations for Small Scale Operations

This information should give you a good foundation for production setup in a small scale operation. Use the worksheets included to build your own capacity and planning charts. As you evaluate options from vendors, you can use these worksheets to ensure that you have addressed all of the operational needs. We also recommend that you read the next section on large scale operations. Although you may start small, you may have the opportunity to expand your organization, and some of the large scale approaches to setup may be useful as you look for more ways to increase productivity.

Large-Scale/Hollywood Operations

Large Scale/Hollywood production covers the greatest variety in setup approaches. Some corporate or non-profit organizations for disc title production may also fit in this category. The common characteristic at this level of operation is the number and the complexity of the titles.

Define Capacity

Table 10.4 compares the effort required to create typical complex discs in large scale operations. Times shown are based on the speed of workstations/clusters and state-of-the-art tools available in late 2006. Advances in multiple processor design, distributed computing, encoder algorithms and the sophistication of authoring tools will reduce the time required, while an increase in the disc sophistication will likewise lengthen production time. During 2007 and 2008, increased testing will be required to ensure player compatibility as more players arrive on the market.

10

Table 10.4 Sample Worksheet, Time Calculation Model for Large Scale Operations

Title Assets/Complexity	Number of Simple	Medium	Complex	Comments
Main Feature	100	120	140	
Bonus/Extra	10	30	45	
Chapters/segments	18	24	34	
Audio Tracks	2	3	5	
Menus (Top and Pop-Up)	19	26	31	
Buttons	46	62	75	
Transitions	5	20	37	
Subtitles	2	4	5	
Language Sets	1	2	3	
Game/Interactive Experience	0	1	2	
Calculations - minutes				
Compression – Video				
2 pass – auto VC1/AVC	1100	1500	1850	Encode times vary depending on the speed of the codec and cluster implementation. For these calculations, we've used ten times real time.
Re-encodes	2200	3000	3700	Re-encoding, depending on the encoder, source video and quality expectations can take about twice as much time as the original encode.
Compression – Audio	220	450	925	Audio compression runs a little faster than real time for AC3 and DTS streams. The HD streams take longer. Benchmark your actual tool.
Title Design/Architecture/Flowchart	93	142	192	Figure about one to three minutes per asset for this step.
Still Graphics	1300	1760	2120	A good rule of thumb is about 10-20 minutes per menu and button
Animation	225	900	1665	Animations will take 20-45 minutes per animation on average.
Authoring	920	1315	1685	Authoring takes 5-15 minutes per menu, button, transition

continues

Table 10.4 Sample Worksheet, Time Calculation Model for Large Scale Operations
(continued)

Title assets/complexity	Simple	Medium	Complex	Comments
Programming	0	480	2400	Programming of custom modules or new feature can take anywhere from a day to weeks.
Multiplex	55	75	92.5	Multiplexing of audio, video and subtitle streams take about half of real time.
Record Test Disc	55	75	92.5	Takes about half of real time
QC – Viewing plus Nav	465	705	950	Each audio, video and subtitle stream needs to be reviewed. Also, figure about 5 minutes per minute, button, and transition for verifying the navigation
Spec Verifications	550	750	925	This step can take hours. It can run as long as five to ten times real time of the video.
Pre-mastering AACS	85	105	122.5	This is for adding the AACS protection. It generally means making a new encapsulated version of the multiplexed streams.
Pre-mastering CMF	55	75	92.5	Creating the CMF image (cutting master format), will take about half of real time.
Totals				
Total minutes	7323	11332	16812	
Total hours	122	189	280	
Total Days (8 hours/day)	15	24	35	

Notes:

1. Game design, implementation and testing times have not been included.

2. Time for other activities such as 3rd party QC, check discs and usability testing have not been included.

3. These times assume all tasks proceed linearly with no concurrency or re-work.

4. Subtitle creation is not included. Time to mux into stream is included.

10

From the time requirements detailed in Table 10.4 and the throughput and capacity numbers required of your production, you will see that an optimized process is essential for an efficient, highly productive operation. Production steps can be arranged in various ways that influence the number of concurrent in-process titles and the throughput of an operation.

Table 10.5 Advantages and Disadvantages of Different Production Workflow Architectures

Workflow	Advantages	Disadvantages
Sequential	1. Streamlined 2. Minimal equipment and staff 3. Minimal schedule and plan production. Each task simply leads to the next.	1. Specialized tools and staff may have idle time. 2. Any delay or hang up in the development will cause all next steps to be delayed. Nobody can proceed until the bottleneck is resolved.
Parallel	1. Turn-around time for a title is minimized because multiple tasks are happening concurrently. Turnaround time is the length of the critical path items. 2. Equipment and staff can be better utilized by balancing the allocations of resources.	1. Requires more organization for scheduling to ensure that all resources are optimally utilized. 2. When requirements change midstream, then the workflow from beginning to end is impacted.

A process may be defined in such a way that tasks are completed sequentially, as they are in a small scale operation. Some advantages may result, but if one title gets stuck on a step, every concurrent project has to wait for the bottleneck to be resolved.

Your team may create dummy assets — video, audio, subtitles, popup and always-on menus — to promote authoring as soon as the project/title flowchart has been defined. As assets are finalized, they can replace the dummy asset placeholders. In this way, the authoring task and testing can begin early for optimal risk management.

Dummy assets may be segments of black video, one, five and ten seconds in length, in the various codecs that you use for your projects — one each for AVC, VC1, and MPEG2. You may also use a stock copyright notice, splash logos, or bars and tone. For audio, you can easily create a set of audio segments with silence and tone to match expected lengths of video, in the standard codecs used on your projects. Your team can set up standardized projects to address standard disc functionality for your target market —

- APS - auto play and stop discs
- APL - auto play and loop

- APM - auto play, display menu at end, or when menu key is pressed
- POPA - play one, play all menus
- MOMA - many play one items and many play all items
- Simple, medium and complex titles for Hollywood movies with the appropriate start-up sequence and navigation structure. These can/are tailored to particular clients and can even be tailored to meet certain genre' or brands of titles

You can also establish standard program code modules, ready to use with different graphics —

- Button push event
- Pop-up transition
- Setup menu with x audio, y subtitles and the appropriate on/off states.

TIP Create project templates with 'dummy' placeholder assets, for streamlined re-use. You can create templates for a number of typical production elements.

Process Definition — Workflow for Large Scale Production

For large scale production, it will be to your advantage to boost efficiency by ensuring that each step creates the best assets or input for the next step (see Figure 10.5). Because of the size of assets and the complexity of projects, reworking assets is undesirable — time consuming and risky. Well-defined hand-offs between each task are dependent on well-defined filenames, format and quality, meeting the needs of the next step. Minimize conversions, copying/moving and asset manipulation.

NOTE Making a "working copy" of the assets for a next generation format disc may mean copying up to 50 Gbytes to a "working" area, taking one to two hours of time. Simply leaving "heavy/big" assets on a shared network resource can save days over the course of a month of projects.

The process should facilitate rapid re-delivery of assets, and avoid the need for reprogramming or rebuilding a title in the authoring environment. Select tools that are designed to make it easy to replace older assets with newer versions, so that replacing an encode or updating the text of a button does not become a big deal.

Figure 10.5 Typical Workflow for Large Scale Operations

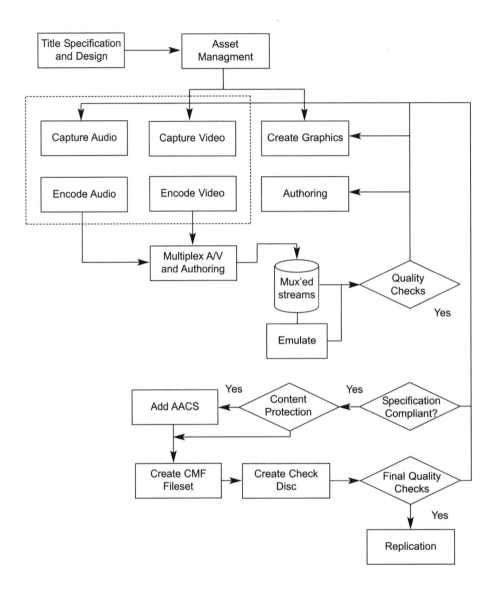

Tools

Given the tools available at the end of 2006, large scale setup requires an investment of over one million US dollars. Just as with standard definition DVD technology, these costs will drop significantly over the next one to four years.

Figures presented in Table 10.6 do not include improvements to a facility. Computing clusters will need special power, cooling and networking setups to support compression tasks. High definition monitors, projectors, and sound systems have much larger physical space requirements than standard definition DVD equipment, so additional floor space is needed, compared to an operation focused on SD DVD.

Table 10.6 Sample Worksheet, Set Up Costs for Large Scale Operation

Setup – Hollywood	Qty	Each	Total	Notes
Total		**$700,645**	**$1,289,190**	
Hardware Subtotal		**$462,945**	**$704,590**	
HD VTR	2	100,000	200,000	Depends on whether you have a D-5, HDCAM SR or DVCPRO-HD deck.
HD Monitor	3	20,000	60,000	Typically you'll want a HD CRT based display. You may need other display technologies, e.g, projection, LCD, Plasma.
Capture Card	2	995	1,990	In this example, this is the price on the Decklink SDI/HD-SDI capability.
Workstation – Capture	2	25,000	50,000	Needs to be fast enough to keep up with the encompressed streams
Compression Blade Server	1	150,000	150,000	In this example, this is for a 32 dual core processor with 4 Tbytes of storage.
RAID	1	50,000	50,000	Figure on 4-40 Tbytes
Workstation – Graphics	4	2,000	8,000	
Workstation – Authoring	10	4,000	40,000	
Workstation – Mux/Premastering	2	3,000	6,000	
Spec Verification Workstation	2	4,000	8,000	
"SneakerNet" disk drives – 100-250G	10	150	1,500	These are disks which can be used to copy and carry data to/from clients and vendors for delivery and review of assets.
Workstation – QC	2	2,000	4,000	
QC Viewing Station	2	2,300	4,600	
HD Demo Room	1	94,500	94,500	
BD and HD DVD Players	6	1,000	6,000	
Hardware Emulators	5	4,000	20,000	*continues*

10

Table 10.6 Sample Worksheet, Set Up Costs for Large Scale Operation
(continued)

Setup – Hollywood	Qty	Each	Total	Notes
Software Subtotal		**$217,200**	**$563,600**	
Production Suite (Apple or Adobe)	3	1,200	3,600	
Authoring	10	50,000	500,000	Current pricing on this class of software is about $50k per seat.
Compression – video	1	110,000	15,000	
Compression – audio	1	25,000	25,000	Estimated prices for Dolby and DTS encoding workstations for the large scale operation.
Specification Verification Software	2	10,000	20,000	Current pricing is running $5-20k per seat for each verification software.
Network		20,500	21,000	
1G Switch	2	500	1,000	
Fiber Channel	1	20,000	20,000	You'll need the fastest network you can get for your capture and encoding workstations.
Total		**$700,645**	**$1,289,190**	

The calculations used in Table 10.6 are based on three to five concurrent titles. You can adjust requirements according to your projected production. This table is also included as a practical worksheet for your convenience on the disc that accompanies this book.

People

In a large operation, the number of people involved is proportional to the time-length and complexity of titles. The organization chart in Figure 10.2 gives an overview of the distribution of large scale operation roles. With the exception of an Assets Management staffing requirement, the team for creating the next generation disc titles for a large scale operation will be identical to the team needed for small scale prosumer/corporate production, including the following skill sets —

- **Assets Management** You will need an individual or group to receive assets and to inventory, inspect and prepare assets as necessary. Skills required include audio/video post production, graphics basics such as Photoshop, Microsoft® Office products, and meticulous detail orientation. Assets Management is your key to avoiding garbage-in-garbage-out.
- **Graphics** One or more designers fluent in the Adobe Creative Suite and possibly the Apple Production Suite. The complexity and number of titles in progress concurrently will determine the number of people needed.
- **Authors** Your team needs someone who is good with details and has some experience in standard definition DVD. Depending on the authoring tool, this person may find basic proficiency with Adobe Photoshop useful.

- **Programmers** If you are creating for the advanced content modes, your staff will include people who are fluent in Javascript™ and XML for HD DVD and Java™, MHP/GEM for Blu-ray.
- **Compressionists** The best kind of compressionist knows how to see, hear and recognize encoding artifacts. The compressionist is also knowledgeable about various codecs, and can configure codecs for the best quality, within the given time and bit budget constraints.
- **QC** This person is responsible for evaluating the compression quality, and the integrity and robustness of the disc's navigation. This person is detail oriented and meticulous in executing test scripts and quality checklists

Conclusions and other recommendations

This information should give you a good foundation for production setup in a large, Hollywood scale operation. Figure 10.6 reflects how a properly staffed and operating large scale production environment can perform over time, in terms of capacity and throughput. As you implement the processes, be sure that you have measurements in place for identifying and evaluating —

- Time required for each task or set of tasks — How long it takes to encode a VC1 from capture to the final QC'd version
- Number of titles completed
- Hand-off errors between steps

Keeping these measurements in place and up-to-date makes it possible for your team to improve the process most effectively. At this early stage of the development of high definition production, count on investing time, money and energy to create your first titles. By 2008, setting up for production should be easier and cheaper, and you'll be way ahead of the pack.

Figure 10.6 Capacity and Throughput Chart

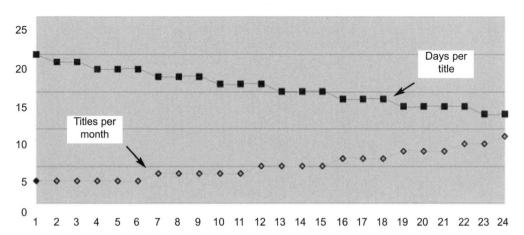

Total number of Titles authored

Chapter 11
Project Design and Planning

This chapter addresses how to set up a project — detailing the user interface, deciding how the title will be assembled, and developing a plan for how all of the activities required will be performed, by when and by whom. Critical to the process is the scheduling of tasks and the documentation that needs to flow from each step of the project to ensure that everybody on the team has the information that they need, when they need it.

The capabilities of the next generation discs are much more extensive than DVD and planning for a project can be much more complex. The amount of planning required depends on the tools used, the functionality requirements of the project, the skill set of your team and the scale of your operation.

Other factors that need to be addressed include the fact that you have much more tightly coupled interaction between compression, graphics, authoring and, depending on the disc mode, programming. Project design and workflow need to be adequately robust so that elements can be modified without having to restart the project from scratch. You know that menu texts will change. Logos will change. The client will request new features and add or delete content. The design needs to be flexible enough to handle the iterations and changes.

Basically, you create a clear specification for the client detailing exactly how the disc works. Here is an overview of the steps that must be considered in the planning and design phase of project development —

- Establish the objective of the disc
- Establish the flow of the disc
- Define interactive aspects of the disc. How will the user interact with this disc?
- Define the graphic elements and how they will be programmed in the title
- Define the bit budget for all of the elements of the disc. A simple bit budget calculator is included on the disc accompanying this book.
- Identify any functions or sections of the disc that need to be prototyped. Are these new ways of accessing video, audio, subtitles on the disc? Will you need to either validate the usability or the feasibility of these features/functions. Some situations that may be prototyped include:
  - How will the user return from the picture in picture function?
  - How fast can text be generated that displays current timecode or location? Will you need to render these features as PNGs or as composited video?

Project Design

There are several ways to approach producing a high definition DVD title. For example, you can start with an idea and then develop graphics and programming iteratively as the title unfolds. This approach may work well for a project with limited scope, loose schedules, or an unlimited budget. This approach is great for initial projects where you are exploring the capabilities of the medium and you have limited the scope so that the project can be completed. However, more often than not, the client wants to know what they are going to get beforehand and they want to know exactly when the title will be done. Oh, and they want to know how much it will cost.

To make the latter dynamic work for everyone, it is imperative to do all of the planning, scheduling and budgeting upfront, so as to minimize the surprises. For this, you need to have a defined, documented design for the project.

Project Objectives

The great thing about next generation DVDs is that they can be used for many applications, whether for education, communication or entertainment. The first step in designing and planning a project is to define the objective of the title. Here are some questions that should be answered that will help to make it clear why the disc is being made —

Target Audience   The audience dictates the expectations and needs of the disc title. For example, the disc could be used by those getting certified in sign language or for a sales force learning about a new product line. Or, this could be a disc targeted at the family watching the latest animated Hollywood film or to teenagers with the latest summer adventure film.

Target Usage Scenario   Look creatively at how the disc can best address where the disc will be used. The disc could be played on a laptop or a home theater or projected in a classroom or used as part of a presentation or museum installation. For example, in a classroom setting you might want to have an attract loop that can exit when it is time for the lesson to begin. At the end of the main program for this type of disc, it might hold on a screen saying "Discussion" before proceeding to a quiz. The user is operating the disc using a remote. This is wholly different than operating a computer interface with a mouse, and certain considerations must be accounted for.

Target Playback Platform   Even though a high definition disc can hold high definition content, can the setup play it? For a classroom, will there be a 5.1 channel setup? The audio mix for a 2 channel output might be different for a classroom or an auditorium than for a laptop or a home theater setup. The menus for a presentation might not need to be as big or animated or entertaining as the menus for a Hollywood disc. Also, many of the players have an Internet connection. If so, you will need to decide if the connection should be available and useful for a title. And, what would happen if/when the Internet connection is not available. In that instance, two possible answers are the disc could simply not play or the disc could play in some restricted mode.

Expectations of Interactivity   How interactive does this disc need to be to meet the project objectives? Does the user expect to have access to cool features like picture in picture, director's commentary and extra content, such as deleted scenes? Will there be games and/or quizzes where scores will be kept and recalled at later dates?

Performance requirements and considerations   Unlike DVD where you might end up with a two to three second delay if the player was processing some input, with the next generation players, it is very possible to experience 30 to 60 second delays. Delays of that length will be less than conducive to a pleasant, interactive experience. You will need to make the disc entertaining or use sleight of hand to make the delays tolerable or, if you are up to the challenge, maybe even enjoyable.

Functional External Specifications

You need a specification for what the disc will do, known as the external specification, and how the disc will do it, referred to as the internal specification. This section is about getting to the external specification.

Clients may have different expectations of what the title will do. And, they probably have some misconceptions about what can be done with the new format discs. It is very important that you define upfront, in as much detail as possible, the expectations for a title.

Here are some tools and approaches that will be helpful for getting everyone to share common expectations for a project —

Develop primers, providing educational background on the technology.   The goal of a primer is to create a context for describing what the title will do and to provide a set of terminology that will help to describe what is desired. You want everyone on the same page when talking about popup menus, normal, selected and activated button states, transitions, load times, codecs, secondary audio and video, picture in picture, et cetera. For example, *DVD Demystified Third Edition* is always a good starting point for reading about the technology that will be used for the project.

Create a flowchart that details how the disc will work.   Some people grasp information graphically, so this can be a good approach. There is a lot of detail that can be communicated in a flowchart. Keep it elegantly simple. Over time you will develop a style that works for you and your clients. The flowchart can be used in all phases of the project. Initially, it is a way to document how the disc will work. It is also helpful in the creation and programming of the interactive assets of the discs. QC can also use the flowchart to create a comprehensive test plan.

Generate a functional outline for disc activity.   A complimentary approach to flowcharts is to use a detailed functional outline that covers first play sequence, menu layout, which remote control buttons are functional at each part of the playback, what happens at the end of the chapter, et cetera. These may be very detailed and explain nuances in the title requirements that you cannot communicate in a flowchart.

Cite existing titles that demonstrate concepts and applications.   Use examples of other projects that you or others have done to show and get agreement with your client about how a feature/function of a disc will work. Some people need to see the feature in action before they understand what it is and how it will work for their project. You can also use the examples to note ways that you don't want a title to work. Maybe the feature on the example did not work the way you thought it would and you now have a better way. Examples show the way to go and/or the way to avoid.

Build prototypes depicting intended functionality. If a function is not clearly understood or it is not clear how to achieve a certain function, you may need to build a model for that part to show it clearly. A couple of ways to accomplish this would be to use Macromedia® Director or Flash®. This may seem like overkill but that will more exactly model the user interface short of actually building the disc. Another approach that is commonly used is to create a video, usually in Adobe® After Effects® and output to Quicktime®, that shows how the menus will be drawn, depicts button transitions and highlight states, and how the navigation is supposed to work. The potential problem with these approaches is that it may be easy to create a user interface that looks really cool but there is no way that it can practically be authored or the performance or implementation costs are unacceptable. However, by prototyping the feature or function early in the process, everyone can see how it will work and you will have a better estimate of the effort required for the full implementation.

Evaluate the different encode selections. There are many audio and video codecs to choose from, for use on a project. You may need to do test encodes with the client's material to ensure that you have the right audio and video codecs to meet the needs of a title. This is also where you can refine the bit budget requirements for a title. You can determine how many bits/second the video really needs.

Functional Internal Specifications

The internal specification addresses these items —

Define the file formats for all handoffs of assets. If you have not worked with a client or vendor or subcontractor before, get a set of test files and test them in your process to ensure that they will really work (e.g., correct frame rate, frame size, color depth, sample rate, et cetera). The test files do not need to be the full delivery. Often just five to ten seconds of the video, audio, and/or subtitle file or a portion of the graphics can confirm that the final delivery will be something that you can easily plug-in and run with.

Define how changes will be made to the project. The specifics of how changes can be made to a project depend on your authoring software. You need to plan that your menu graphics will change, and that you will get new video and audio. Subtitles will need minor tweaks or redeliveries. Functionality of the title will also change, either to fix a bug or to add a new feature. Some of the items to define include: filename conventions (e.g., same filename for each delivery within a folder with the date of the delivery), how the change is to be delivered (email attachment, FTP, DVD-R, HDD, WAM!NET®, on-line repository or other media).

Determine the compression requirements. Create and use a bit budget to plan the project. The next generation DVDs have a much larger capacity but the assets are also much larger. The compressionist will need to know the in/out timecodes, input file formats, chapter points, and the codecs to use for audio and video. The next section includes an example bit budget.

Establish graphic parameters. Create the internal specifications for the graphics person or group that details the raw assets they can expect (e.g., logos, models, screen grabs, et cetera.), a wire diagram or other way of specifying the button layout and interaction. Also, this specification should detail any requirements for localization, guidelines on the use of lower thirds, and colors. This will also detail the format that is needed by authoring for the tool that they are using.

Create Authoring requirements. The author and/or programmer will need to have a detailed specification that takes the encodes from compression and the graphical elements from graphics and implements the connections and functioning of the disc.

Formulate a project schedule. Establish and agree on a schedule that includes key check-points (e.g., Viewing QC, menu design reviews, Navigational QC, check discs and master image to replication). If you can, identify how you will handle problems and have contingencies. Have a plan for when the plan isn't working!

Asset Management

Asset management is much more critical in next generation discs than with standard DVD. For example, in a DVD menu, you may only have two to three assets to load into a project (e.g., background (still or video), highlight state and, maybe, audio). With next generation discs, each button has 3 sets of assets — Normal, Selected and Activated. This means that a menu with 10 buttons could now have at least 30 assets instead of just two.

Naming Conventions

Choose a naming convention for assets that is logical, consistent, and assists those preparing and assembling the title. A good naming convention is one of the most important things to establish, for titles as well as for your operation. Poorly naming assets are often the root cause of a defect in a disc.

Here are some examples that help identify what an asset is by knowing how to read the filename —

`MM_PU_N.PNG`	Main Menu, Pop-Up, Normal State
`MM_PU_S.PNG`	Main Menu, Pop-Up, Selected State
`MM_PU_A.PNG`	Main Menu, Pop-Up, Activated State

For a project requiring individual buttons, you would have a file for each button state and for its button click, if needed —

`MM_PU_B1_N.PNG`	Main Menu, Pop-Up, Button 1, Normal State
`MM_PU_B1_S.PNG`	Main Menu, Pop-Up, Button 1, Selected State
`MM_PU_B1_A.PNG`	Main Menu, Pop-Up, Button 1, Activated State
`MM_PU_B2_N.PNG`	Main Menu, Pop-Up, Button 2, Normal State
`MM_PU_B2_S.PNG`	Main Menu, Pop-Up, Button 2, Selected State
`MM_PU_B2_A.PNG`	Main Menu, Pop-Up, Button 2, Activated State
...	
`MM_PU_Bn_N.PNG`	Main Menu, Pop-Up, Button n, Normal State
`MM_PU_Bn_S.PNG`	Main Menu, Pop-Up, Button n, Selected State
`MM_PU_Bn_A.PNG`	Main Menu, Pop-Up, Button n, Activated State
`MM_PU_B_Click.wav`	Main Menu, Pop-Up button click sound for all buttons

11

Depending on the assets, you may want to include in the filename, the frame size, frame rate, main video, special feature, bitrate, encoder type, number of channels, channel layout, screen position, et cetera. One other option is to put this in a database or use a metadata file, in .xml file format, to hold asset description information.

Graphic Design

The graphic design is even more intertwined with the rest of the process in the new disc formats. With popup menu graphics, the graphic designer must include some additional design constraints. Also, for many of the authoring tools, there is a set way that the graphics must be prepared so that the author can more easily load them. If the graphical elements are not prepared just right, the authoring step will be required to do the preparation. The problem with doing that preparation in authoring is that it must be done each time there is a new delivery. It is often easiest to include the preparation steps as part of the graphic design workload. Changes can be implemented more quickly, the graphic designer maintains the artistic control and the author can focus on the programming of the disc.

Considerations for menu graphics deliveries include —

- What does the authoring tool need? For example, with the Sonic Designer tool for preparing graphics for Scenarist® BD, you may need to prepare elements with especially named layers.
- The location of where the graphic elements fit on the screen needs to be communicated to the author/programmer. These can often be included as part of the filename. Location is best referenced to upper left corner of the graphic (e.g., 0,0 is upper left corner).

The graphic designer must also address these issues in the design —

- Developing a plan for how the graphics will be sliced up to create the buttons, transitions and backgrounds.
- Define the buffer usage requirements —
  - For Blu-ray Disc™, the designer must manage the 16 Mbyte menu space buffer. When preparing advanced content for Blu-ray, with a tool such as Sony's Blu-Print™, it is necessary to plan how the 16 Mbytes IG graphics buffer is being used. Blu-Print sees flattened PNGs (8 bit). A set of PNGs is delivered for each menu or transition. The menu set includes all of the button graphics with one PNG for each states — Normal, Selected and Activated. The drawing of the buttons and the definition of the transitions is defined in the authoring process. The problem to watch for is the usage of the buffer. You want to reduce the size of the PNGs by minimizing the dimensions of the sliced PNG and using an optimized palette so that you can fit all of the IG layers graphics in the 16 Mbyte buffer. Trial and error is one approach but it means that the author loads the graphics, slices up the buttons, programs the button interactivity and the transitions, and then finds that when multiplexing the project that the IG is over

full. So, that workflow approach is not very efficient. Another technique is to slice the graphics before loading into Blu-Print to ensure that the graphics will fit in the 16 Mbyte buffer. This calculation is best done in the design process by the graphic designer, although it may also be done by an author as soon as the graphics are ready. If there are any buffer use issues, the graphic designer can resolve them sooner and the author is not re-authoring the project upon each delivery of the graphics.

■ For HD DVD, the size of the graphics plays a big part in the performance of the title. Define with authoring the graphics element requirements. For example, settop players can take as long as .75 second to load 1 Mbyte of data (packaged in the .aca file). This means that a 16 Mbyte file will take 12 seconds to load. If this data is out of multiplex, then the user will be waiting for 16 seconds for something to happen. Without planning, it is easy to make files bigger than this and the load times will be correspondingly long. Work with the author to see if it is possible to multiplex the data with the opening material (e.g., FBI Warning, opening trailers, logos, splashes, et cetera). It may be necessary to create a special video segment that simply tells the user that the "Program is Loading".

These issues and their solutions are discussed further in the chapters on graphics and authoring. With planning in the design phase of the project, these potential problems may be creatively solved so that the user has a great interactive experience.

Compression

The compressionist needs a few things to be successful. The most obvious is a good bit budget. The second thing is a good list of assets including in/out timecodes, a list of chapter points and clearly labeled tapes or files.

The goal of the bit budget is to make sure everything is going to fit on the disc and everything to be presented is within the peak bitrate constraints of the format. If you end up having to re-encode due to exceeding the budget, it can be very painful and time consuming. Table 11.1 presents the disc limits for single layer and double layer for each format. Calculating the peaks is a little different than DVD. Now, you need to include all of the streams.

TIP   Get the bit budget defined upfront and encode only once, but leave some breathing room. Pick a number. Four percent has worked well for many projects. It is really great to find that you have filled the disc to within seven megabytes of capacity. It is really sad when you find that you are seven megabytes over and the deadline is in 30 minutes.

It is okay to leave gigabytes open. Often, bit budgets will specify video bitrates of 26 Mbits/second just because there was room when 18 Mbits/second looked just as good.

Table 11.1 Maximum Disc Capacity

Capacity	HD DVD 15 (single layer)	HD DVD 30 (double layer)	Blu-ray 25 (single layer)	Blu-ray 50 (double layer)
Size (bytes)	15,000,000,000	30,000,000,000	25,000,000,000	50,000,000,000
Size (bits)	120,000,000,000	240,000,000,000	200,000,000,000	400,000,000,000
Allow overhead (4%)	600,000,000	1,200,000,000	1,000,000,000	2,000,000,000
Total Available Bits	119,400,000,000	238,800,000,000	199,000,000,000	398,000,000,000
Maximum bits per second	30,240,000	30,240,000	48,000,000	48,000,000
Maximum Video bps	30,240,000	30,240,000	40,000,000	40,000,000

Table 11.2 presents a bit budget for a sample project. The total size of this sample disc is 27 Gbyte. So, if you are using HD DVD, you would need to create an HD DVD 30 (dual layer) disc. In your authoring tool, you would need to specify the layer break. If the layer break is in the middle of a video clip, you will need to encode that section of video so that the bitrate is low enough during the layer change to not create a pause in the video.

Table 11.2 Example Bit Budget

Element	Duration mins.	secs.	Data Rate bits per secs.	Size (in bits) bits per secs. x number secs.
Main Feature Video	90	5,400	18,000,000	97,200,000,000
Main Feature English 5.1 Audio – DD	90	5,400	640,000	3,456,000,000
Main Feature English 2.0 Audio – DD	90	5,400	192,000	1,036,800,000
Main Feature English 5.1 Audio - DTS	90	5,400	3,018,000	16,297,200,000
Main Feature Spanish 5.1 Audio - DD	90	5,400	640,000	3,456,000,000
Subpicture tracks (IG/PG for BD and SP for HD DVD)	90			
Subtotal (main feature audio and subpicture)			22,490,000	121,446,000,000
Motion menu video[a]	3	180	20,000,000	3,600,000,000
Intro Splash logos[a]	0.5	30	20,000,000	600,000,000
Trailers - video	2	120	24,000,000	2,880,000,000
Trailers - audio	2	120	1,536,000	184,320,000
Extras - Interview - video	20	1200	18,000,000	21,600,000,000
Extras - Interview - audio	20	1200	192,000	230,400,000
Subtotal (non-main feature elements)				29,094,720,000
Total of all assets except video				150,540,720,000
Bits available for Video			Max Bitrate	
HD DVD 15[b]	90	5400	5,655,689	30,540,720,000

continues

Table 11.2 Example Bit Budget *(continued)*

Element	Duration mins.	secs.	Data Rate bits per secs.	Size (in bits) bits per secs. x number secs.
HD DVD 30[c]	90	5400	16,566,533	89,459,280,000
BD 25[d]	90	5400	9,159,126	49,459,280,000
BD 50[e]	90	5400	46,196,163	249,459,280,000
Maximum video bitrate with allowances for audio				
HD DVD 15			8,673,689	
HD DVD 30			13,548,533	
BD 25			6,141,126	
BD 50			43,178,163	

[a]Use 20 Mbits/sec including audio.   [d]Capacity in bits 200,000,000,000

[b]Capacity in bits 120,000,000,000   [e]Capacity in bits 400,000,000,000

[c]Capacity in bits 240,000,000,000

Notes

1. This material will not fit on an HD15 disc. We are over by more than 30 Gbits before the video is added. To make this fit on a single layer disc, we would have to eliminate the extra feature material.

2. This sample does look ideal for an HD30 disc. The allowable bitrate would provide enough bits for a decent quality encode with either AVC or VC1. MPEG2 would be a little on the low side for quality.

3. For a BD25, this will fit but the allowable bitrate would be low for most applications.

4. The allowable bitrate exceeds what is allowed by the specification for a BD50 disc. The maximum video bit rate for Blu-ray is 40,000,000. So, although the total bit rate of 46,196,163 bits per second is within specifications, the video bit rate would have to be reduced to 40,000,000 bits per second on less.

Project Flowchart

Flowcharts can be used to represent how a disc functions (the external specification) and how the title will be built (the internal specification). Flowcharts are a very helpful means to communicate project design to the whole team (designer, client, authoring, and QC). See Figure 11.1 for a sample flowchart.

TIP

There are many tools you can use to create a flowchart, including the AutoShapes tool in the Microsoft® Office Suite. Also, Visio® is a very good tool for creating flowcharts. There are other options that work quite well, such as OpenOffice's drawing tool (OpenOffice is multiplatform — Windows®, MacOS®, Linux, Sun Solaris™ and FreeBSD® — and is a free download available at www.openoffice.org). All of these drawing tools support the ability to draw boxes with connectors that move with the boxes. This is very handy when you need to change the flowchart and move things around on the page.

Figure 11.1 Sample Flowchart

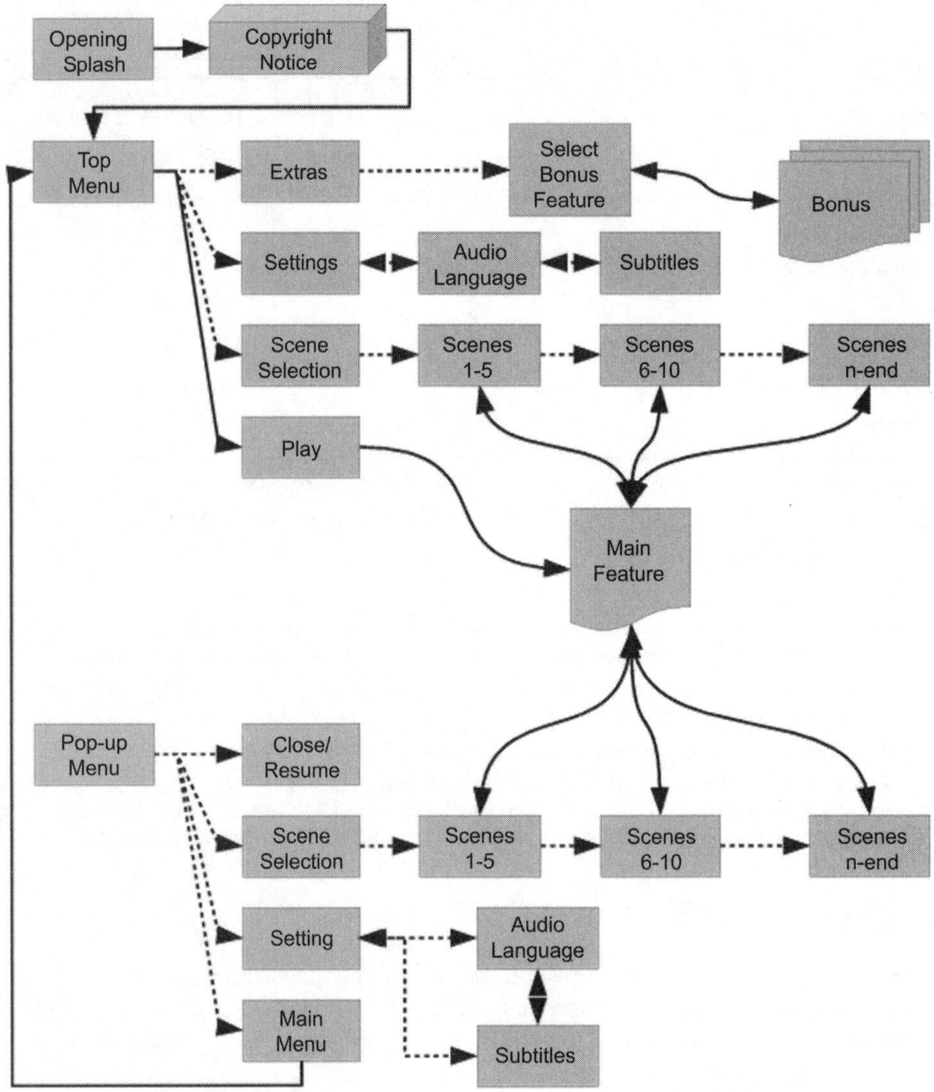

The flowchart should consistently represent all of the interactions that a user can have with the title. The example shown in Figure 11.1 uses dashed lines to represent transitions between menu and menu elements. And, the dashed arrow represents that the graphic should scroll up as it is fading in and if the arrow is a double ended arrow then the graphic will scroll down as it fades out. The solid lines represent no transitions.

Further, you may add as much detail on the flowchart as you want. You can specify items such as —

- Menu structure, including which buttons are available on the menu and, when selected, whether a User Operations is locked out or permitted
- Audio tracks selected

- Graphic element filenames can be placed next to the boxes on the flowchart
- Length of transitions
- Network connectivity
- Persistent storage usage

User Interactivity

Because the new format discs allow much more flexibility than DVD and depending on the disc mode selected, you will have differing amounts of flexibility in what can be done with the user interface.

The basic way that users have to interact with the disc is the remote control. Figure 11.2 depicts archetypical remotes for HD DVD and for Blu-ray. Each player manufacturer has the ability to layout the remote in any manner they deem appropriate for how they see the market. What this means is that you have quite an opportunity and quite a challenge to make the disc usable by your target audience.

Figure 11.2 Typical HD DVD and BD Remote Controls

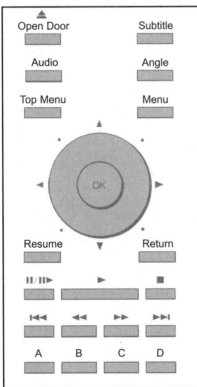
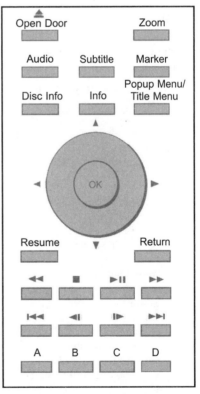

Notes: The differences between the remotes are subtle but significant.

1. These controllers are typical representatives for each of the format players, but the actual controllers will vary from model to model.

continues

11

Figure 11.2 Typical HD DVD and BD Remote Controls *continued*

2. On the HD DVD controller there is a Top Menu button versus the Disc Info button on the BD controller.

3. BD has a PopUp Menu button. On HD DVD, you have to use another button, usually the Menu button.

4. There is no Resume button on the BD controller.

5. Both controller have application defined buttons labeled A, B, C, D. You get to define their function in your title. For BD the buttons are colored, but the colors may vary from country to country. In the U.S., they are red, green, yellow and blue, respectively.

6. The controller is often used for controlling the CD and DVD functions of the player as well as buttons to control a TV. These representations are simplified to include just the High Definition DVD buttons.

Either in a flowchart or in a detailed outline of the disc specifications, it is imperative to address the following items and considerations —

- Opening sequence (e.g., copyright, studio splash, trailers), show the allowable UOPs for button operations and whether a user can go to the main menu, popup menus and submenus (also show when these functions are available), default buttons, and transitions.

- First play, play main feature (with notice of how to use the disc, e.g., press menu button for main menu and pop menu button for popup menu), multiple menu objects, lyrics, reference, PiP, network access, and for persistent storage.

- For each area of the disc, you need to specify which of the user operations are allowed. For example, during the first play with the logo splashes, you would show if you are allowing the user to skip ahead to the next segment in the opening sequence or to menu button select to the top menu (see Table 11.3).

Table 11.3 Available User Operations for Blu-Ray and HD DVD

Blu-ray Check List	HD DVD Check List
Menu Call	Menu Call (Title)
	Menu Call (Root)
	GoUp() in TT_DOM
	GoUp() in Menu space
	GoUP() in case of GoUP lin has 'FFFFh' value
Skip to Previous Point	PrevPG_Search
	TopPG_Search
	NextPG_Search
PG textST Stream Change	SP_Stream_Change
Title Search	PTT_Search
Pause On/Off	Pause On/Off
Secondary Video Enable/Disable	
Resume	Resume
Still Off	
Secondary Video Stream Change	FP_SP_Stream_Change
Play FirstPlay Title	
Forward Play (Speed)	Forward_Scan
Secondary Audio Enable/Disable	
Stop	Stop
Backward Play (Speed)	Backward Scan
Secondary Audio Stream Charge	FP Audio Stream Change

continues

Table 11.3 Available User Operations for Blu-Ray and HD DVD *continued*

Blu-ray Check List	HD DVD Check List
Chapter Search	Chapter Search
Primary Audio Change	Primary Audio Change
PiP PG textST Stream Change	
Time Search	Time_Search
Angle Number Change	
Text Subtitle Style Change	
Skip to Next Point	
PG textST Enable/Disable	Video_Presentation_Mode_Change
PopUp On/Off	
	Menu_Audio_Stream_Change
	Menu_SP_Stream_Change
	Menu_Angle_Change
	Video_Presentation_Mode_Change

Authoring and Programming

Here you will take the external specification, create an internal specification if not done previously, and implement all of the functionality for the title. Depending on the intended disc mode and the authoring tool set you may need to create a software framework and a library of standard modules. The framework and library should be created with the goal of being able to reuse them from title to title.

For the advanced modes of each disc format, you will want to define common modules/objects for controlling navigation, generating effects and giving a general framework to a project. Reusable modules will increase the authoring and programming efficiency as each title will not have to be created from scratch. And, over time, the quality of the titles produced will increase because these modules are already debugged.

In these early stages of the development of the next generation disc formats, it will be best to use an iterative approach to authoring and programming. You will design, prototype, test, refine, test with users and with players to find how well a project works. Then you may need to redesign and try again, repeating the steps again and again.

WARNING
There are speed differences between settop players and computer players. Computer players are faster than their settop brethren. You want to test early and often on both settop and software players to ensure that you are implementing a title in a manner that yields a pleasant, desirable experience on both. Not all players are created equal. The specifications for Blu-ray and HD DVD mandate only a minimum performance level (if/when performance is specified).

Quality Control

Using the flowcharts and the rest of the specifications for a title, Quality Control will verify that the disc has been created per the client's requirements and the video, audio and subtitle streams are meeting the client's expectations.

11

In the project plan, allow enough time for checking the title. This may mean that even when a small change is made to the disc, a full check is required. Understanding the internal specification of how the title is built may allow smarter spot checks but depending on the tool used to create the disc and the scope of the change, even a simple last minute change may necessitate an extensive checking of the disc.

Scheduling

There are many activities that may be done concurrently. However, although this can shorten the schedule, you need to plan on the dependencies to optimize throughput and turnaround time. The times required for each of the tasks can be considerable and need to be factored into the project plans. Re-work or tweaking of a title must be planned for in both the workflow and in the overall schedule. Keep in mind that it will be much easier to miss a deadline with the next generation discs.

Chapter 10 described sequential and parallel workflows and although the sequential workflow is more straightforward the project will take longer. Yet, with the parallel workflow, the challenge is making sure that the work is coordinated and completed.

Figures 11.3 and 11.4 list a set of tasks for a sample project. Each of the tasks have an estimated duration for how long the activity will take. In determining the task estimates, it is important to include time for —

- Preparing assets for a step or task (although, ideally, the previous task prepared the asset(s) in the exact format that is required for the next task).
- Testing to ensure that the results are good when input to the next step.

Checkpoints are a good way to ensure that you will complete and deliver a project on time. For example, here are some good checkpoints to include —

- Assets Received — the cleaner this checkpoint is the easier the project will proceed. If you have all the assets received according to the disc title specification, it will feel like you are almost done.
- Client Creative Approval — client approves the graphics, audio and navigational design aspects of the disc.
- QC sign-off — this date signifies that the disc has passed a viewing QC (all of the encodes for audio and video, and the subtitles) have been viewed and are approved. Also, the disc title has been checked to ensure that all of the buttons work and go to where they are supposed to in the manner they are supposed to.
- Approved for Replication — all fixes have been implemented and approved. Some anomalies are now features of the disc (they won't be fixed on this disc). Specification compliance has been verified. Copy protection (e.g., AACS and BD+) has been applied and the check disc has passed.

The goal in the parallel workflow is to have concurrent activities so that the overall project is completed as quickly as possible. Comparing the two sample project workflows, you can see that you can potentially complete a project weeks earlier by doing many of the tasks in parallel.

Figure 11.3   Sample Task Schedule — Sequential Workflow

Item	Task Description	Duration
1	Specifications agreed	1 day
2	Assets received	1 day
3	Compression	5 days
4	QC viewing, 1st	1 day
5	Re-encodes	3 days
6	QC viewing, 2nd	1 day
7	Graphics design	5 days
8	Graphics - client approval	2 days
9	Authoring/Programming	10 days
10	Multiplex all final assets	4 hours
11	Navigation QC, 1st	1 day
12	Fixes for authoring	1 day
13	Fixes for graphics	1 day
14	Fixes for encoding	1 day
15	Fixes for subtitles	1 day
16	Final QC	1 day
17	3rd party QC/Client approval	1 day
18	Specification verifications	1 day
19	Pre-mastering (CMF)	2 hours
20	AACS	2 hours
21	Check discs	4 days
22	Final QC/Approval	1 day
23	Okay to ship	0 days

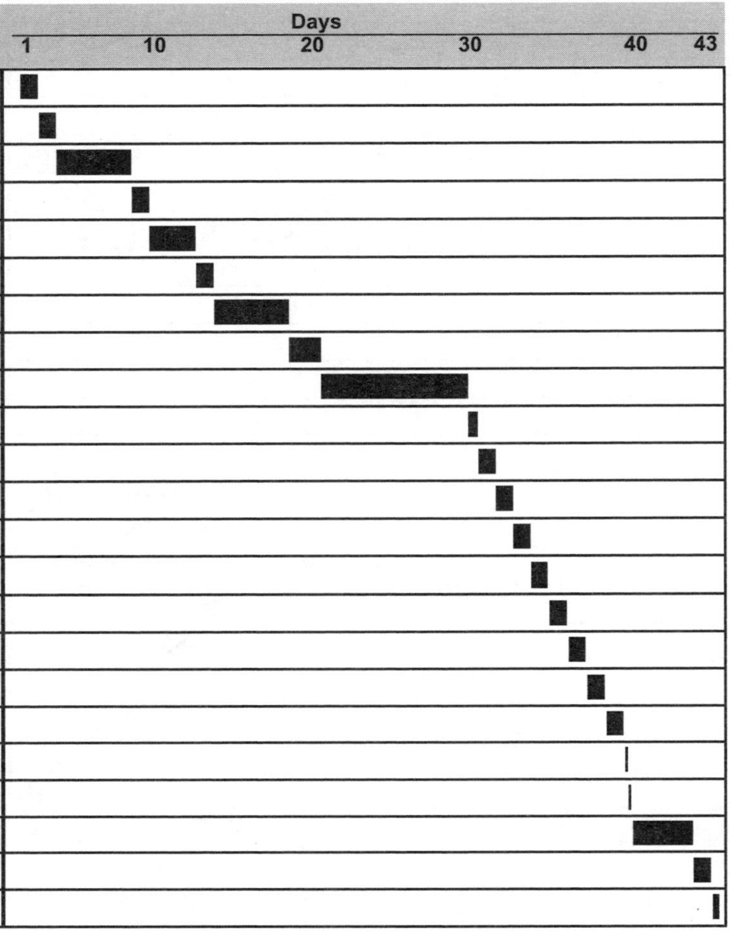

Figure 11.4 Sample Task Schedule — Parallel Workflow

Item	Task Description	Duration	Days
1	Specifications agreed	1 day	
2	Assets received	1 day	
3	Compression	5 days	
4	QC viewing, 1st	1 day	
5	Re-encodes	3 days	
6	QC viewing, 2nd	1 day	
7	Graphics design	5 days	
8	Graphics - client approval	2 days	
9	Authoring/Programming	10 days	
10	Multiplex all final assets	4 hours	
11	Navigation QC, 1st	3 days	
12	Fixes for authoring	1 day	
13	Fixes for graphics	1 day	
14	Fixes for encoding	1 day	
15	Fixes for subtitles	1 day	
16	Final QC	8 days	
17	3rd party QC/Client approval	3 days	
18	Specification verifications	1 day	
19	Pre-mastering (CMF)	2 hours	
20	AACS	2 hours	
21	Check discs	4 days	
22	Final QC/Approval	1 day	
23	Okay to ship	0 days	

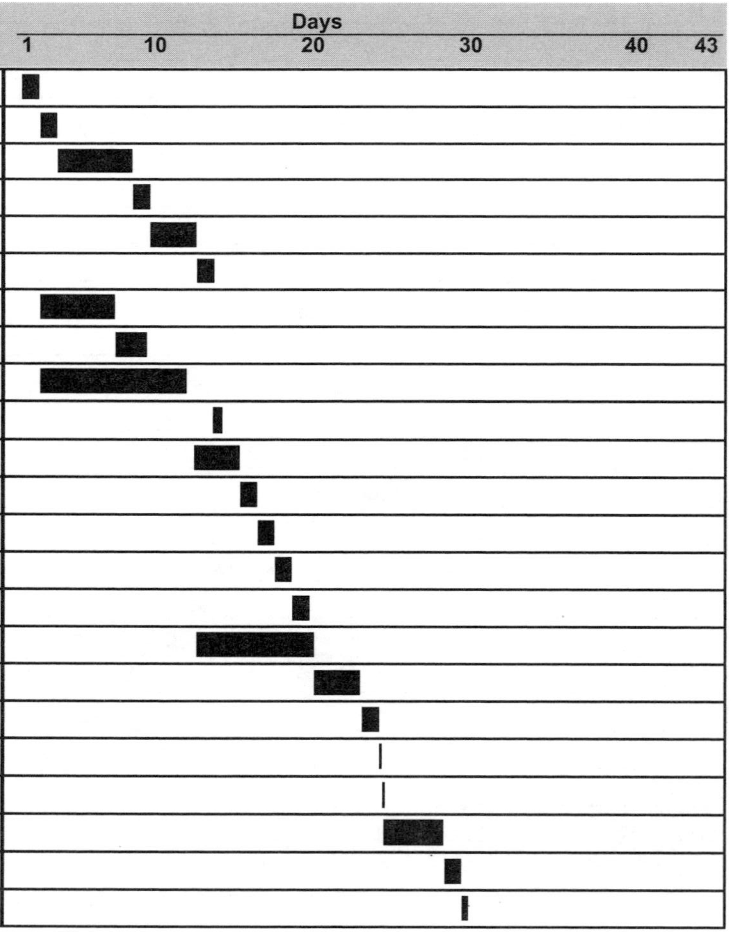

Ready, Aim, Fire

There comes a time in every project when you've got to shoot the engineer and start production. - Unknown

When all the planning is done (and it is never done), it is time to begin production of the title.

One of the client expectations is the ship date. So, off you go. With the planning in place and a way of tracking your progress, you'll have a better chance of maintaining your sanity and meeting your schedule.

You may need to change the schedule due to whatever unforeseen problems show up but with a plan you will have a starting point for the deviation.

Chapter 12
Graphic Design

The graphic design of a video disc makes the greatest impact on the impression of the end user. This is where the coolness factor is applied. The graphic design for these next generation discs goes beyond what can be created in Photoshop®, After Effects®, Flash® or whatever graphic tool(s) are used. Graphic presentation for HD discs is a combination of the images created with the design tools and the enhanced authoring techniques that are available for the disc formats. The art is in optimizing what aspect of the image design happens in the graphics and what happens in authoring.

Now, there are many more options in authoring than simply turning on and off two bit sub-pictures. Buttons can now be separate graphics for the Normal, Selected and Activated states. Also, each button state can be not just one graphic but an animation sequence. Depending on the authoring environment, the programming tools, the authoring team capabilities and the specific format mode being designed for, the graphic designer can create a user experience that is really awesome.

The graphic design workflow for creating a disc is shown in Figure 12.1, at the end of this chapter. The first step is to have the client provide a design wish list for the disc. From this list, the designer develops several concepts. The designer may need to take those concepts to the author(s) for a review of their feasibility in terms of implementation difficulty — resources and techniques to use, can it be done, how long will it take. This is a creative give and take negotiation with the client, graphic design and authoring.

Some of the effects that can be achieved in programming include pans, fades, scaling, cropping, and rotation, as well as many others. Though, some of the effects may not be practical because of limitations of the authoring environment, time constraints on writing code, player performance and, most often, buffer management issues.

Once graphics and authoring have determined what they can implement within the budget and time frame as well as the design wish list, the design can be pitched to the client. With approved detailed design requirements, the graphic designer can create the graphic elements and animations for use by the author. Depending on the size of the operation, someone may pre-qualify and stage the assets for the author. Rejection of assets in the pre-qualification step is most frequently caused by buffer size overflow, positioning, missing assets, clarification on naming of assets or a missing or ambiguous navigation path. Further, the author may find that even with all of the planning that the assets will not work. Most often, this is caused by buffer overruns. Resizing, re-slicing and simplification of the design are the options for fixing these problems.

The Quality Control (QC) stage checks the disc for positioning, graphic artifacts (e.g., banding), functionality (e.g., do all the buttons link to where they are supposed to and do the transitions work as designed), and performance (e.g., do the buttons display in a reasonable time and are transitions smooth, is the disc interaction responsive and intuitive).

The client ultimately approves a check disc for replication. Initial titles may require several iterations to get the actual implementation that meets everyone's expectations. Due to time and budget constraints as well as limitations of the authoring tools (and skills) and the hardware implementations, the design requirements may need to be refined.

In the HD formats there are two types of menus, static/full screen and popup over audio/video/subtitle streams. The popup menu is available in all disc format modes except HD DVD Standard Content. A better term may be popover, as the main video and audio can play uninterrupted while the presentation graphics (PG) and the interactive graphics (IG) are displayed. These popover graphics are either —

- PNGs (Portable Network Graphics)
- Series of PNGs or MNG (Multiple-image Network Graphics)
- Rendered graphics (vector or bitmapped) that have been programmatically generated.

Figure 12.2 Graphic Model Comparisons for HD DVD and Blu-ray Disc™

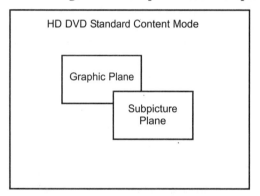

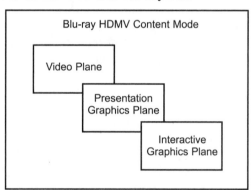

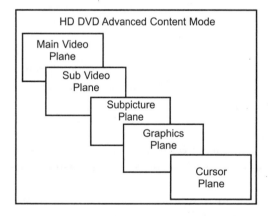

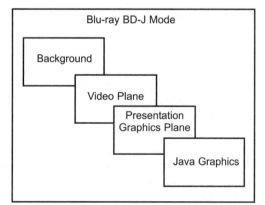

The graphics planes (PG or IG) may be loaded with objects that have X, Y and Z coordinates, a palette, and an alpha channel. Figure 12.2 depicts the design models for graphics in each of the next generation disc modes.

With DVD, menus are created with a background, which could be video, and a subpicture for button highlights. Button highlights are selected from four colors, one of which you cannot use. For the next generation discs, these restraints have been removed.

Also in DVD, there is menu space and video space. In video space, the menu button highlights use the subpicture area that is used by the subtitles. So, the choice is subtitles or menus buttons.

For HD DVD Standard Content mode, the planes from back to front are:
- Graphics plane contains either video or a still graphic
- Subpicture plane contains either a subtitle (in video space) or a subpicture graphic for menu button highlights, and colors are mapped via either a 2 bit index to a color palette or an 8 bit color value

For HD DVD Advanced Content mode (i.e. HDi), the planes from back to front are:
- Main Video plane contains the main video stream
- Secondary Video plane contains the secondary video stream
- Subpicture plane contains elements from the subtitle stream
- Graphics plane contains elements drawn by markup and API's
- Cursor plane contains image file for the cursor. Size, position and alpha level controlled by APIs.

For Blu-ray, HDMV mode, the planes from back to front are:
- Video plane contains the video stream
- Presentation Graphics plane contains subtitles and frame accurate graphics
- Interactive Graphics plane contains menus and interactive graphics

For Blu-ray, BD-J mode, the planes from back to front are:
- Background plane - contains a still frame graphic or solid color
- Video plane contains still frame images and video
- Java™ Graphics plane typically contains menus and interactive graphics
- Presentation Graphics plane contains subtitles and frame accurate graphics

When designing and producing graphic elements for the next generation formats, here are some things to consider —

■ **Full HD resolution is 1920 × 1080.** This is a higher resolution than most computer users have today. Hurrah. But this means you need to be very cautious about simply reusing DVD assets for high definition. If you try this, the graphics are going to look fuzzy when scaled to a higher resolution. You may need to re-create, or at the very least re-render, your assets at the higher resolution. Remember that if your assets were originally designed for 16:9 anamorphic, you'll need to stretch them to square pixels. Figure 12.3 is an illustration of the FBI logo in HD resolution and at DVD resolution. On the left is the source at 442 × 435 pixels. On the right is the graphic that worked fine for standard definition at a source resolution of 220 × 246 pixels. The point is to use the highest resolution source for your graphics. It will show up at these resolutions.

Figure 12.3 Comparison of High Definition versus Standard Definition Graphics

■ **Title Safe and Action Safe Areas.** The current consumer HDTVs and displays have been observed to have an action safe area of about 2 to 3 percent of the screen size and a title safe area of about 7.5 percent of the screen size. But, to be conservative, retain the accepted safe area guidelines of five and ten percent respectively (see Figure 12.4). You may want to review the installed base of displays in coming years and re-assess this guideline but for now it is best to stick with it. Also, always consider the application of the disc and its intended use. For example, if the disc will be for a museum exhibit that will be projected on a 100 inch screen with no title or action safe restrictions then these guidelines do not apply. If you are preparing a general entertainment title for a large audience where they may have any type of display then you will probably want to go ahead and be conservative regarding the action and title safe areas. A sample template is included on the accompanying disc.

Figure 12.4 Title and Action Safe Areas for 1920×1080 Screen Size

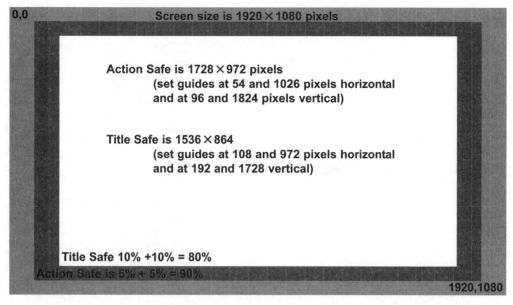

0,0   Screen size is 1920×1080 pixels

Action Safe is 1728×972 pixels
    (set guides at 54 and 1026 pixels horizontal
    and at 96 and 1824 pixels vertical)

Title Safe is 1536×864
    (set guides at 108 and 972 pixels horizontal
    and at 192 and 1728 vertical)

Title Safe 10% +10% = 80%
Action Safe is 5% + 5% = 90%

1920,1080

- **Font Size** HD often means big screen. The old rules for DVD said keep the fonts larger than 20 point. Good practice says to try your fonts and graphics on the target displays. Display technologies have their own artifacts and you do not want to be aggravating your end users by not having addressed how graphics will be displayed. That being said, you can use much smaller fonts than with DVD. Although, 4-6 point fonts are now possible because of the increased display resolutions, you may need to actually use larger fonts because of the distance that the viewer will be from the screen. It all depends on how the disc will be used.

- **Buffer Management is Critical**. For HD DVD, the buffer for program and all of the graphics is 64 MB. For Blu-ray HDMV mode, 16 Mbytes is available for the graphics. With BD-J mode, you have 45.5 Mbyte for graphics and code. You can load and unload material from these buffers but there are a couple of considerations:

  - 1st generation HD players load data from disc at the rate of approximately .75 MBs per second. So, to load 16 MBs can take up to 21 seconds and for 64 MBs it is taking about 84 seconds. Loading the graphics can appear to the user as if the player has stopped. As we are sure you agree, 20 to 80 seconds can seem extremely long time to wait.

Figure 12.5 depicts a simple popup menu construction, and Figure 12.6 illustrates the elements for the menu depiction.

Figure 12.5 Simple Popup Menu Example

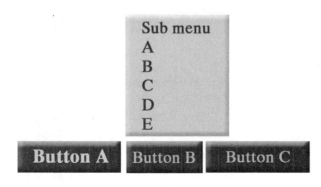

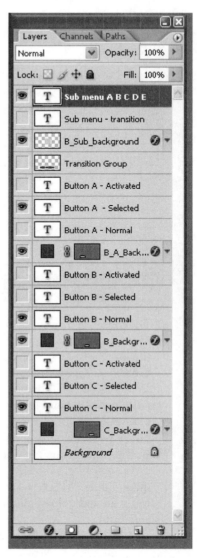

Figure 12.6 Example of Menu Layers and Elements
(Screenshot of Layers/Channels/Paths Panel from Adobe® Photoshop CS)

Please note, in the simple popup menu example depicted in Figures 12.5 and 12.6, the following:

- rectangular button area
- buttons with alpha layers
- three states for each button: Normal, Selected and Activated

Table 12.1 is a sample buffer bit budget calculation for the simple menu depicted in Figure 12.5. This example is for Blu-ray HDMV mode, but the same type of calculations can be

made for discs in HD DVD Advanced Content mode and BD-J mode. Basically, the size of each graphical object is calculated by height times width times bit depth times states times frames. The buffer size required equals the sum of all object sizes. This table is fairly simple in that it contains only a few objects.

Table 12.1 – Example Buffer Bit Budget Calculation for Blu-ray HDMV

Button/ Graphic Element	Height	Width	Bit depth	States	Total Size Frames	6,392,160 Buffer Usage (bytes)
A	120	400	8	3	1	144,000
B	120	300	8	3	1	108,000
C	120	400	8	3	1	144,000
Main_PopUP	120	1138	8	1	24	3,277,440
B1	240	400	8	1	24	2,304,000
B1_A	72	384	8	3	1	82,944
B1_B	72	384	8	3	1	82,944
B1_C	72	384	8	3	1	82,944
B1_D	72	384	8	3	1	82,944
B1_E	72	384	8	3	1	82,944

The first generation of HD design and authoring tools do not help this step. Future tools should have a meter widget to help the designer and the author monitor the projected buffer usage. Just as with bit budget calculators for figuring disc utilization, one should look to see if any homegrown tools have been developed to help with this important aspect of graphic design. For example, Sony's Blu-Print™ authoring software provides a utility called PNGSizeCalculator.exe. Check the enclosed disc as we may have included a utility. In the end, you may end up wanting to create your own tool that is optimized for your workflow.

Understanding buffer management and how to optimize it in the graphic design will help you maintain an efficient implementation. If the graphics are not designed to fit within this constraint, everyone will be involved in the re-work steps and the design concept is at risk of being compromised. In other words, get this right and the design gets implemented as the designer planned. If the buffer management is not addressed, the final implementation will not look like what the client specified.

One of the biggest challenges of implementing interactive graphics for the high definition formats is the perceived limitations of the playout buffer. One of the early CD-ROM titles, *Sherlock Holmes, Consulting Detective*, took a very creative approach to the then-existing limitations. This was back when video on a computer meant image sizes of 160×120 pixels at 10 fps and large computer displays had resolutions of only 640×480 pixels. One of the clever slights of hand that this title used was to have a static background graphic with rectangular areas of the display defined for motion such as the flicker of a candle or the moving of the actor's lips or the swinging pendulum on the cuckoo-clock. The total of all of this motion fit into the 160×120 pixel "buffer" and the effect was a really cool "full screen" motion video experience. The buffer size limitations are a creative opportunity. Don't limit your imagination.

12 Design and Planning

In DVD, menus for any single VTS need to fit within 1G and menu graphics are stored on the disc as compressed files. For HD, the good news is that the disc is much larger. Also, you can define the graphics for individual objects (e.g., each button state). Now for the bad news, there is still a limited buffer space for the graphics. Table 12.2 reflects the buffer space limits by disc mode.

Table 12.2 Comparison of Blu-ray and HD DVD Buffer Sizes

Format: Mode	Buffer restriction
Blu-ray: HDMV	There is a 16 MB buffer that holds all of the button objects and transitions. This buffer is for the graphics as they are uncompressed. In other words, how big the graphics are in pixels times the bit depth.
Blu-ray: BD-J	For BD-J, the 16 Mbytes aremeaningless. The limitation is The buffer is 45.5 MBytes. However, these are 32 bit graphics, not just 8 bitgraphics. Hence, they'll take up 4 times the space.
HD DVD: Standard Content	Full 1920x1080 graphic (still or motion) with the subpictures for highlights. Same as DVD, except the 1GB limit for each title set has been lifted.
HD DVD: Advanced Content (HDi)	Multiple application files (the ACA file which contains the graphics and the code) can be active concurrently as long as they all fit within 64 Mbytes. You can load/unload ACAs.

 TIP As you compare design alternatives you will have to consider the buffer size and how graphics can fit into them. There is one buffer per title set. Therefore, you can split your graphics into different title sets but you need to consider the user experience as the titles load. Use .75 MB per second as the approximate rate of loading graphics. You can also use the in-mux technique to load the interactive graphics while you are displaying a splash or copyright warning or a "please wait, loading..." graphic, with cool entertaining video playing.

During the planning and execution of the graphic design, a number of issues need to be addressed —

- What is the target authoring environment?
  - HD DVD Standard Content
  - HD DVD Advanced Content (HDi)
  - HDMV - Blu-ray's Movie Mode
  - BD-J
- How do the graphics need to be delivered?
  - Layered Photoshop file with layer per button state
  - Layered Photoshop file with layer group per button state (for the Sonic® Designer plug-in)
  - Set of PNG files, sliced from overall graphics. Typically for advance authoring modes of BD and HD DVD, there will be one PNG for each button state.

- Flash project with graphics and interactivity defined.
- What is the performance of the target player?
    - How long will it take to load the graphics?
    - When will the menu graphics be loaded?
    - Will the graphics be multiplexed with the audio/video/subtitle streams or will they be loaded separately, ie "out of mux"?
- What is the player's capacity?
    - Size of buffer
    - Size of persistent storage

Implementation

Probably the best way to create the graphics for the user interface is to first create a storyboard that depicts how the user interface will work.

① Storyboard

② Define what will be done with graphics versus what can be programmed

③ Create graphic elements in Photoshop or similar graphics toolset. Slice the graphics and estimate the buffer size impact and how it fits with your budget.

④ Create animations in your favorite tools such as After Effects or Flash
- Create QT movie to be used as reference for client approval and for authoring/programming guide
- Render to PNGs

⑤ Slice the graphics into the individual PNGs (one PNG for each state). This step depends on the authoring tool. It is not required to slice the graphics for Sony Blu-Print, Sonic's Scenarist® Designer and for HD DVD Standard Content mode.
- What is the pixel size of your PNGs?
- What is the bit depth of the PNGs?
- Have you reduced the pallete depth? Caution: Palette reduction for PNG-8+ isn't easily done in Photoshop or Fireworks. One palette is shared over multiple images. Blu-Print and Scenarist Designer solve this problem for you. But you need to ensure that it still looks good.

 TIP   TIP: Plan on creating Actions in Photoshop to automate many of the repetitive steps that you will do when creating your menu elements, such as —
- Set Action and Title Safe Guides
- Slice
- Save with slice attributes in filename

As shown in Figure 12.7 at the end of this chapter, please note that each button is a separate graphical element in a separate file. As such, depending on the authoring tool, each button needs to be placed individually on the screen and each button gets an instruction for left/right navigation routing (i.e., the arrows). The oddly shaped squares in Figure 13.7 are

12

dummy buttons used for implementing logic for transitions. Specifics for preparing graphics will vary with each authoring tool and with the conventions established for each authoring/graphics team.

Sonic Solution's Scenarist Designer, a plugin for Adobe Photoshop, may be used to aid the project workflow by automating the creation of individual files and allowing the files to be auto-loaded into the authoring tool (Scenarist BD). A key advantage of Scenarist Designer is that this automated loading addresses the problem of matching the shared palette that Blu-ray HDMV mode requires.

Graphics Basics for HD DVD

Standard Content

Uses the same approach as DVD except that the graphic files are now created to HD size, 1920×1080 pixels. Subpictures can be either 2 bit indexed graphics as in DVD or they can be 8 bit values.

Advanced Content (HDAC)

All of the buttons and graphics are objects which are positioned on the display via XML file instructions and JavaScript™. Filenames for handoffs to authoring is critical, as the author needs to know:

- Position in X,Y coordinates — upper left corner is 0,0 and lower right is 1920,1080
- Timecodes for event-triggered graphics
- Button sounds
- States defined for normal, selected and activated
- Any transition information (e.g. QT animations can be provided as examples)

Graphics Basics for Blu-ray

Standard Content, HDMV

Note graphics are limited to 8 bit (256 colors) with an indexed palette plus 256 alpha values (PNG-8+). These alpha values add a powerful dimension to yours design. All graphics and code for each title must fit within the 16 MB buffer. Effects that are available include scrolls, wipes, cuts, fades (transparency changes) and color changes.

The BD-ROM Player can present either a Text Subtitle stream or a Presentation Graphics stream, but it will not present both at the same time.

Blu-ray uses the same coordinate system as the HD DVD Advanced Content mode.

The BD authoring systems abstract the actual player implementation in different ways. The advantage is that authoring is easier — the disadvantages include not being able to get to all of the raw features.

Sony Blu-Print

Define the graphics as a multiple layer Photoshop file with a layer for each button state

and for the background. When the buttons are drawn in Blu-Print, the graphics are sliced and the groups of buttons and their relationships are defined. Changing graphics is as simple as loading a new multiple layer Photoshop file. Blu-Print has all of the button areas, so that slicing can be done or redone and any new generation of palettes can take place in Blu-Print.

Sonic's Scenarist BD

This authoring system requires every button and its three states (normal, selected and activated) to be positioned in the tool. This works for the initial loading of the project but if a button graphic should change, which will require a new palette, all of the button graphics have to be re-assigned or the palette for the PNGs is not correct. To overcome this feature of their abstraction, Sonic has a different workflow that includes a Photoshop plug-in. Instead of creating buttons, defining placement and assigning all of the graphic files to each button state in the authoring tool, Sonic has this work done as part of the preparation of the graphics. While this can mean more work for the graphic designer, there are a couple of advantages. Each button is structured as a set of layers (that have the normal, selected and activated states). First, the designer is aware of how the graphics will be sliced and can help optimize the buffer size (there is a built-in feature that calculates the buffer size so the designers can optimize the design while still in the design phase). Second, because the buttons are defined in named layers, the authoring links the named graphics to buttons and their interactivity. When the button changes, say to fix/change text or a graphic treatment or the position on the screen, when the PSD file is reloaded into Scenarist, all of the authoring is preserved. This has some very significant implications. The buttons could change position, text and graphics and because button_1 is still button_1, any action/interactivity that was programmed for button_1 still exists. So, you could create a popup menu with buttons for top_menu, Scenes, Extras and Setup. You could author that functionality for a disc. Then change the PSD for a new disc, replace the content in the project and now you have a new disc with the same functionality of your template disc. So, a little bit more work upfront for graphics, but it can save significant time in authoring and QC. The new disc works just like the template because the interactivity has not changed. Another benefit is that the graphics designer is able to specify more explicity how graphics will look.

Advanced Content, BD-J

All graphics can be 24 bit with an alpha channel. Buttons and other interactive items are defined via objects. Objects can be positioned on the Interactive Graphics plane by their X, Y and Z coordinates. The Z coordinate allows objects to overlap and for the order of the overlap to be defined.

Summary

The tools, techniques, playback platforms and even the specifications are all very immature as this book is going to press. These are the early days of these formats. The best in state-of-the-art user interfaces is changing every few months.

Over the next one to two years, tools and techniques will be developed so that creating great graphics will be much easier and faster. For now, this is the time when what can be done with the formats is being defined. How hard can graphic designers and the authors push these "limitations"? Stay up to date with us as we watch in the years ahead!

Figure 12.1 Graphic Design Worflow

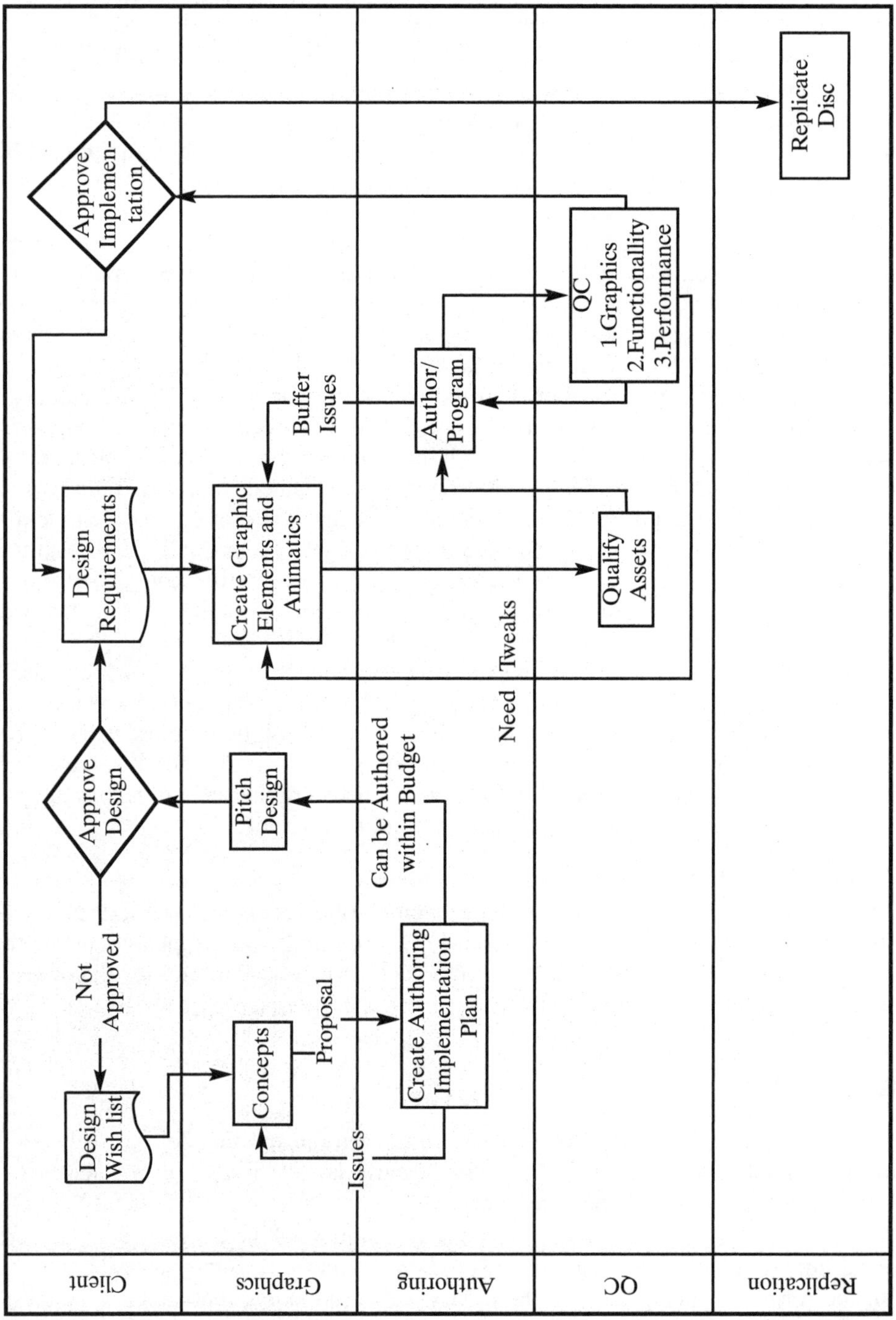

Figure 12.7 Screen Shot of Sliced Graphic from Sonic Scenarist BD Tool[a]

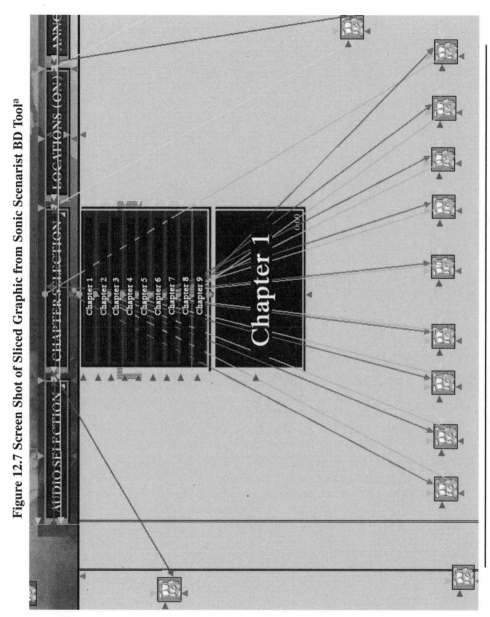

[a]Note that each button and button state is a independent graphic file sharing a common pallet and placed at the specific x and y coordinate. In the case of Scenarist BD, the individual files are listed in the data tree in the lower left hand side of the screen.

Chapter 13
Compression

This chapter focuses on the art of compression. The compressionist uses tools and techniques to reduce the file size and bitrates of the audio and video streams to fit within the constraints of the high definition discs specifications while minimizing any perceptible artifacts. Tables 13.1 and 13.2 summarize the impacts of high definition video and audio and the challenges for the compressionist. Although the discs can now store three to ten times more than DVD, the requirements of high definition video have increased by a factor of six and for audio, using the highest resolution audio, by a factor of over 40. Depending on the program length on a disc, there may be enough space on the disc. But looking at the maximum total bit rates allowed for each format we see that for DVD, the maximum is 10.08 Mbps vs. 30 Mbps for HD DVD and 48 Mbps for Blu-ray™. So, the bitrate maximums for the total streams is only three times more than DVD for HD DVD and only about five times for Blu-ray. This means that to get the same quality that you expect on DVD, we need higher levels of compression.

Table 13.1 Video Compression Ratio Comparisons to DVD

Format	Mbps	Ratio
HD DVD Video	29.4	3.0
HD DVD total max bitrate	30.24	3.0
Blu-ray – video	40.0	4.1
Blu-ray – total max bitrate	48.0	4.4
DVD Video	9.8	1.0
DVD – total max bitrate	10.08	1.0

Table 13.2 Audio Compression Ratio Comparisons to DVD

Codec	DVD Max Bitrate	HD DVD Max. Data Rate (Mbps)	Ratio to DVD	BD Max. Data Rate (Mbps)	Ratio to DVD
Linear PCM[a]	1.536	13.500	8.8	27.600	18.0
Dolby® Digital[b]	.448	.448	1.0	.640	1.4
Dolby Digital Plus[b]		3.024	6.8	4.736	10.6
Dolby TrueHD[b]		18.000	41.1	18.000	41.6
MPEG-1 Audio[c]	.384	.384	1.0		
MPEG-2 Audio[c]	.384	.912	2.4		
DTS™	1.509			1.509	1.0
DTS-HD High Resolution™[d]		3.024	2.0	6.000	4.0
DTS-HD Master Audio™		18.000	12.2	24.500	16.2

continues

13

Table 13.2 Audio Compression Ratio Comparisons to DVD *continued*

ᵃFor comparison to DVD, Linear PCM sample frequency of 48 kHz, 2-channel audio is compared to the maximum now allowed for HD DVD and Blu-ray

ᵇDolby Digital comparison is for maximum allowed data rates for each of the formats. The base DVD Dolby Digital codec is used for comparing the new codecs available to HD DVD and Blu-ray.

ᶜThe highest bitrate available for each format is used for comparison.

ᵈThe DTS comparison uses the highest bitrate available for each format is used for comparison.

When you are compressing either the audio or the video, ideally you don't want to introduce any artifacts into the encoded stream. In reality, any lossy compression introduces artifacts. This chapter is about minimizing those artifacts. You want the source and the compressed stream to be indistinguishable to the end user on their playback platform. Sometimes, and this is touchy area, you'll want the compressed stream to sound or look better than the original. There is a fine line between restoration, which is an art in and of itself and compression.

NOTE At high definition resolutions, defects or imperfections in the source film or video are now very visible. The defects were largely masked on DVD due to the smaller frame size and the detail that the codecs are able to expose at the higher data rates. Not only are banding and grain now visible, but so are out of focus scenes and all of the wrinkles and warts on the actors and actresses. You may have material that you do a great job of compressing that gets rejected because "it doesn't look good". New film transfers or pre-processing may be required to get a master that is good enough for High Definition DVD.

Audio

For high definition discs, there are many more choices of codecs than with standard definition DVD. With DVD, the audio might only take 10% of the disc space and the bit rate allocation. Dolby TrueHD at a peak of 18 Mbps, uses between 38% and 63% of the available maximum data rate (for Blu-ray and HD DVD respectively). These "fat" audio streams create more need to plan and co-ordinate the audio and video compression efforts to ensure that a good bit allocation trade-off has been made.

When compressing or encoding audio, be sure to check —

- Has the audio been conformed so that all of the audio for a disc works together well?
- Do all audio levels match?
- Do all language track audio levels match?

■ Are the commentary track levels appropriately loud, so that as users change tracks, they are not forced to adjust the volume constantly?

■ Menu audio is usually lower than main audio but it depends on the disc.

■ Button sounds should be at an appropriate levels compared with the main feature, particularly if a popup menu is used. Note that with HD DVD, the author is able to programmatically fine tune the levels.

■ Does the bitrate meet the bit budget?

■ For advanced lossless audio codecs, are data rate spikes within limits and/or co-ordinated with the video bitrate at the frame level, to prevent buffer overflow?

■ Make sure that you create an audio stream with the right parameters for the target format.

  ■ As noted in the chapter on codecs, different formats have different requirements. For example, DTS HD High Resolution, and Dolby Digital TrueHD, each encode a stream differently for Blu-ray Disc than for HD DVD, even though they have the same information in the stream. Be sure you encode the stream for the correct target HD format (HD DVD or Blu-ray Disc).

■ Is the audio synchronized with the video?

  ■ Use the same start time and handles (if the video starts at 00:00:58:00, then the audio must start at 00:00:58:00)

  ■ The end time can be the same, or depending on the authoring tool, it can have a longer handle of a few seconds

  ■ Use embedded timecode, if the encoder and the authoring/multi-plexing tools support it

  ■ Make sure that the timecode for the audio and the video are the same type (e.g., 29.97 fps, non-drop frame).

TIP

Make sure your file systems and networks can handle the large files that you'll be dealing with. Record data to optical discs (DVD, Blu-ray and HD DVD recordables) formatted to UDF 2.1 or later. Format hard disk drives for NTFS. Mixed platform networks should be checked to ensure support for files larger than 4 GB. Feature length, encoded high definition audio files can now easily be over two GB. ISO-9660 formatted DVD-ROMs only support files less than 2 GB. Early implementations of network file sharing between Mac OS/X and Windows truncate files at the 2 or 4 GB size. SneakerNet with hard disk drives formatted with NTFS file system works. FAT32 formatted hard disk drives will work for most audio but files are limited to less than 4 GB. Feature length, high resolution, uncompressed, multi-channel, PCM files will be larger than 4 GB.

13 Video

The biggest factor affecting video quality is the encoder. A well implemented encoder will have more effect on quality than which codec is used. With over 10 years of experience, MPEG-2 has been the codec of choice for many the first shipping Hollywood discs. Compressionists know the parameters and the developers of the encoders have been quick to expand their encoders to handle high definition DVD video.

Initially, the VC-1 and AVC encoders had the capability to generate video streams with equivalent MPEG-2 quality at bitrates that were at least 10% less than MPEG-2. Over time, as the encoder developers and the compressionists have more experience, it is possible that the bitrates for equivalent quality encoded video will be 50% of the MPEG-2 bitrate.

The basics of approaching video compression are —

- ■ Does the encode fit the bit budget?
- ■ Does it look good?
- ■ Is it compliant (does it work in the project)?
- ■ When is it due?

Here are the features that should be included in any of the professional level encoders (regardless of the codec) —

- ● Target format: HD DVD and/or Blu-ray
- ● Set frame rate and size, e.g., 23.976 fps at $1920 \times 1080p$
- ● Bit rate – target average, maximum and minimum rates
- ● GOP structure —
  - ● I, B, and P intervals
  - ● Pyramid B structure for AVC
  - ● I frame interval – fixed, maximum and minimum.
  - ● Dynamic placement of the B frames depending on content.
- ● Quantization
  - ● Dynamic (automatic)
  - ● Manual
  - ● Constant or variable bitrate
- ● Quality priority
- ● Bitrate priority
- ● Number of Passes

Depending on the codec, the encoder may expose other parameters and algorithms. Although you may want to have access to all of these details, when initially learning an encoder and codec, it is best that they are not exposed. Some of the algorithms in the codecs, although they seem like they will be helpful, are interesting but in practice turn out to be a waste of time. Let the art of compression be your guide.

Some encoder capabilities will increase the ability of the compressionist, improving both quality and productivity —

- ■ Support for multiple processors or computers. Some encoders can be configured to run a cluster of computers greatly reducing the time it takes to do each pass of an encode.

■ Present a quality graph of the encode showing PSNR (Peak signal-to-noise ratio vs. time)

■ Present a bitrate graph of the entire video stream (bit rate vs. time). This is useful for seeing the average bitrates of the GOPs (Group of Pictures) and identifying where there may be issues for buffer underflow or in setting the layer break.

■ Access to the quantization matrix — display and allow adjustment

■ Pre-processing including debanding and various filtering options

■ Compare source with encodes and versions of encodes onto one or more monitors. Also, allow split screen capabilities (horizontal, vertical, diagonal, side by side and butterfly). Some monitors will also display the difference between two signals. These comparisons should be viewable at speed and frame by frame with forward/reverse stepping capabilities.

Pre-processing can greatly increase the quality of an encode, with serious impacts on time and disk space. Video that is 1920 × 1080 pixel frame, 23.975 fps, 10bit, 4:2:0 uncompressed requires 370 Mbytes/second. A 90 minute feature will take about 2 Tbytes of disk space. Processing this file means moving a lot of bits. Even with today's fastest RAID's, just reading this file takes at least real time for the video. Having the pre-processing built into the encoder and not as a separate step will speed up the overall encoding process.

Each encoder has capabilities that allow for improved ease of use or for automatically getting higher quality encoding. Here are some of the features to look for in an encoder —

● Automatic Scene Detection

● Number of Passes

● Quality priority mode — VBR with constant quality

● Adaptive Quantization — automatically adjusts I, B, P quantization for maximum quality

● Dynamic GOP Structure — automatically adjusts number of frames in a GOP. You should be able to specify the maximum number of frames and minimum number of frames. If I frames are too close, the bitrate can become too high to multiplex with the other streams

● Layer break support

● Multiangle

● Seamless branching

Each encoder has its unique capabilities and personality. Not all features of each codec are available on each encoder. Each encoder may implement the features differently with different degrees of effectiveness. Just as with Standard Definition DVD, High Definition DVD will probably require multiple encoders that allow the best encodes of all types of video scenes. Both VC1 and AVC encoders for Next Generation DVDs are at the moment immature in their feature set implementations. Expect that the encoder implementations get much better, easier to use and faster as they are used more and have the experience of the early encodes integrated back into them.

13

Examples of some problem type video segments and how to work with them —

- Music video — light flashes less than just a few frames apart with scene detection set to "ON" will sometimes cause the encoder to create GOPs that are too short.
- Banding that is an artifact of going from 16 bit to 8 or 10 bit.
  - May need pre-processing to deband the input — happens in animation and computer generated sequences
  - High frequency noise
  - Fine detail can be lost with the deblocking settings

Film grain technology (FGT) can be used to lower the bitrate and preserve the feel of a feature. This is only available for AVC encodes on HD DVD format discs. All HD DVD players include the FGT capability.

Getting Started with a New Encoder

Each encoder and codec has features and capabilities that can at first impressions often overwhelming. Here is the pragmatic approach to learning how to get the best results from an encoder and codec —

- Review your bit budget for peak bitrate and average bitrate of the video.
- The average bitrate for the video stream includes the requirements for audio and subtitles as well as secondary audio and video. To ensure that you'll have space for re-encodes, set the target bitrate in the encoder at about 90-95% of this average bit rate.
- Determine the audio bitrate requirements for peak birate. Set your peak bitrate to the maximum allowed by the format minus the peak bitrate required for the audio stream.
- When using VBR, the maximum bitrate should be at least 15% higher than the average bitrate setting. If the maximum and average bitrates are too close, the quality will be worse than a CBR encoding at the average bitrate setting.
- Set the appropriate frame size and frame rate with inverse telecine as required by the specifics of the source video.
- Set all other settings in the encoder to default.
- Encode some typical difficult scenes with these parameters.
- View these scenes to see if they are of acceptable quality. If these scenes have no objectionable encoding artifacts, you are ready to do a complete encode with these settings.

The above approach is most efficient for getting a base encode. It is your reference point for further refinements. Basically, if the encode looks good and it fits on the disc, then it is good. It is compressed enough. If there are scenes where these settings did not result in acceptable results, then you need to go into the tool box of codec capabilities and encoder features to get more compression so that it looks good.

Advanced Encoding Techniques

When the simple approach doesn't produce acceptable results, then you need to use the advanced capabilities of the codecs and the encoders. Each encoder has recommendations for how to proceed. The encoders will often have a way that works well for that encoder.

Getting the best encode for a particular video is an iterative process. Figure 13.1 shows an over all workflow where there are many iterative loops. You will want to set up your encoding setup (hardware, software, and network) so that you can make test encodes very quickly. To get the best encode means doing many, many test encodes.

Figure 13.1 Compression Workflow

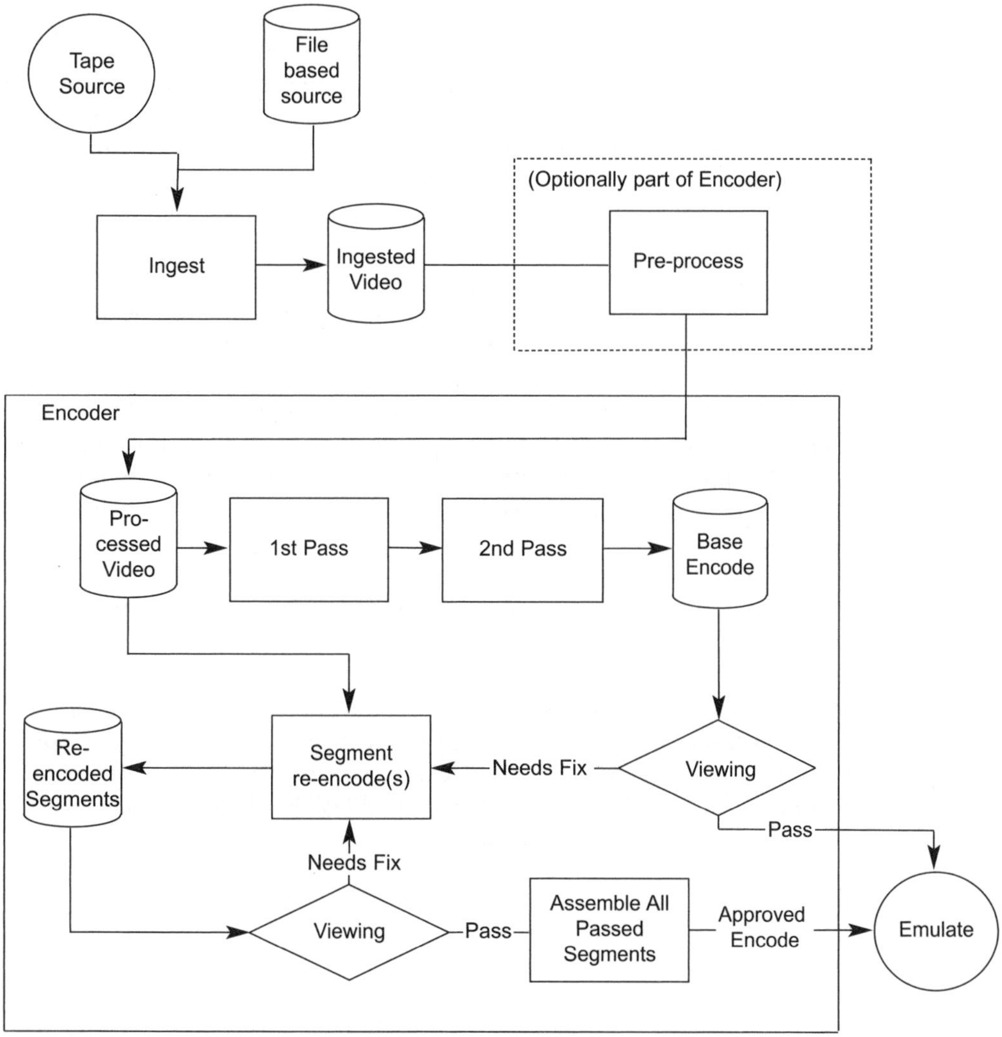

While different encoders and different hardware implementations vary, the following techniques result in quality encodes in the shortest amount of time. They apply whether you have a 128 CPU cluster or you run on a single core CPU. Just scale your implementation to fit your projects.

- Set up a batch of encodes of the same scene with different settings where you've changed 1-2 parameters for each encode. This will allow you to review the encodes all together and see which ones look the best. This helps you more quickly builds your experience/expertise in determining what effect the different parameters really have.

- Use measurements such as PSNR when you can. Often these are small or subtle improvements that you are looking for. So, if you have a base-quality PSNR of 42db, then an encode that moves you from 39.1 to 39.4db is a big deal. A quantitative measurement helps track results and ensures that you find and use the best resulting encode settings. Even if you are using a subjective judgment of an encode, try to give that test encode a numeric value, such as "This encode was a 7 on a 10 point scale." You may have tens of test encodes that you are trying to review. Quantifying the quality of each of the encodes will help you more quickly sort out the results.

- Some of the encoding algorithms are very time consuming. Remember to calculate how long the entire encode will take with this setting. You may be able to just use the setting for a set of scenes. You may also need to compromise to meet your schedule.

- Some of the algorithms will have negative effects when combined with other algorithms. Conversely, the result of one algorithm may over shadow the impact of another algorithm. Log the results of your test encodes to be able to quickly and more efficiently get the best encodes and to build your expertise in using a particular encoder and codec.

- Segment encoding is essential to getting the best encode for a full feature length video. The best settings for one scene may totally mess up another scene. The best encoders let you encode segments, and choose and concatenate the best segment for the final full encode.

- Because this is an iterative process and the encoding times can be quite lengthy even on the fastest available hardware, advanced encoding can take days if not weeks to get a good final encode. Set expectations accordingly. Even with all the interactive cool features of the new high definition DVDs, for most uses of these discs, the video is the content that people want to see.

TIP   Plan on intermediate in-process backups of a full encoded stream. Although this may mean large amounts of disk space, you get a couple of benefits. First, if you have a hardware or software or operator glitch that causes a loss of work, all is not lost. Second, you may find that with a deadline approaching, the encode you had a couple of days ago now looks acceptable (okay, maybe it is looking better and better).

Summary

The new audio and video codecs allow a new level of freedom for the compressionist to encode at relatively low bit rates and yet have high quality audio and video. The challenges include audio codecs that currently are not supported by the A/V receivers in the market (you can encode something that no one can play back), video codecs that are very bit efficient but that consumers may have difficulty playing back on computers because the computational requirements for the decoder exceed what most computers can accomplish today. Also, the initial VC-1 and AVC encoder implementations are immature and don't always create streams that are specification compliant or compatible with the authoring system or players. Perceiving artifacts for the new codecs requires a new eye. The artifacts are more advanced (i.e. weird) compared to MPEG-2. For a lot of applications, MPEG-2 at 24 Mbps will look just as good as a VC-1 or AVC and it will take you a fraction of time and expense of the new advanced encoders.

For some applications, the new codecs will absolutely be required. For example, VC-1 or AVC will allow 45 minutes of 24 Mbps, very high quality, high definition video on a DVD-9. The popularity of all of the extra content for DVD may also be true for this next generation of discs. The art of compression is what will make it happen.

Chapter 14
Authoring/Programming

This chapter is intended to provide an overview about the different authoring and programming procedures used on HD DVD and Blu-ray Disc™. Since both formats have already been discussed earlier in this book, this chapter will focus on different techniques to achieve certain functionality. We are aware that the implementation of the format specifications is software dependent. Every software application will behave differently. And some may not provide the low-level access required to achieve the functionalities we'll talk about. With that in mind, this chapter will talk about general techniques and not software specific details.

HD DVD Standard Content

Authoring for HD DVD Standard Content (HD DVD SC) is very similar to authoring for current DVD. Meaning, it should be very easy for everyone with DVD authoring experience to start working within this new format. The authoring does not require any scripting or programming skills, because it exposes all possible commands and functionality within the environment of an authoring application. Some of the enhancements to HD DVD SC like the availability of more General Parameters (GPRMs), more allowed commands per PGC (Program Chain), Button or Cell, don't really add new functionality and are already explained in an earlier chapter. However, we would like to focus on a few examples that actually provide new or improved capabilities compared to current DVD, such as Multi-Language Authoring, Accessing Resume Information and improved techniques to set audio and subtitle streams.

Multi-language Authoring

Using current DVD, there is only one System Parameter (SPRM) storing the Player Default Language. However, this is not very sufficient attempting to author a title with multiple language menu sets. In theory, it was possible to create multiple menu PGCs specifying the language for each of those. The player should automatically jump to the PGC with the matching language set. Unfortunately, in practice, this only works when there is a menu language that matches the Player Default Language. In case the player is set to a language that is not contained on the disc, each player behaved totally different, leaving an unacceptable user experience.

On HD DVD, this mechanism has been improved. Although there still is the System Parameter specifying the Player Default Language (SPRM 21), there is also a new System Parameter (SPRM 0) that can be used to specify the Current Menu Description Language. SPRM 21 still has to be set in the Player Setup Menu, and cannot be modified by the author.

14

But SPRM 0 can be changed by the author to make Multi-Language Authoring a much easier task. Now the disc can contain a so-called option card listing all available menu languages to choose from. Since the First-Play PGC is now also a "true" PGC, which can hold content, the option card could be placed within the First-Play PGC making it the first menu presented upon disc playback. Once the user selects the desired menu language, SPRM 0 would be updated to reflect this choice.

Now it is very easy to navigate the appropriate menu language. Each menu set would be stored within a language specific PGC, and the value of SPRM 0 would be used to jump to the correct menu language. This way, the player setting could still be used as a starting point, for instance to reflect the availability of this language by highlighting this button on the option card. But in case the Player Default Language is not available on the disc, it would not cause unexpected playback behavior, because the user can update SPRM 0 with the desired language.

Another great new feature is the option to present multiple angles and audio streams within a Menu PGC. This is perfect for Multi-Language titles as well. For instance, some of the current DVD titles use language during the menu loops to tell people to "push a button". Another use case for such a feature would be a disc for visually impaired, making the button choices audible. Implementing such a feature on a title with multiple menu sets was rather complicated. Each of the menus would have to be placed in a different PGC with the respective language audio. This not only complicated the programming, but also impacted the bit budget of a disc, because the background video used for the menu would have to be stored multiple times on the disc. With HD DVD SC, this becomes very easy. Each PGC could simply contain multiple language specific audio streams and also different video angles for the background.

Access to Resume Information

The concept of Resume compared to current DVD has not really changed in the sense that there is still only one set of resume information available for the player. However, there are a couple of improvements providing control over when this information gets updated. On DVD, whenever a jump from Title Space to Menu Space was executed, the resume information, specifying where the content got exited, was updated. The player would use this information to return to the exact same location when the user called the resume command from Menu Space. Unfortunately, this didn't always allow for the desired feature. Let's say the user exited the feature jumping back to the menu. Then some bonus game was played and the user would now want to resume the feature from the menu. This use case would not be possible on DVD. With some tricks using dummy PGCs, the author would be able to redirect to the feature instead of the game PGC, but it would not be possible to resume the feature, only a re-start would work.

On HD DVD SC, this mechanism has been improved. When exiting a PGC in title space (i.e. feature title), a so-called Resume Permission Flag is checked to find out whether the resume information shall be updated or not. This would make the implementation of the use case described above very simple. As illustrated in Figure 14.1 below, the feature title permits the Resume Permission Flag, resulting in an update of the resume information before displaying the menu. Executing a title jump to the game will start a new title in which the Resume Permission Flag is prohibited. The result is that exiting this title, will not update the resume information but go straight to the menu. If the user now performs a Resume Call, the resume information would still specify the location where feature was exited and return to this position.

Figure 14.1 Resume Process Example

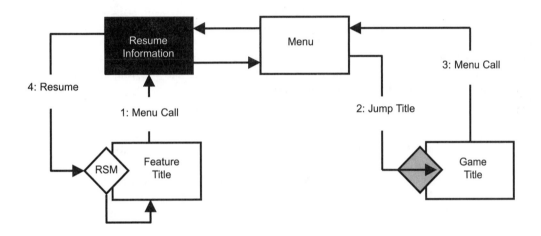

However, before jumping to the resume position, the player jumps to a resume command area that is associated with the PGC of the feature. If there are any commands in this area, they will be executed before playback resumes. This area could be used to make sure that all the appropriate audio and subtitle streams are set. It could also be used to branch out of this title altogether in case something else should be played.

NOTE   All commands in the Resume Command area will be executed before playback resumes. However, in case there is a break command present in this area, processing of the commands will be stopped and playback resumes immediately.

Setting Audio/Subtitle Streams

For every DVD author, setting audio and subtitle streams is a very familiar task. Due to the format only allowing one command per button and not having a dedicated resume com-

mand area, changes to the stream configuration had to be immediate. This means that the audio and subtitle streams were set in the Pre-Command of the Feature PGC. Once this was passed, jumping out of the feature and resuming playback, the Pre-Command area would not be passed through. Hence, any updates to the stream configuration would be ignored leaving the previously set configuration. The only work around on DVD was to jump to the Post-Command area of the menu and setting the appropriate streams. In this case, resuming the movie would have the changes applied.

For HD DVD SC, there are now two different ways of doing this, each of them being much easier than the DVD work around. Since HD DVD SC allows for up to eight commands in a button, setting the stream could be part of the button command. This way, the user would select a specific audio or subtitle stream, which could be applied immediately without any jumps to other command areas necessary. Another way to apply the changes would be to set the streams in the Resume Command area. This method is very similar to the one used on DVD, where the stream settings were applied in the Pre-Command area. However, now the settings would also be applied in the Resume Command area, knowing that all commands in this area will be executed before playback is resumed.

TIP

The new feature of multiple commands per button (up to eight) can be used for many other issues as well. Whenever there is a need for multiple instructions to change settings as described above, this feature is helpful. In addition to that it can also be used to modify multiple parameters when a single button is selected making the navigation of the title more interesting.

HD DVD Advanced Content

Compared to DVD authoring, HD DVD Advanced Content (HD DVD AC) is a totally different concept. It's better compared to programming for web applications, because it's using XML and ECMAScript, meaning that Web programmers are typically familiar with both languages. ECMAScript is the official standardized version of this scripting programming language, which is also loosely referred to as JavaScript™ or Jscript (although there are differences between ECMAScript and JavaScript or JScript). However, it should not be mistaken for Java™, which is an object-oriented programming language and not a scripting language. For regular DVD authors used to software applications with nice User Interfaces, this new format will require a different way of thinking and skill set. However, this chapter will provide an overview on where to best get started.

In terms of programming environment, there is really not much needed for HD DVD AC. The only "special" software required to create titles is used for multiplexing and formatting. Creating the code for the applications on the other hand doesn't require any special toolsets.

It could literally be done using Notepad or any other Text Editor. However, it is recommended to use a more sophisticated XML or JavaScript Editor, because they also contain verification mechanisms to avoid little mistakes, such as not closing tags.

Playlist

The first thing a player will load is the playlist file. It defines the general structure of a disc linking the multiplexed EVOBs together with the advanced applications. It also configures the player and provides information about the media attributes used on this disc. To manage all this information, the playlist is divided into three parts — Configuration, MediaAttributeList, and TitleSet. Table 14.1 shows an example of the Configuration and MediaAttributeList portion of the playlist.

Table 14.1 Example Configuration and MediaAttributeList of a Playlist File

```
<Playlist xmlns="http://www.dvdforum.org/2005/HDDVDVideo/Playlist"
        majorVersion="1" minorVersion="0">

        <Configuration>
                <StreamingBuffer size="2048"/>
                <Aperture size="1920x1080"/>
                <MainVideoDefaultColor color="108080"/>
        </Configuration>

        <MediaAttributeList>
                <VideoAttributeItem index="1" codec="MPEG-2"/>
                <AudioAttributeItem index="1" codec="DD+" channels="2"/>
        </MediaAttributeList>

        <TitleSet>
        ...
        </TitleSet>

</Playlist>
```

The Configuration portion is responsible to setup the player correctly. Attributes like the size of the streaming buffer and aperture, or the color of the background plane can be defined in this section. The MediaAttributeList on the other hand defines all the attributes of the media being used on the disc, for instance whether a video is compressed with the MPEG-2, AVC or VC-1 codec. This way, playing back a title will reference the associated attribute and instantly know how to correctly configure the available decoders.

Table 14.2 Example of the Title Set portion of a Playlist File

```
<TitleSet timeBase="60fps">
        <Title id="main " titleNumber="1"
        titleDuration="00:01:24:00" displayName="Hello World"
        onEnd="top_menu">

                <PrimaryAudioVideoClip titleTimeBegin="00:00:00:00"
                titleTimeEnd="00:01:24:00" dataSource="Disc"
                src="file:///dvddisc/HVDVD_TS/TITLE001.MAP" >

                        <Video track="1"/>
                        <Audio track="1" streamNumber="1"/>

                </PrimaryAudioVideoClip>

                <ApplicationSegment titleTimeBegin="00:00:00:00"
                titleTimeEnd="00:01:24:00" zOrder="0"
                src="file:///dvddisc/ADV_OBJ/menu.aca/menu.xmf"
                sync="hard" language="en">

                        <ApplicationResource priority="1" multiplexed="false"
                        src="file:///dvddisc/ADV_OBJ/menu.aca"
                        size="20000000" description="archive"/>

                </ApplicationSegment>

                <ChapterList>

                        <Chapter titleTimeBegin="00:00:00:00"/>
                        <Chapter titleTimeBegin="00:01:19:15"/>
                        <Chapter titleTimeBegin="00:01:08:15"/>

                </ChapterList>

                <TrackNavigationList>

                        <VideoTrack track="1"/>
                        <AudioTrack track="1" langcode="en:00"/>

                </TrackNavigationList>

        </Title>
</TitleSet>
```

The Title Set portion as shown in Table 14.2 describes the general layout of the disc itself. It specifies each individual title with the total duration of it and the content that's being played together with the In- and Out-times of each clip. This way, one title can reference multiple clips. It also specifies the number of audio, video, and subtitle tracks that are present in each clip, and assigns the associated media attributes to them. Additionally, it contains the chapter list of all available chapters and defines where to jump to when the title is finished playing. But most importantly, it assigns advanced applications to each title.

TIP
There can be multiple playlist files available in a player. For instance, one playlist is contained on the disc while an updated playlist resides in persistent storage having updated content. The playlist is named VPLST###.XPL with "###" being a number from "000" to "999". The player will always use the playlist with the highest number in the filename. So whenever some content is updated, the playlist only needs to get a higher number assigned.

An advanced application is the menu set that's being presented while some video is playing back. Basically, all the elements needed for the menus set are contained in so-called Advanced Content Archives (ACA). While creating the actual ACA is the easiest part (it is similar to creating a ZIP file) and getting all the menu and effect sound elements assembled is also not a very big problem (this process has been discussed in previous chapters), we will focus on the programming of the disc. In order to make a disc work with Advanced Content, there are three basic elements involved: manifest, markup and script.

Manifest

The manifest is a basic description of the application. It is named manifest.xmf and an example is shown in Table 14.3 below.

Table 14.3 Manifest File Example

```
<Application xmlns="http://www.dvdforum.org/2005/HDDVDVideo/Manifest">

        <Region x="0" y="0" width="1920" height="1080" />

        <Script src="file:///dvddisc/ADV_OBJ/menu.js"/>
        <Markup src=" file:///dvddisc/ADV_OBJ/markup.xmu" />

        <Resource src="file:///dvddisc/ADV_OBJ/markup.xmu" />
        <Resource src="file:///dvddisc/ADV_OBJ/main.js" />
        <Resource src="file:///dvddisc/ADV_OBJ/mainMenu_Btn1.png" />
        ...

</Application>
```

Basically, the manifest file describes the region and dimensions of the application. In our example, the application will cover the entire screen, but it would also be possible to only have an application running in one particular area of the screen, for instance a logo animation for a corporate title in one of the corners. The manifest also specifies all the assets (images, sounds, fonts, etc.) required, defines which script and markup file shall be used for this application and in which order the scripts will be executed.

Markup

The Markup defines the general layout of the menu pages. It consists of a <head> and <body> section just like HTML web pages. The <head> contains styling and timing information whereas the <body> creates various buttons and other graphical objects. Similarly to HTML pages, the objects are grouped in divs – a division or section of the document. This method is useful to separate the contents of a div from the rest of the page.

Table 14.4 Example of a <body> Section of a Markup File

```
<root xml:lang="en" xmlns="http://www.dvdforum.org/2005/ihd"
         xmlns:style="http://www.dvdforum.org/2005/ihd#style"
         xmlns:state="http://www.dvdforum.org/2005/ihd#state">

  <head>
     <styling>
           ...
     </styling>

     <timing>
           ...
     </timing>
  </head>

  <body>
     <div style="divStyle">

        <button id="HelloWorldButton1" style:width="296px"
           style:height="77px" state:focused="true" style:opacity="1.0"
           style:backgroundImage="url(mainMenu_Btn1.png')
           url('mainMenu_Btn1_s.png') url('mainMenu_Btn1_a.png')"/>

     </div>
  </body>
</root>
```

As shown in Table 14.4, buttons will be defined in the <body> section of the markup. Multiple buttons and other graphical objects will be grouped in div sections. This allows the author to control groups of objects with a single command. For instance, a sub menu can be moved or turned on and off with a single command avoiding the necessity of changing attributes for every individual object. A <button> can contain multiple URL's. It makes it easy to implement all image files for the different button states, such as normal, selected, and activated state. The image can then easily be changed by calling for the specific style:backgroundFrame as shown in Table 14.5 below for each of the button states.

Styling can be implemented in two ways, so-called in-line styling and referential styling. The <styling> attributes defined in Table 14.5 below are implemented as referential styling. Meaning, they are generally defined and have an ID, so that they can be applied (or referred to) from any place in the XML file. For instance, the div in Table 14.4 uses this method to retrieve the attributes by referring to the ID. It is typically used to avoid redundancies in the code. For instance, if multiple objects have common attributes, referential styling can be very useful. The individually different attributes are better implemented using in-line styling such as the button defined in Table 14.4.

Table 14.5 Example of a <head> Section of a Markup File

```
<styling>
      <style id="divStyle" style:position="absolute" style:x="100px"
      style:y="100px"/>
</styling>

<timing clock="page">
      <defs>
            <set id="ButtonNormal" style:backgroundFrame="0" />
            <set id="ButtonFocused" style:backgroundFrame="1" />
            <set id="ButtonSelected" style:backgroundFrame="2"
            state:focused="true"/>
      </defs>

      <par>
            <cue use="ButtonFocused" select="id('HelloWorldButton1')"
            begin="id('HelloWorldButton1')[ state:focused()=1] "
            end="id('HelloWorldButton1')[ state:focused()=0] "
            />

            ...

            <cue use="ButtonSelected" select="id('HelloWorldButton1')"
            begin="id('HelloWorldButton1')[ state:actioned()=1] "
            end="id('HelloWorldButton1')[ state:actioned()=0] ">
                  <event name="ButtonChapter1Event"/>
            </cue>

      </par>
</timing>
```

14

In addition to the styling information, the <head> section also contains timing information. As described in a previous chapter, timing sections consist of three separate clocks — the title clock, the application clock, and the page clock. In our example we're using the page clock for timing, because the application is not tied to the video timeline. If the application would have to be triggered based on time code events from the video, the title clock would be more appropriate.

There are also multiple segments within the <timing> section. The general definitions such as which image should be selected for the individual state, will be described in the <defs> section, whereas the actual timing areas are <par> and <seq>. Cues within the <par> element will be processed in parallel whereas the cues within <seq> element will be processed sequentially. Meaning that only if a process is finished, the next one will be executed. This method can be used to finish the rendering of an animation before executing an event, i.e. jump to another title. As shown in Table 14.5, each cue references the general definitions defined in the <defs> section. For instance, the information about which image should be displayed as the current backgroundFrame is retrieved from there. In addition, every cue defines a "begin" and "end", or alternatively a "begin" and "duration". The method used for querying the different states or timing events is called XPath. There are a lot of different options on how to query things using this language.

NOTE In order to handle user input from the remote control, so-called virtual key events are used. The specification defines multiple VK_Key events that can be listened to both markup and script to control the disc.

Script

As shown in Table 14.6, once the button in our example has been selected and activated (state:actioned()=1), an event named "ButtonChapter1Event" is triggered. This event calls an Event Listener with the same name in the associated script file (menu.js), which was loaded from the Manifest file.

Table 14.6 Script File Example

```
function ButtonChapter1EventHandler(objEvent)

{
Player.playlist.titles[ "main"] .chapters[ 0] .jump("00:00:00:00",false);
}

application.addEventListener("ButtonChapter1Event",ButtonChapter1EventHa
ndler,true);
```

First of all, one important step using ECMAScript in addition to XML is to register Event Listeners (addEventListener). This step is easily overlooked, but it is crucial to make sure the events with this name in the XML file will indeed be triggered. Once the EventListener receives a notification of such an event (i.e. after the button was activated), it calls the function defined in the EventHandler. In this example it will call the function "ButtonChapter1EventHandler".

In our example, this function will start the playback of the video. It calls the player's playlist and references the title to be played - in this case the title called "main". It can also specify which chapter or time it wants to start playing.

TIP   In order to make the programming easier, there are schema files available for all the XML documents that help creating and verifying the code. Additionally, a list of all available API's for scripting will make things easier. All these documents are available from the DVD Forum (www.dvdforum.org).

Markup vs. Script

As it is with almost everything, there's always multiple ways to achieve the same things. This also applies for HD DVD AC functionality — both markup and script can achieve similar results. Particularly implementing disc logic, using XML is generally easier than script, but it is also more limited. Advanced Logic typically requires using script functionality as it is more versatile. Now one may argue that having two ways for implementing disc logic may be a great thing providing multiple solutions to the author. On the other hand, having logic split between two places is undoubtedly calling for problems. The way to best approach this matter is to try and use each of these two methods for what they're best for. Markup should be used for the user interaction and generic elements while logic issues should be dealt with in script due to the more advanced capabilities. The idea behind this approach is simply the nature of production. It is very common that menus are created with a specific design in mind and the functionality is based on this. Keeping the menu layout as a generic component contained in markup with all the functionality and logic separated in script provides two main modules for a title. This way the design can easily be changed whenever necessary without having to touch the script at all. If the logic would be kept in markup as well, this scenario would not be possible, because every design revision would also break the logic. With this approach, the division is very simple. All markup elements have certain events implemented that are called, and the script implements listeners for these events to handle the logic associated with them.

Blu-ray HDMV

The authoring process for Blu-ray HDMV is similar to Standard DVD authoring in the sense that it also requires an authoring application to get access to all the parameters available. Due to this, it is of course very dependent on the specific authoring software whether it provides the full low-level access as defined in the specification or applies additional abstraction making it easier to author titles, but also hiding some of the functionality. One other difference to current DVD authoring is the fact that HDMV uses very different terminology

14

and adds a lot of new functionalities that is not available at all in DVD offering a much richer user experience. Some of the new features we'll talk about are popup menus, menu effects and animations, and a browse-able slideshow. Additionally we'll look into new concepts like subpaths and resume functionality.

Popup Menus

For HDMV, two kinds of menus exist – Top Menu and Popup Menu. The Top Menu is very similar to the Menu Space in Standard DVD. It is a designated area where menu elements are presented and the user has the chance to browse through various menu pages to either change settings or jump to other disc content. The popup menu on the other hand is a new concept, which doesn't exist on current DVD. It allows the user to browse menu pages while the feature keeps playing in the background. This way, the user can change settings like audio or subtitles without having to leave the feature. As the name suggests, popup menus don't have to always be present while the feature is playing. Only when the user wants them to be present, the popup menu button on the remote can be pressed and the menus show up. Pressing the same button again makes them disappear. There is also the option to specify a timeout for Popup menus. If no navigation happened for the specified time, the popup menus will automatically disappear making it very user friendly.

NOTE Some remote controls may only have one menu button implemented making it impossible to return to the Top Menu once the feature plays back. With that in mind, it might make sense from a content authoring perspective to always include a "Return to Top Menu" button on the screen.

As described in an earlier chapter, the way menus are constructed in HDMV differs quite a bit from Standard DVD authoring. HDMV contains things like Epochs, Display Sets, Palettes, Pages, and Button Overlap Groups. All these elements are new to an author and will most likely take some time to get used to.

Basically, the smallest piece of a menu is a button. And just like with current DVD, each button can have three states – normal, selected, and activated state. Also, each button has a set of commands associated with it, and can additionally have a click sound and different animations per state. The Button Overlap Group (BOG) can contain up to 32 buttons, but only one button within a BOG shows at any given time. This means that for regular menus, a BOG will typically have only one button, with all possible states defined. One exception, however, may be a menu using more enhanced menu sets. For instance, implementing a checkbox indicating which audio stream is currently selected would be realized using this approach. An audio menu with multiple audio streams would consist of two different buttons per BOG. Each audio option would be contained in its own BOG. But one button would show the checkbox marked as "checked", and the second button would be marked as "unchecked". Scripts control which button will be displayed per BOG.

There can be 256 BOGs per page. However, BOGs cannot overlap each other, adding some more complexity to the menu authoring. Usually this requires to "cut down" menus into many little pieces to make the menus work. Each authoring tool treats this differently, some create the "cut outs" themselves, others require the graphics to be prepared separately beforehand. An application example is shown in Figure 14.2. The first image shows only the main menu whereas the second image shows the display expanding the submenu options.

Figure 14.2 Application Example for an HDMV Menu

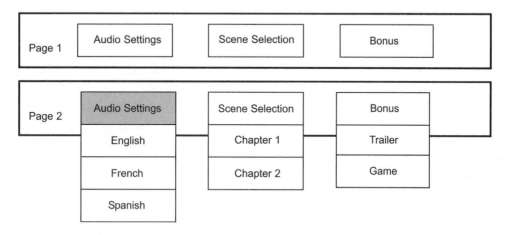

Since BOGs cannot overlap each other, this requires cutting down the images of a menu set. Having different images for each button state is easy to understand, as only one of them will be displayed at any given time. But the background image will also be dissembled into many small images that will or will not be loaded depending on the current page being displayed. How this would look like for a background image is shown in Figure 14.3.

Figure 14.3 Example of "cut out" Images for an HDMV Menu

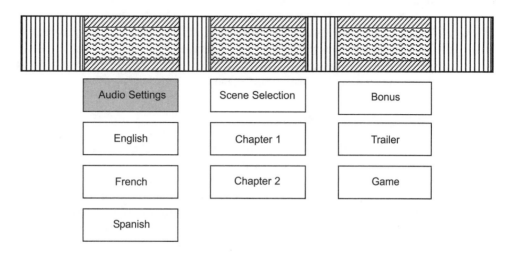

14

A page is an interactive display, meaning the display of the full menu page including all menu elements. Each page is limited to 256 colors (including alpha). Using this limitation creatively may result in very nice looking menus. One thing to keep in mind though is the fact that certain graphical elements may be used on different pages, making the calculation of the color palette much more complicated. Because nobody would like to see color shifts on certain graphical objects when changing from one page to another. For that reason, one color palette is often used for the entire project, making it much more challenging for the design group to create decent looking menus.

TIP

Avoiding color transitions, shadows and gradations will help reducing the number of colors within a palette.

Only one page can be active at any one time. This principle is used to create multi-page menus as shown in the application example in Figure 14.2 above. Each possible "version" of the menu would be contained in its own page. For example, the main menu display is defined in one page, the audio sub-menu is defined in a second page and all other sub-menus would be defined in additional pages. In order to figure out which page to display requires some calculation and programming. It is recommended to have one "dummy" page for each menu set that only contains one invisible auto-activated button. This dummy page will always be entered first from the MovieObject using the SetButtonPage command. Then the auto-activated button will execute the necessary code to determine which page should actually be displayed.

NOTE

Each page has a UO Mask defined. This means that user operations from the remote can be enabled or disabled on a page-by-page basis. For instance, within an audio menu, the "audio change" button on the remote might be disabled, but in all other menus it might be enabled.

A group of pages with the same duration composes a Display Set and there are up to 4096 Display Sets allowed per epoch, which is the highest level of graphics components. Because an epoch defines a decoder state, and the decoder state will not be preserved between epochs, one menu set will always be within one epoch.

Menu Effects and Animations

Each page can also have an in-effect and out-effect defined. The in-effect will be played prior to loading the page and the out-effect will be played before the page transitions into the inactive state. Supported effects include scrolls, wipes, cuts, fades (transparency changes) and color changes. In order to realize these effects; multiple pages are rendered for every frame of the animation. By defining different cropping areas for every image, scroll, wipe, or cut effects can be achieved. At the same time, fade or color changes will be applied by changing the color palette for every page of the animation. However, one downside of this is the fact that menu animations are very memory intense. As explained in a previous chapter, the memory limitation for all graphical elements is 16MB. Although each page uses the same source image and applies different cropping or palette parameters, it will still be decoded multiple times into the buffer, taking up a lot of space. This makes it very challenging for an author to implement many transitions on a menu set, especially if the graphics are rather big.

Also, there are a few other implications adding some more complexity to menu effects. Although there are multiple commands possible after activating a button, some commands will terminate this process, including PopupMenuOff, SetButtonPage, and various link or jump commands amongst others. So how could the following usage scenario be realized? We want to link to another chapter while the feature is playing, but before we link, we want to play an out-effect and have the popup menu turned off. Implementing all of these commands subsequently in the button command area will only play the out-effect and terminate the processing, leaving the popup menu "on" and the link to the other chapter will never be executed. Now there are a few creative ways to work around these limitations. One solution would be storing the selected chapter number in a GPR and jump to the exact same title. This would play the out-effect and immediately after disable the popup menu (jump title commands disable popup menus), turning it off. On re-entering the same title, the GPR would be used to jump to the specified chapter. The disadvantage of this method is that re-entering the same title may result in the player re-loading the IG stream. It is recommended for players not to do that, but in the end, it is implementation dependant and may result in longer loading times. To avoid this, another solution would be to use other hidden pages and buttons that carry the individual commands and would be executed one at a time. However, the authoring for such a case will be much more complicated, but at least there would be no additional loading times.

In addition to page effects, menus can also have an animated sequence of graphics for buttons. This would enable buttons to be animated with a sequence of different images for each button state (normal, selected, activated). And there is the possibility to implement button click sounds as well.

NOTE	Button animations and button click sounds will be immediately terminated when another button navigation is executed. For instance, having a menu with multiple buttons and each of them contains a sound. Navigating through all buttons will immediately stop the sound as soon as the next button becomes selected.

Subpath Concept

A previous chapter already described the general concept of a subpath as an additional presentation to the PlayList associated to the main path. While main paths contain PlayItems, subpaths contain less surprisingly SubPlayItems. Subpaths are typically used for out-of mux assets like audio for browse-able slideshows as described below, asynchronous interactive graphics menus, text-based subtitles, or additional audio streams. But they can also be used for in-mux or out-of mux picture in picture applications.

WARNING	The BD specifications only allow for one non-multiplexed IG stream. In case there are multiple non-multiplexed IG streams on the disc and the streams change, this can cause interruptions to the video. Depending on the specific player, the AV interruption may be a couple of minutes.

Basically, the concept of subpaths allows for various features to be implemented. For instance, separate audio/video files can be linked together – either synchronous or asynchronous. For instance, another picture in picture window can play additional information about the movie while the main feature keeps playing (and the audios are mixed together). The picture in picture can either be synchronized to the video in case the content is only relevant for a specific scene, or can be asynchronous and be launched whenever the user pushes a specific button on the menu. This concept can also be used for downloadable content, additional commentaries or audio languages would be linked to the main path. And despite the fact that the files are physically located in different area (i.e. disc and persistent storage), the user would not notice, because the experience is seamlessly linked together.

Browseable Slideshow

One other type of subpath is the browseable slideshow. This kind of feature is something that everybody involved in DVD authoring was looking for. It allows presenting still images with audio playing in the background. In Standard DVD, this was possible, but only as a time-based slideshow. Meaning the timing of the slides was predefined. So while audio was playing, the still images changed after predetermined times. However, this has never been the perfect solution. Instead, it was desired to allow the user to decide when to switch to the next still image while audio should play without interruption.

A browseable slideshow does exactly what was desired. The time duration of an image is user defined and may be infinite. It stays on the screen unless the user decides to go to the next one. The way this is implemented is using a main transport stream containing the still images. Each still image is one PlayItem and PlayMarks will be used to navigate to the next or previous slide. The audio will be played from a SubPlayItem. As a result, jumping between the PlayMarks will not affect the audio at all.

Resume Concept

Similarly to DVD, there is only one resume information available in each BD player. Additionally, every Movie Object contains a so-called resume intention flag indicating whether the resume information for this particular title should be stored or not. In case the resume intention flag is set, the resume information will be saved into backup registers whenever a Menu Call is executed from a title. After browsing through some menus and possibly updating some disc configurations, the user could call the resume, and the movie would continue playing where it was left off. However, as soon as a Title Search, or Jump command is executed, the backup registers will be initialized and the resume information is lost. Unfortunately, this prevents the usage scenario in which the user would exit the feature, play a game in a bonus title and then resume feature playback.

The reason for the lack of a better resume functionality is the more advanced control available with BD-J. Using this format, resume information can be stored into multiple variables allowing for much better resume functionality. This also removes the limitation of only one resume information per player, using BD-J multiple resume information could be stored in different user-defined variables, allowing to track where every video on the disc was exited and resuming it whenever it is jumped to again.

BD-J

Programming for BD-J is undoubtedly the least comparable of the new formats to DVD. It can only be compared to software programming. This statement alone will probably raise concern for everyone having experience with software development, because the time and effort it takes to achieve an acceptable result can be tremendous. On the other hand, everybody using software, particularly media-related software, will agree that there are always bugs and unsatisfying quirks in each software application, raising even more concerns. The reason for this is simply that there is never enough time to actually test these applications and ensure the functionality. Given the tight timelines under which most of these titles have to be created, this situation doesn't look any better, if not worse. So why is it expected to be any better?

Looking at it from a distance, the situation is less dramatic than it seems. First of all, a lot of Standard DVD titles have common features implemented, so each title cannot be compared to different software applications. There are a lot of similarities between different titles that can be leveraged. Basically, programmers will create a basic framework that can be used on every title, getting them involved step-by-step, learning new things with every subsequent release. And each new title will improve the framework being used and stabilize the playback performance on the various platforms.

From a technical perspective, using an object-oriented programming language with Java for this multimedia application compared to a scripting language certainly provides a much more powerful platform. It is also a system that has a proven track record in this area, because it is used in millions of TV Settop boxes based on the Multimedia Home Platform (MHP). So in the end, the decision to go with Java was not totally out of the blue, but it definitely requires a different skill set in terms of content programming.

In terms of programming environment, every content author has its own preferences. However, those who plan to directly write Java code will probably want to at least use an Integrated Development Environment (IDE), because it also allows for code syntax verification and makes it easier to keep on target.

Basic Architecture

The underlying technology used on BD-ROM is shared between HDMV and BD-J. Meaning that the layout of the disc can share titles from both modes. Some titles may contain HDMV content, others BD-J. An Index table administers which title contains which content. As a result, the basic elements of a disc such as multiplexed audio, video, and subtitle files, PlayLists, and Clips are shared across the disc. The only difference between both modes is that instead of a MovieObject controlling the commands and playback for a BD-J title, it is a BDJObject (BDJO). Figure 14.4 illustrates in which way the BDJO controls the playback of the disc. The BDJO links two important elements together, the PlayLists containing the audio and video elements (stream files associated to clips) and the JAR files containing the menu elements (graphics, sounds, fonts, and other resources together with the class files containing the programming logic).

Figure 14.4 Basic Control Hierarchy of a BDJO

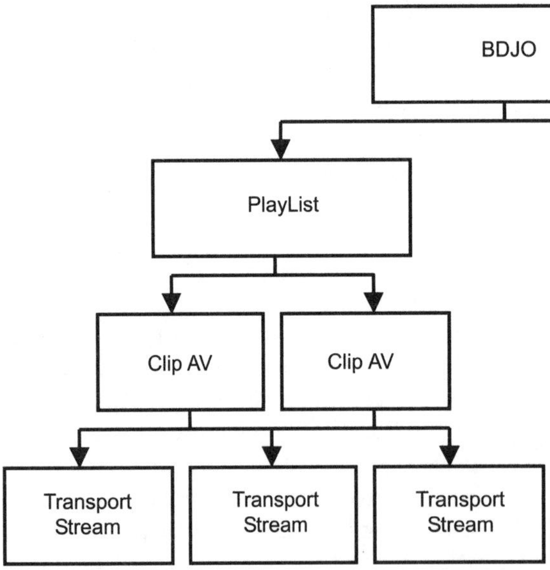

Besides the actual disc assets (both AV and menu assets), the BDJO also contains all the administrative information about each BD-J title. For instance, the Content ID, Provider ID, and others are contained within the BDJO. This information is necessary for certain authentication purposes in order to access dedicated areas in persistent storage or allow for network applications. In addition to this, one other very important detail is contained in the BDJO file – the initial class. This is necessary for the player to know which JAR and class file to access in order to start loading the application. Although we're not attempting to discuss the programming in full detail, the following pages will touch on the most important concepts and steps necessary to create a BD-J title.

Concept of an Xlet

Once the initial class has been identified, the player will look for the main component to launch the BD-J application. This component is called Xlet, which is implemented by the initial class as shown in Table 14.7 . However, before even initializing the Xlet, a few other programming resources, such as packages and variables have to be imported and setup. The different packages that are used for BD-J (i.e. JavaX, Havi, Blu-ray, etc.) have been discussed in a previous chapter and every programmer used to the Java language is familiar with this concept. Also, a few parameters will be initialized in this area. For instance a scene or tracker could be created and the screen width would be defined before even starting anything else.

Table 14.7 Example of Initial Class

```
package com.xxx.bluray;

import java.awt.*;
import javax.media.Player;
...


public class TestMenu extends Component
       implements Xlet, Runnable, UserEventListener, ControllerListener {

protected Graphics2D g;
protected HScene scene;
protected MediaTracker tracker ;
protected int width = 1920;
...
```

The Xlet itself is similar to a Java applet, but it is specifically designed for media playback with limited memory resources and bandwidth restrictions. It can be paused and resumed to support the features required by media applications such as TV settop boxes or Blu-ray players. Table 15.8 outlines the basic states of an Xlet and the corresponding methods.

Table 14.8 Example Xlet Outline

```
public void initXlet(XletContext ctx) throws XletStateChangeException {
    this.xletcontext = ctx;
}


public void pauseXlet() {

}

public void destroyXlet(boolean unconditional) throws
    XletStateChangeException {

}

public void startXlet() throws XletStateChangeException {
    //Adding user event listeners
    userEventRepo.addAllArrowKeys();
    userEventRepo.addKey(HRcEvent.VK_PLAY);
    userEventRepo.addKey(HRcEvent.VK_ENTER);
    userEventRepo.addKey(HRcEvent.VK_POPUP_MENU);
    EventManager.getInstance().addUserEventListener(this,
        userEventRepo);
    ...


    // Starting the main thread, which calls the run() method
    mainThread = new Thread(this);
    mainThread.start();
}
```

As shown in Table 14.8, the Xlet has four state-changing methods – initXlet, startXlet, pauseXlet, destroyXlet. The initXlet is for the Xlet to initialize itself in preparation to get started. It shouldn't contain any time consuming tasks. The main focus for this method is to get ready in a reasonable amount of time and remain in paused state until started. The startXlet can hold all the shared resources like adding listeners for key events of the remote control and can also start a thread and enter an active state by calling the run() method of the class. The pauseXlet can only be called while the Xlet is in active state. The Xlet should minimize the resource usage, and one way of doing this might be to terminate threads before entering into paused state, as this would release all shared resources. The destroyXlet method will terminate the Xlet and can be called from the loaded, active, or paused state. It will release all resources and save preferences or states before terminating. The destroyXlet also has a boolean value associated to it called "unconditional". If this value is true, the Xlet will be destroyed regardless of how this method will be destroyed. In case it is false, the Xlet will throw an exception. For instance, it could result in the methods to not be destroyed and stay active until a later request.

Implementing Graphics Environment

After the Xlet is created, initialized, and started, it will launch the run() method. Within this method, we're trying to accomplish the next task, which is to have graphics (e.g. buttons) appear on the screen. For that, we first have to create an environment, and a so-called HScene, which is part of the havi.ui package typically used for user interfaces. Once we created this, the HScene can be added to the screen. Table 14.9 shows an example of such an implementation.

Table 14.9 Example of Graphics Environment Implementation

```
public void run() {
    try{
        // Create Graphics Environment, define initial parameters
        GraphicsEnvironment ge =
            GraphicsEnvironment.getLocalGraphicsEnvironment();
        GraphicsDevice gs = ge.getDefaultScreenDevice();
        gc = gs.getDefaultConfiguration();

        setSize(width, height);
        setLocation(0, 0);

        // Creating an HScene and adding it to the Sceen
        scene = HSceneFactory.getInstance().getDefaultHScene();
        scene.setVisible(true);
        scene.add(this);
        setVisible(true);

        requestFocus();
        g = (Graphics2D)getGraphics();

        ...
```

As illustrated in Table 14.9 a LocalGraphicsEnvironment, DefaultScreenDevice and DefaultConfiguration have to be instantiated. The size and location of the environment has to be defined as well. As the example shows for most parameters, attributes can be either real integer values or globally defined parameters, such as width and height that were defined earlier in the code for the entire class. Then the HScene has to be created and added to the scene. Additionally, the scene has to be made visible and request focus to gain control.

Loading & Drawing Images

Now that the general environment for graphic displays is created, we can focus on loading individual images and managing them as resources. In order to do that, a so-called MediaTracker will be created to manage all the media files added to this scene. Once this is done, as shown in Table 14.10, the images itself can be loaded.

14

Table 14.10 Example of Loading and Drawing Images

```
// Creating Media Tracker to add images to screen
    tracker = new MediaTracker(scene);
    g = (Graphics2D)getGraphics();

// Creating an array to hold all button images
    button = new Image[ 2] ;
    button[ 0] = Toolkit.getDefaultToolkit().
            createImage(TestMenu.class.getResource("btn1.png"));
    button[ 1] = Toolkit.getDefaultToolkit().
            createImage(TestMenu.class.getResource("btn2.png"));

// Adding all button images to the tracker
    for (int idx = 0; idx < button.length; idx++) {
            tracker.addImage(button[ idx] , idx);
    }

// Wait for all images to be loaded
    try{
            tracker.waitForAll();
    } catch (Exception e){
            e.printStackTrace();
    }

// Draw images with initial states on screen
    g.drawImage(button[ 0] , 100, 100, button[ 0] .getWidth(null),
            button[ 0] .getHeight(null), null);
    Toolkit.getDefaultToolkit().sync();
```

It is usually smart to add all or at least groups of images to an array in order to better manage them. For our example, we created an array for two buttons, and load these two buttons (btn1.png, btn2.png) respectively. Generally speaking, there are multiple ways to load resources. They can either be loaded through the classloader (from within a JAR file) or mounted individually. In our example we're loading the images through the classloader, and they are in the same location as our class file itself. Once the images are added to the array of buttons, they are also registered to the Media Tracker. When doing this, it is typically recommended to implement the code to wait for all images to be registered before drawing them onto the screen. This way it can be ensured that all resources (images) are available before executing the next operation — the drawing on screen. Finally the objects will be drawn to the screen using their native resolution and a defined location. Once all this is done, the screen will be synchronized and updated.

Focus Management and User Events

Unlike other authoring environments, BD-J is very flexible, which also means that most processes have to be defined by the programmer. One of those processes is the management of the button focus and the resulting events. The content author needs to always be aware which button is selected and when entered, what action should be triggered. Although this seems more complicated, it also allows way more advanced capabilities with regards to the user experience. The focus is not dependent on any object areas or other influences. Instead it is a method designed and fully controllable by the content author. To know when to change focus and trigger events, the content author relies on user events received from the remote control. Table 14.11 provides an example of how such user events are received and how they are used to control the disc playback.

Table 14.11 Example of User Events

```
public synchronized void userEventReceived(UserEvent e){

    // Track User Key Events
    if(e.getType() == HRcEvent.KEY_PRESSED){
        switch(e.getCode()){
            case HRcEvent.VK_ENTER:
                playVideo();
                break;
            case HRcEvent.VK_DOWN:
                buttonFocus = buttonFocus + 1;
                focusManagement();
                break;
            case HRcEvent.VK_UP:
                buttonFocus = buttonFocus - 1;
                focusManagement();
                break;
        }
    }
}
```

First of all, the user events have to be registered as UserEventListeners. Each individual Virtual Key (VK) that should be recognized has to be added to the UserEventListener. Since we've already done this at the beginning of this example project (Table 14.8), we only have to implement the behaviors if such an event occurs. As shown in Table 14.11, there are three different Virtual Keys implemented that result in some action — VK_ENTER, VK_DOWN, and VK_UP. In case the UP arrow key gets pressed, the value of the local parameter buttonFocus is decreased by value one. Similarly, in case the DOWN arrow gets pressed, the value of buttonFocus increases. In both cases another method called focusManagement() is called as well. In case the ENTER button is pressed, the playVideo() method is called.

Since all these user events only listen to dedicated buttons on the remote, at this stage it

14

is irrelevant which button or object on screen is actually selected. The UserEventListeners only recognize that a user event occurred and will then typically change some parameters and call other methods in which the content author then determines which button was selected and which event should be triggered. Table 14.12 shows an example of the code required to manage the button focus.

Table 14.12 Example Code for Focus Management

```
public void focusManagement() {

    if (buttonFocus > numButtons){
        buttonFocus = 1;
    } else
    if (buttonFocus == 0){
        buttonFocus = numButtons;
    }

    switch(buttonFocus){
        case 1:
            b1State = 1;
            b2State = 0;
            break;
        case 2:
            b1State = 0;
            b2State = 1;
            break;
    }

    // Redraw images on screen based on the changed button states
    g.drawImage (button[ b1State] , 100, 100,
            button[ b1State] .getWidth(null),
            button[ b1State] .getHeight(null), null);
    g.drawImage(button[ b2State] , 100, 400,
            button[ b2State] .getWidth(null),
            button[ b2State] .getHeight(null), null);
    Toolkit.getDefaultToolkit().sync();
}
```

Since the user events are very generic, it has to be determined whether increasing or decreasing some values would cause invalid values. For instance, according to our code in Table 14.11, having button number 1 focused and pressing the UP arrow button would decrease the buttonFocus value and result in value 0. However, for such a value, no corresponding button exists that could receive the focus. Such cases have to be caught by the code, defining if in such a case either nothing should happen, or another button should be selected. In our case, the buttons would loop - pressing the UP arrow button would highlight the bottom button. Once we have determined which button actually receives focus, this change also has to visually apply to the screen. Meaning, the images for some buttons (in our case all buttons since we only have two), have to change from normal to selected state. Table 14.12 shows how this is being done using a switch based on the value of the buttonFocus. In the end, the buttons are re-drawn to the screen. The difference this time is that we don't call for specific buttons, instead we use a parameter to identify which image from the button[] array will be loaded.

Video Playback

To finalize the example project, the user event ENTER has to be executed and should start the playback of the video. For this, the method called playVideo() is called, which is shown in Table 14.13 below.

Table 14.13 Example of Start Video Playback

```
// Create datasource and define media location to play
DataSource ds = Manager.createDataSource(new
    org.davic.media.MediaLocator(new
    BDLocator(bd://0.PLAYLIST:00000.MARK:00000)));

// Create player with datasource
player = Manager.createPlayer(ds);
player.addControllerListener(this);

//starts playback of the video
player.start();
```

First of all, a new datasource has to be created defining the location on the media that should be played back. In order to realize this, a so-called BDLocator is created, which specifies the exact content to be played back. For instance, in Table 15.13, it defines PlayList "00000" and Chapter "00000" to be played. Once a BDLocator is defined, an actual player has to be created and added to the ControllerListener. Once this is done, the player.start() method can be initiated and the video would start playing.

Chapter 15
Quality Control (QC)

As you produce new format disc titles, quality checklists will help you deliver increased quality and reliability, without creating a bottleneck in your production. Several different quality control (QC) tools and approaches are available for achieving continuous improvement. Final assembly will mean ready-to-roll, with no need for extensive testing at the end to ensure that the media really works.

The complexity of the new formats requires a new commitment to quality. The new discs and their playback platforms look more like computers and consumers will not accept titles that crash and hang up. An extra trip to the kitchen for another beer while waiting for a reboot doesn't count as interactive functionality.

Quality control is both a distinct set of steps separate from the actual work and a way to build a process that builds quality products. Quality is assured in a product not by inspecting it but by building it into the product. Each step in the process is set up so that it uses quality assets and generates good quality assets for the next step in the process. This means that there should be no reworking or "massaging" of assets to get them to work.

Regardless of the size of your operation you want to set up a quality program for your titles so that you only produce good high quality discs. Each step in the process is specified so that the tasks can be performed per specification. When a problem is found, a fix is found for not only this particular problem but how to fix the process so that the problem cannot happen again. Any problems need to be discovered at the earliest steps in the process. The objective of a good quality control process is to ensure that you only create good quality assets. The process should be set up so that you cannot put square pegs in round holes. Each person in each step knows what is good input and what is good output.

Here is an example of how the process can be adjusted when problems arise. At one compression and authoring facility, clients require both Blu-ray and HD DVD titles from the same source assets. An audio stream had been prepared initially for HD DVD and then loaded into the authoring tool used for creating Blu-ray discs™. The audio stream worked through simulation and they were even able to multiplex the stream. But, when playing the stream on the hardware player, the audio stream didn't play. When the specification verification tool was run on the project image, the verifier software reported that the stream had an illegal header. It took a bit of looking at the assets to realize what had happened. Was it a player problem, a multiplexing problem or a bad stream? Running a specification verifier tool on the final image reported that the audio stream had a bad header. The problem is that although both formats used Dolby® Digital Plus, Dolby Digital Plus has been specified to have a different data structure in the header for each format. So the compression and authoring facility then re-created the audio in the correct format for Blu-ray and everything worked. But the troubleshooting, fixing and re-multiplexing of the project, re-running the verification tool, creating new check discs and re-checking the operation on the players, took two days.

Often fixing the immediate problem is the end of the quality checking. This facility then took the next step which was to fix the process. The audio department now has a separate check box on their checklist to indicate the disc format (Blu-ray versus HD DVD) as well as the audio codec to be used. The file naming convention was also updated to include the disc format. The asset inspection step now has a tool that allows the reading of the header to verify the proper format for the target platform. When the author gets the assets, they note the file name and know that it has been checked for the right format. Also, the authoring tool was updated to ensure checking the header. All of these little steps now ensure that the process will not allow putting "square pegs in round holes" (see Figure 15.1).

Figure 15.1 Square Pegs Get Rejected, Round Pegs Go Through to Next Process

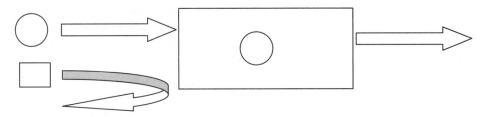

Setting Up for Quality

Your operation and the format(s) and authoring modes that you are supporting will necessitate unique solutions. Each of the authoring modes has its own challenges and is susceptible to different types of problems. A good way to approach setting up your operation for quality is to look at each process step and determine which quality checks can be put in place so that only good assets move on to the next step. Here are some things that need to be addressed —

- A quality measurement for each step:
  - **Go/no go** For example, does the file have the correct filename? Are all menu assets present in the delivery from the graphics design group? Are the file sizes for the assets matching the bit budget for the project?

 TIP Assets delivered on DVD may have problems being read if file sizes are over two GB. This is important to note for audio as, even compressed, HD format audio files for a full-length feature may now be two and one half — three GB. So, if you see a file size reported on the Windows platform that is 2,147,483,648 bytes, be suspicious. That file has probably been truncated.

- Quantitative measurements: Each quality attribute should have a metric. Either use an industry standard measurement or if there is not one, create one that allows a quantitative, objective measurement that is repeatable by different people.

- Use examples, written and documented cases, for judging acceptable quality. For example, video banding may be in less than 20% of the screen and, if present, must have at least 10 color bands in that screen area. For audio, no audio pops can be greater than -3db from the background level. Video must look acceptable on certain types of display technologies. Ideally, for video, there should be PSNR (peak-to-peak signal-to-noise ratio) levels and other metrics that are quantifiable and describe repeatable, objective criteria that technically describe any artifacts and let pass those that are acceptable to the end user.

Not all potential problems can be identified when a process is initially implemented. You can set up your quality process to improve with time. As new problems arise, you can use these as opportunities to fix the process and, as required, put in new checklist items to ensure that a problem will not happen again. Further, identify ways to streamline the process so that problems cannot happen again.

Before release for shipment of a title, the title needs comprehensive testing in two areas — *Visual QC* and *Navigational QC*. Visual QC needs to review all streams on the disc. This means that if you have three angles of a video then the disc needs to be viewed at least three times, once for each angle. Table 15.1 is an example of how all streams can be viewed. The minimum number of viewing passes is the maximum number of any one stream type. In the example in the table, there are five subtitle streams so it will take five complete viewings to review all streams for the disc.

Table 15.1 Sample Scheme for Visual QC

Stream		Viewing Number				
		1	2	3	4	5
Angle 1		x			x	
Angle 2			x			x
Angle 3				x		
Audio 1	English	x				
Audio 2	French		x			
Audio 3	Spanish			x		
Audio 4	Commentary				x	
Subtitle 1	English	x				
Subtitle 2	French		x			
Subtitle 3	Spanish			x		
Subtitle 4	Commentary				x	
Subtitle 5	English SDH					x

15 Quality Checking Techniques

There are several methods that can be used to ensure quality. As the creation process proves itself through consistently delivering quality results, the techniques below can be used to maintain the high level of quality yet increase the efficiency of the operation. Here is a selection of some of these methods that may be used to achieve a high level of quality in your titles —

- **100% Visual QC** Every video, audio, and subtitle stream is viewed in its entirety. Every navigational branch is verified to operate correctly by running the disc through every possible button push. This can be time consuming. For more complex interactive discs, this may not always be either practical or possible. Some of the techniques below will be required to be used in conjunction with a 100% Visual QC to get the high quality that is required.

- **Sampling** Check a few key spots in the streams, such as, the 30 seconds on either side of a chapter marker. And check the audio level, audio and video quality, audio and subtitle sync and that the chapter marker is properly placed with respect to the audio and the video and the subtitles. This technique yields the best results when you have a solid process. The spot checks that are being done with the sampling are validating that the process is generating good results. The premise of this technique is that if the process is solid and being followed, then the results will be solid and of a known consistent high quality.

- **Inspection** You can run a spell checker on subtitles. Also, subtitles can be inspected either visually or with the aid of a software tool to ensure that subtitles do not cross chapter markers (which causes the subtitle to stop being displayed). For the ECMAScript/XML or Java™ code, review the code for complete patterns (e.g., buttons one through seven all have code to process actions) or check that the logic always covers the "if all else fails" case.

- **Specification Verification** Various tools are available to verify that the data is prepared within specification. Verification software will check either elementary streams, multiplexed streams or completed CMF images of titles. As the compression and authoring tools become more mature and tested, these verifications may not be required but until then verification is needed to ensure that your title will play on all specification compliant players. Syntactical verification of the program code may also be needed. Sony, Mitsubishi, and Panasonic offer format verification tools for the Blu-ray format and Toshiba and Microsoft offer a set of tools for HD DVD. Specification verification is appropriate for mainly the high profile, high volume titles (e.g., Hollywood titles). HD DVD Advanced Content can use XML Schemas to check that Playlists, Manifests and Markup are well-formed and correct. BD-J apps should be checked for invalid function calls.

- **Measurement** Review bitrates for audio and video to ensure that there are no spikes above certain thresholds that may cause problems in multiplexing or playback. Also, look at the audio waveforms for pops or clicks or other artifacts that may have happened in encoding. Professional audio tools can be used to check for these audio defects. There are measurements for video quality such as PSNR (peak-to-peak signal-to-noise ratio), MSAD (mean sum absolute difference of the color components), Delta, Blurring and Blocking. Some of the tools that can help with these measurements are made offline such as Moscow State University's "MSU Video Quality Measurement Tool" (http://www.compression.ru/video/quality_measure/video_measurement_tool_en.html) or the commercial tools from such companies as Tektronix with their Cerify product (http://www.tektronix.com) which measures the quality of an encoded stream or K-will (http://www.kwillcorporation.com/products/VP21H.html) with their automated video quality and measurement system which is a hardware-based system that allows an A/B comparison of HD SDI signals.

■ **Navigational QC**   Using both experience and the title flowchart and functional specification for a title, a test plan should be created that ensures that all logical branches of a disc are tested for operation (see Table 15.2). This test plan should then be performed on the target platform player(s). For a limited application title such as one for a presentation or exhibition, testing on one player may suffice. Extensive testing on all players available may require the use of a third party that has a bank of players.

Table 15.2 Navigational QC Test Plan Checklist

Title: Away with the Wind		**Disc Revision:**
Brand/Model, serial number and FW revision		061126A
Display Sharp, LC-42D62U		**Date of Viewing:**
A/V Receiver None		20Dec06
Disc Player Samsung, BD-P1000, serial number, 060822		
HeadphonesSpeakers Sony, MDR-V700DJ		

First Play Splash

☐ Confirm Correct UserOps allowed and locked out — let play through

☐ Correct logos, trailers, etc per title specification document

Transition to Top Level Menu

☐ "Play" button is highlighted after first play

☐ Up, Down, Left and Right – function and follow specified button navigation convention. All buttons tested and in sequence

☐ Confirm all buttons have correct highlight graphic and area

☐ "Play Movie" button should begin playing main feature with correct angle, audio and subtitle

☐ Test Activate action of every button on menu — Confirm correct target

☐ Confirm audio and video sync on all targets

Continues

15

Table 15.2 **Navigational QC Test Plan Checklist** *continued*

Top Menu	Scene Selection	Setup	Bonus	
				Top Menus (return from feature or bonus)
				Up, Down, Left and Right – function and follow specified button navigation convention
				Confirm all buttons have correct highlight graphic and area
				Activate all sub-menu items – ensure correct target
				Return from lower level menu should show that menu button highlighted
				Transitions to lower level menus or target content played as per specification
				Confirm audio and video loop correctly

Top Menu Bar	Scene Selection	Setup	Bonus	
				Pop-up Menus – while video is playing
				Up, Down, Left and Right — function and follow specified button navigation convention
				Confirm all buttons have correct highlight graphic and area
				Activate all sub-menu items — ensure correct target
				Return from lower level menu should show that menu button highlighted
				Transitions to and from lower level menus or target content played as per specification
				Button sounds per specification
				Return to Root menu functions with "Play" button highlighted

	Feature Video
	Confirm User-Ops available (per specification):
	Popup Menu, Top Root Menu, Stop, Pause, Rewind, Fast-Forward, Previous Chapter, Next Chapter, Angles, Audio, Subtitles
	Audio, angle and subtitle surfing should be reflected in Setup Menus (top and popup)
	Return from Top Level menu should resume playback
	Audio, video and subtitles should be correct stream and should be in sync

Note:

This is a fake/non-existent film title so as to not imply anything with current or future projects at Technicolor or other C&A house.

Visual Inspection Checklists

Eash asset type should have a checklist. The checklist items are used to ensure that the earlier process steps were performed correctly. These lists are used to ensure that both this project is good as well as that the process was followed and that the process is generating good results.

Subtitle Checklist Items

- Timing/synchronization
- Misspellings
- Positioning
  - Not covering in-video-text or key objects in the frame
  - Left, right, top, button to signify who is speaking
- Correct font and typography (e.g., style such as italics, bold, and underline as well as kerning, ligature or the equivalent for eastern fonts)
- Quality of rendering

Menu Graphics Checklist Items

- Buffer calculations are complete, meeting the bit budget
- Filenames meet compression and authoring facility requirements
- Files line up as required
- Spelling correct
- Animations used for authoring/programming reference cover all button states and transitions

Audio Checklist Items

- Pops, clicks, dropouts
- Lip sync or sound effects are off
- Levels are not consistent (either in the feature or with other audio material on the disc)
- In/out points clip or downcut or upcut the audio (e.g., jumping to a chapter comes in mid-sentence or more commonly mid-word or sound)

Video Checklist Items

- Video levels are consistent (clipped whites and blacks)
- Compression artifacts such as blocking, excessive smoothing/softening, banding

15

- In/out times clip action in video or start/stop too early/late
- Chapter markers are placed so that there are no flash frames (one or two frames of a previous scene) or clipped audio (either phrases, words or notes, depending on the context)
- PSNR measurement meets compression quality expectations

TIP   Displays vary in the ways they show artifacts from encoding. Although you like to see your work on the best display possible, review encodes using consumer displays that actually don't look as good as a studio quality CRT monitor.

Navigation QC Checklist Items

Navigation QC focuses on the interactivity of the disc. Compared to standard definition DVD, the next generation discs are more complex. Here are some of the items that need to be checked —

- Button navigation (e.g. left/right/up/down)
- Correct destination of button
- Highlight states are correct
- Transitions display completely
- Button sounds
  - present
  - level appropriate (or per specification)
  - not being clipped
- User operations are permitted or restricted as per specification
- All delays in response meet or exceed client defined expectations
- Intuitive interface for the target market (e.g. the user doesn't get lost or confused)
  - Button highlights are easily identified within two seconds of when a menu or submenu is displayed
- Standard remote control works
- Computer interaction with mouse works as would be expected
- No hardware/player errors — all failures or unexpected conditions should be handled gracefully. Restarting the disc or unplugging/rebooting the player is not acceptable.

ROM Content Checklist

Although the new formats support extensive capabilities for interactivity, there may be ROM content on the disc. PDFs and computer application software are two examples that may still be needed. Here are some of the items to check —

• filenames readable by the target platforms

NOTE   All next generation discs use UDF version 2.5. As of December 2006, this format level is not natively supported by either Windows® or MacOS® without additional drivers or file system extensions that most users will not have. UDF 2.5 is supported on Microsoft's Vista™ operating system so systems with that OS should not have a problem reading data files from these discs. Apple® should be updating their operating system as well. Linux and other operating systems have the capability to support UDF 2.5 but it depends on the build. You may need to include a note in product packaging advising users about the file system support of their operating system.

• Check that all files can be read back from the disc on the target playback platform
• Check that any file references in the programs or documents on the disc use relative pathnames. Clicking on all links in the disc applications is a way to check that the links are correct. Please note that each operating system uses different path structures for removable media, so just because it worked on the Windows operating system does not mean that the link will work on MacOS.
• Path/filename length and directory depth is different on UDF 2.5 than the target platform operating systems. Check the longest and deepest files to ensure that they can be read on the target system.

Test Plans

All possible user interaction paths need to be verified. As discs become more complex, trying to test every possible way of navigating a disc becomes very difficult. Testing a disc is best accomplished by —
■ Reviewing the title flowchart
■ Creating a button by button navigation plan that takes and verifies each path or branch in the logic
■ Checklists should be used to verify that each path has been checked
The checklists should allow the tester to document the steps needed to repeat any odd behavior. This is crucial so that the author can repeat the problem, make the fix, ensure that they have really fixed it and then have Navigational QC perform the test plan again to ensure that not only the problem was fixed but that no other problems were introduced.

Test Reports

Thorough reporting is required for all steps of a QC plan (see Table 15.3). Not only do you want to record the version of the title that is being tested but you want to record serial num-

bers, and firmware levels of the entire playback platform (player, display, A/V receiver, speaker/headphones). Also, specify settings of the display, player, A/V receiver, and other components in the platform.

You need to be able to tell that you were actually listening to the DTS-HD™ audio at 3Mbits/second being decoded by the player (and not the A/V receiver) that is being played back in-phase on the correct channels and that you are actually watching the video at 1080p via the HDMI interface on a LCD monitor that has been calibrated to properly display the video.

Table 15.3 Sample Visual QC Log

Title	Disc Revision:	Date of Viewing:
Away with the Wind	061126A	11/27/06

QC Environment		
Display	Sharp, LC-42D62U	
A/V Receiver	None	
Disc Player	Samsung, BD-P1000, serial number, 060822	
Speaker/ Headphones	Sony, MDR-V700DJ	

Video Segment	Timecode	Element Stream (Video/Audio/ Subtitle)	Observation
e.g. Feature	01:15:30:15	V2	Video encode artifact, pixelization in upper left corner
e.g. Bonus: Making of AWTW	00:04:04:00	A1	Drop out in audio for approximately 1 second

The Player Challenge

The next generation DVD specifications give the player manufacturers much more latitude in how features and capabilities are to be implemented when compared to DVD. Performance is often not specified so some players may take over 90 seconds just to turn on and open the disc tray drawer while other players can complete this basic operation in less than seven seconds. And, the player response time to the user pressing a button is variable. So, what do you test for? And, what is acceptable to the users of the disc and to the client?

Unlike DVD players, where only a handful supported firmware updates, all next generation players support firmware updates either via CD or network download using the player's network connection if available. The firmware update can significantly change the way the player works (which is, of course, the point of the updates), but which disc players at which firmware level will your title play on? When the firmware level changes, do you need to do regression testing on the titles or will you rely on the installed base to tell you that the titles no longer work? What are the client and end user expectations? Which features of this disc will work in all players? What is acceptable performance for a title? How long should it take for a button push to respond? Are you going to test, either in-house or via third party, all players that are on the market? What happens if a feature is not supported on a player? Are you explicitly only supporting certain players?

The Case for Third Party QC

Whereas your operation may be focused on creating titles, the third party QC testers are more distant from the titles and can see things that you might not be able to see. They are seeing the title for the first time as an end user would. Here are some of the possible advantages of using a third party testing facility —

- They will have invested in banks of players, often having multiple copies of players and a copy of each of the players on the market. They will also have access to the various player firmware levels and can advise you on which levels to support and therefore test accordingly. Because this is their business, they often have many more players than you would as an authoring and compression house.

- They will have the market data to know the installed base of each player. This information is critical when assessing the impact of a bug or anomaly on a title. This data can tell you whether 50% or .05% of consumers might be affected by any particular player/title anomaly.

- Third party vendors may have invested in automated and/or specialized tools to increase the quality and repeatability of the testing. Automation may also help with speeding up the testing.

- They will have experience with knowing how titles fail. They are good at breaking titles. You want them to find the problems before end users do.

- You should consider using third party QC vendors when you are creating titles that will have high volumes or wide exposure. They will have a plan or approach that meets the needs of a title. You may not need third party QC when the title has a low volume or low complexity or when the target market is very well defined (as in corporate or interactive exhibits).

Criteria for choosing a third party QC vendor include —

- Experience with the title's genre
- Recommendation of the client
- Bandwidth or capacity to complete testing in the required timeframe
- Location and proximity of check disks
- Costs of testing and retesting
- Priorities and reputation for quality, accessibility, convenience and comprehensiveness

■ Quality of reports — How easy are they to read? Do the reports tell you what is wrong and include constructive suggestions? Are recommendations based on speculation or on the actual requirements of this title?

Summary

As you manage quality as a process each step of the process of creating a disc title will provide checks to ensure good input and output for the next step. Quality checklists facilitate the process throughout production.

Monitor process quality as well as product (final disc title) quality and over time you will find that you are building discs with higher quality and reliability faster than you ever thought possible.

Chapter 16
So, You Think You're Done

Releasing a final HD DVD is similar, but not identical, to the process for releasing a standard DVD. However, releasing a final Blu-ray Disc™ requires an entirely new manufacturing process. As this book went to press, there were several replicator options for HD DVD in the United States, but for BD there were only two. Although it is expected that the options for BD replication will expand before too much time passes.

Other than feature films, given the low numbers of next generation discs that may be needed for an HD disc project, it may be prudent to consider standalone duplication products that can perform small quantity runs. As of early 2007, these duplicators were only available for Blu-ray Disc, but it is expected that HD DVD duplicators will be available soon.

Interestingly, the changes that are made to accommodate HD DVD replication have improved the control, quality, and yield rates for standard DVD. As no mechanical changes were required for the next generation HD DVD format, the manufacturing lines can make both standard DVD and HD DVD with only small modifications to the process. This tweaking of the production process had the added benefit of making improvements to the DVD replication process.

The manufacturing changes required for Blu-ray Disc do not accommodate this dual-use approach to disc production. Although current DVD equipment can be re-purposed to BD construction, the same manufacturing line cannot produce both DVD and BD. Thus, a dedicated equipment line is needed for Blu-ray Disc.

The early-on issues of not being able to make dual layer, hybrid or twin formatted discs have dissipated, and there should no longer be any concerns regarding the ability to generate discs in the various flavors of side/layer/type combinations. Naturally, though, there are some general questions that should be considered before choosing a replication facility —

- Is the facility licensed to manufacture next generation discs?
- What is their process for mastering?
- How do they maintain quality control?
- What test equipment do they own — MEI Verifier, Eclipse?
- Are they participating in the IRMA anti-piracy effort?
- How many other titles have they produced in the same format configurations

16 Replication Workflow

Following the selection of a replicator for a project, getting the content files to their location can be accomplished in a few ways. Given the very large file sizes of a next generation disc, many producers are using a networking service to deliver their disc content. Commercial digital content delivery services such as WAM!NET® can be utilized, as well as FTP delivery, depending on the replicator. And, replicators are also taking delivery on hard drives or BD-R/RE discs. However, some replicators are now requiring a checksum error report, regardless of the file delivery method.

The content protection schemes that are available for the next generation discs are the principle alteration in the workflow for replication. Figure 16.1 provides a general workflow overview to the disc replication process for an HD DVD project.

Figure 16.1 Replication Workflow, HD DVD

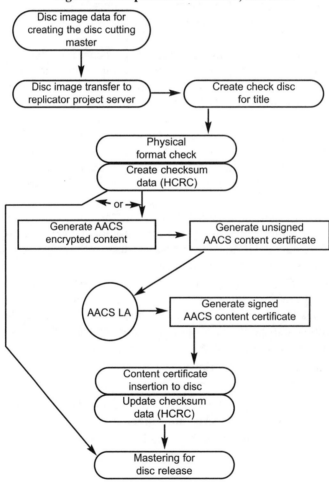

You will note that the AACS LA step is optional for HD DVD, as shown in Figure 16.1. An HD DVD disc does not require AACS.

The replication workflow for a Blu-ray Disc project will proceed in the same manner as that for HD DVD, but AACS is mandatory for Blu-ray. Further, if the BD project includes a form of BD+ protection, then the certificate steps as shown in Figure 16.1 would be duplicated to accommodate the BD+ process. The BD+ step will also require a second encryption for BD-ROM Mark, and content certificate insertion prior to the certificate cycle for the BD+ protection.

Scheduling availability and turnaround times vary from facility to facility. The AACS certification cycle generally takes two days, more or less, to send and return the necessary keys and authorizations. Further information on AACS and BD+ protection schemes is presented in the next chapters of this book.

Disc Format Configurations

The disc content industry has been striving to deliver the latest and greatest image formats to the marketplace, and it really does not matter whether the purchaser is ready with an HD player or not. On one hand, titles in either HD disc formats are merged with standard DVD versions in an array of disc combinations. On the other hand, Warner Brothers' recent announcement of their patented "Total HD" disc offers HD DVD on one side and BD on the other. A variety of names and terms are assigned to the format marriages taking place in the disc world —

- **Twin Format** supports both HD DVD and DVD. The discs may be produced in various side/layer combinations. One form of the format has DVD 4.7 or 8.5 GB on one side, and HD DVD 15 or 30 GB on the second side. Toshiba has announced a single side, two or three layer disc, with DVD 4.7 GB on one layer and HD DVD 15 GB or 30 GB on one or two layers. Or, it could contain DVD 8.5 GB on two layers and HD DVD 15 GB on one layer. These discs may be called "Flippers", "Combination", or "Combo".

- **BD Hybrid** could contain one or two BD layers, one CD layer, and one or two DVD layers. These layer presentations may be in one or two sided arrangements. The BD Hybrid discs are theoretical as there have been no marketing or product announcements, as of early 2007.

- **Total HD** contains both HD disc formats, and was developed by Warner Brothers. Although officially announced in January, 2007, there have been rumors about this development since late summer 2006. As of this writing, the exact side/layer combinations are unclear. Early stories described Total HD as a three layer, double sided disc with a BD-ROM layer and an HD DVD layer on one side, married to a DVD-ROM layer on the other side. A more recent announcement by a prominent industry news service, tragically conflating the format standards, described the Total HD concept as having "the ability to store 15 or 30 gigabytes on the red-laser HD-DVD side and 25 or 50 GB on the blue-laser Blu-ray side." As we all know, HD-DVD is 15-30 GB using blue laser. Ah, well...

16

The following Tables 16.1 and 16.2 provide technical overviews of the HD format specifics for the various disc types.

Table 16.1 Technical Overview for HD DVD and HD DVD/DVD Format Discs

	HD DVD	3X DVD ROM	Twin Format
L0 (Part 1 - Physical)	HD DVD	DVD	DVD
L1 (Part 1 - Physical)	HD DVD	DVD	HD DVD
L0 (Part 2 - Filesystem)	UDF 2.50	UDF 2.50	UDF 1.02
L1 (Part 2 - Filesystem)	UDF 2.50	UDF 2.50	UDF 2.50
L0 (Part 3 - Video)	HD DVD	HD DVD	DVD
L1 (Part 3 - Video)	HD DVD	HD DVD	HD DVD
Transport Protocol	DDP 3.0/CMF 2.0	DDP 3.0/CMF 2.0	DDP 2.10 (L0), 3.0 (L1)
Transport Medium	USB HDD, DMD, W!N	USB HDD, DLT, DMD, W!N	USB HDD, DMD, W!N
Standard Content	EVOB	EVOB	VOB (L0), EVOB (L1)
Advanced Content	HDi	HDi	HDi (L1 only)
Network Connectivity	HDi	HDi	HDi (L1 only)
Copy Protection	AACS	AACS	CSS (L0), AACS (L1)
BCA	Mandatory	Mandatory	Mandatory
BCA Cut Method	Mastered or Serialized	Mastered or Serialized	Mastered or Serialized
Channel Bit Rate	64.8 Mbps	78.47 Mbps (at 3x)	L0 DVD, L1 HD DVD
Track Pitch	.40 microns	.74 microns	L0 DVD, L1 HD DVD
Layers	0.6 mm substrates (x2)	0.6 mm substrates (x2)	L0 DVD, L1 HD DVD
Scanning Velocity	6.61 m/s	10.47 m/s SL 11.52 m/s DL	L0 DVD, L1 HD DVD
User Data Bit Rate	36.55 Mbps	33.24 Mbps (at 3x)	L0 DVD, L1 HD DVD
Numerical Aperture	0.65	0.6	L0 DVD, L1 HD DVD
Lead-In Start Sector	1E 400h	2F 200h	L0 DVD, L1 HD DVD
Lead-In Start Radius	23.3 mm	22.6 mm	L0 DVD, L1 HD DVD
User Data Start Sector	30 000h	30 000h	L0 DVD, L1 HD DVD
User Data Start Radius	23.8 mm	24.0 mm	L0 DVD, L1 HD DVD

(Table courtesy of Eric Carson and Henry Boon Kelly)

Table 16.2 Technical Overview for BD and BD/DVD Format Discs

	Blu-ray (BD)	BD9	BD Hybrid
L0 (Part 1 - Physical)	Blu-ray	DVD	DVD, CD or BD-R/Re
L1 (Part 1 - Physical)	Blu-ray	DVD	Blu-ray
L0 (Part 2 - Filesystem)	UDF 2.50	UDF 2.50	UDF 1.02, ISO, or UDF 2.60
L1 (Part 2 - Filesystem)	UDF 2.50	UDF 2.50	UDF 2.50
L0 (Part 3 - Video)	Blu-ray	Blu-ray	DVD, VCD, Blu-ray
L1 (Part 3 - Video)	Blu-ray	Blu-ray	Blu-ray
Transport Protocol	BDCMF	BDCMF	BDCMF
Transport Medium	USB HDD, BD-R, DMD, W!N	USB HDD, BD-R, DMD, W!N	USB HDD, BD-R, DMD, W!N
Standard Content	HDMV	HDMV	VOB (DVD), HDMV (BD)
Advanced Content	BD-Java™	BD-Java	BD-Java (L1 only)
Network Connectivity	BD-Live	BD-Live	BD-Live (L1 only)
Copy Protection	AACS, BD+	AACS, BD+	CSS (DVD), AACS & BD+ (BD)
BCA	If PMSN Used	If PMSN Used	If PMSN Used
BCA Cut Method	If PMSN, Serialized only	If PMSN, Serialized only	If PMSN, Serialized only
Channel Bit Rate	66 Mbps	No spec yet	No spec yet
Track Pitch	.32 microns		
Layers	1.1 mm + 0.1 mm cover layer		
Scanning Velocity	4.917 m/s (Video at 1.5x or 7.376 m/s)		
User Data Bit Rate	35.965 Mbps		
Numerical Aperture	0.85		
Lead-In Start Sector	B9 200h	2F 200h (guess, no spec yet)	
Lead-In Start Radius	22.512 mm	22.6 mm	
User Data Start Sector	100 000h	30 000h (guess, no spec yet)	
User Data Start Radius	24.0 mm	24.0 mm	

(Table courtesy of Eric Carson and Henry Boon Kelly)

16 HD Disc Packaging

The packaging for the two next generation disc formats is similar to current DVD boxes, but slightly thinner and with rounded corners. Also, HD DVD boxes will be tinted red, and Blu-ray Disc boxes will be tinted blue. The respective format logos will be prominently displayed in a banner across the front top of the boxes. The reduced thickness of the boxes should allow retailers and consumers to save a little on shelf space.

The cover art and inside booklet options are generally the same as those for standard DVD. Disc producers should contact their own replication facility for templates that reflect the specific layout and size needs for disc related artwork.

Chapter 17
Advanced Access
Content System (AACS)

Content protection has always been one of the most important issues for content owners. That is totally understandable, because after spending millions of dollars in producing a movie, they want to make sure the investment is recovered. Protecting this valuable content is very high on the priority list of Hollywood and independent studios.

When DVD was first released, the content protection scheme used was called CSS (Content Scramble System). It encrypted the content with a combination of Title Keys and Disc Keys. The Title Keys were stored in the sector header whereas the Disc Keys were stored in the control area of the disc. Since both areas containing the keys to unlock the content are not directly accessible from DVD-ROM drives, this protection mechanism was supposed to be very secure. However, it only took about two years after format launch until CSS was hacked. The reason for this was a security flaw in one software player exposing keys, which made it very easy to reverse engineer the entire scheme and find all other keys necessary for decryption. Unfortunately, once the keys were exposed, not only one disc was hacked, but also all released titles at that point in time and in the future were affected. The only measure against this attack was to disable all player keys for all playback devices in the market. Since that would most likely have killed the format altogether, the only way out was accepting the failure of CSS, abandoning this system as a serious protection scheme and trying to develop something new for the future.

The lessons learned from the CSS attack proved to be very important for the development of the next copy protection mechanism called CPPM (Content Protection for Prerecorded Media). Although it was generally based on the CSS system, there have been a lot of improvements and modifications making it a much more robust and reliable system. With its improved security, it was first used for protecting DVD-Audio discs. CPPM didn't use Title and Disc Keys anymore as they used to be the source for the attack. Instead, the Disc Keys were replaced with an album identifier, which was stored in the control area of the disc lead-in. This area is not available on recordable media, only prerecorded discs can write data into it. The idea behind this approach was that even if it would be possible to decrypt the content, the keys could not be recorded, making it impossible to create a duplicate copy of the disc.

Also, each player came with a set of 16 confidential Device Keys that are not stored on the disc. Instead, a so-called Media Key Block (MKB) is stored on the disc. The Licensing Entity provides the MKB to the Disc Replicator. Playing back the disc, a number of logical operations are performed between the MKB and the Device keys to calculate a Media Key. The Media Key in combination with the album identifier is then being used to decrypt the protected content of the disc. In addition to this new approach using Media Key Block instead of Title and Disc Keys, a revocation mechanism was added to the format making the system future proof. This means, if one player was hacked, future discs would use a different MKB disabling the Device key of the hacked player. As a result, the logical operations between the

Device keys and the MKB would calculate the wrong Media Key preventing the disc from playing back.

When the Hollywood studios started their mission effort in the Next Generation Optical Disc formats, the protection of their content was more important than ever. The video material was in high definition, with a much better quality, making the material even more valuable. To make sure that history doesn't repeat itself and the concerns of all participants were being addressed for the new formats, eight companies (IBM, Intel, Microsoft, Matsushita, Sony, Toshiba, Walt Disney, and Warner Bros.) formed the AACS (Advanced Access Content System) group to develop the next generation copy protection system.

Objectives

As one can imagine, with three industry sectors (studios, consumer electronics, IT industry) being involved, there are very different objectives for the new system. Content Owners were looking for robustness and renewability, making sure that both audio and video content are protected in a future proof manner. The other parties were primarily interested in one solution suitable for both CE players and computer implementations. On top of the individual interests, the solution had to be transparent for legitimate consumers.

The result of the AACS group now promises a seamless, advanced, robust, and renewable method for protecting audio-visual entertainment content hoping to address all the problems found with CSS in the past. The protection system provides a format neutral platform targeting all next generation formats including HD DVD and Blu-ray Disc™. It can be used for prerecorded and recordable media.

When AACS was designed, the founders also considered new business models for content providers, distributors and device manufacturers. For instance, new distribution management options have been added for Content Providers extending the high definition audio and video delivery by offering e-commerce capabilities on next generation discs. As a result, usage scenarios are much wider than they were with DVD. In addition to the local playback of the disc, electronic distributions with secure copy, home media servers, and portable devices are all covered using AACS — different forms of distribution being protected by the same system.

Ecosystem

From an organizational point of view, there are mainly four different entities involved. Figure 17.1 describes the following scenario: the Content Providers create, own, and distribute copyrighted material and define the usage conditions for every individual title. The Replication facilities (Licensed Content Producers) encrypt and produce the media according to the usage conditions defined by the Content Provider. Adopters build licensed software or hardware products that will be able to play back AACS encrypted media. Holding all these strings together is the AACS Licensing Agency (AACS LA), which administers the specifications, defines licensing terms, and issues keys used for encryption.

Figure 17.1 AACS Ecosystem

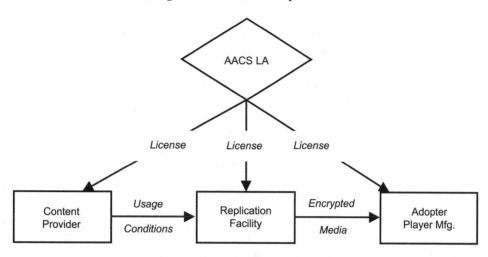

 TIP

Authoring facilities do not have to be an AACS licensee. They should, how-ever, verify that their clients are licensed before applying AACS encryption.

Another important required part to process AACS is the so-called ISAN number. It is the International Standard Audiovisual Number that identifies audiovisual works. Basically, this system is voluntary and provides an internationally recognized, unique, permanent reference number for every version of a movie. Instead of referencing a movie by its title, which may be different in other languages, this system provides a unique reference number across all countries. It is also very helpful for anti-piracy purposes as it verifies the title registration. However, the issuance of an ISAN number is not a process of copyright registration and also does not declare any ownership rights. The Content Provider can obtain an ISAN number for each title by registering the works at http://www.isan.org. The ISAN number is a required for AACS processing, hence the Content Provider should make this number available to the Authoring Facility. The AACS disc cannot be finished without the ISAN number.

 NOTE

The ISAN number is a 24 digit hexadecimal number divided into three parts. The first 12 digits specify the root (i.e. title of the movie), the next 4 dig-its define the episode and the last 8 digits specify the version of this title.

Technology Overview

Although there are different format specific implementations for HD DVD and Blu-ray, both formats use the same underlying common definition of this technology. The encryption for AACS is based on the robust Advanced Encryption Standard (AES) with 128-bit keys.

For the Key Management, the Media Key Block Technology is used. Also, a big emphasis was put towards an enhanced robustness for proactive software renewal with revocation. Due to the new usage scenarios that were incorporated into AACS, an enhanced drive authentication was necessary to allow advanced operations such as content distributions to home media servers. Also, network connectivity was added to enable e-commerce applications.

In order to allow all these usage scenarios, a number of identifiers are necessary for the discs. Table 17.1 below provides an overview of the most important identifiers.

Table 17.1 Identifier for AACS Protected Discs

Volume ID	It is a unique number for each Volume (i.e. replication run). The Volume ID is stored on Prerecorded media and cannot be copied to recordable media.
Media ID	It is unique for each recordable media. This way, content can be bound to a specific recordable disc.
PMSN	The Prerecorded Media Serial Number (PMSN) is a unique identifier for each individual disc. It is used for Network-based transactions, i.e. Managed Copy transactions (described later in this chapter). The PMSN is optional for both HD DVD and Blu-ray and would be stored in the BCA area of the disc.
Provider ID	This number is used for the segmentation of the Persistent Storage inside the player. Each content provider has a specific area in Persistent Storage that it can read and write data to and from. Discs from other content providers cannot access this area.
Content ID	This is a unique number that identifies the content on a remote server in an online transaction. It contains the ISAN number of the specific title.

Media Key Block

Similarly to the CPPM protection (described above), each AACS compliant device contains a set of Device Keys provided by AACS LA. The keys can either be unique for an individual device or used for multiple devices. Device Keys are highly confidential. AACS LA also issues a Media Key Block (MKB) to the Replicator to include on each replicated disc. All compliant devices can calculate a Media Key using their specific Device Keys and the MKB on the disc.

The Media Key will then be used to decrypt the Title Key, which is used to protect a given Title (or multiple titles) on a disc. This means that it is the most vulnerable key, because once this key is known, the content can be decrypted without any other keys needed. With that in mind, this encryption chain is used - the Title Key is encrypted with the Media Key, which is calculated with a combination of the Device Key and the Media Key Block.

Revocation

There are a number of methods to ensure renewability of the system in case an attack was successful. One of the biggest targets for an attack is the computer environment, because it is not a "closed system", but uses a drive and a software player (host) at least two separate components. In order to play back an AACS protected disc in such an environment, both components have to perform a Drive-Host authentication. Basically, each drive must be

authorized by AACS LA and receives a Drive Certificate and Drive Private Key in order to perform the required drive authentication. The same mechanism applies for host applications; they receive a Host Certificate and a Host Private Key to perform authentication. Before disc playback can start, the drive and host have to verify that both counterparts are indeed AACS compliant and have a valid certificate signed by AACS LA. Additionally, both parts check that they are not on one of the Revocation Lists.

The Host Revocation List (HRL) and Drive Revocation List (DRL) are part of the Media Key Bock that is stored on the disc itself. Both lists contain information about which hosts (software players) and drives are revoked. AACS LA provides the lists when they issue the MKB for a given disc. The Type and Version number of the Host Revocation List and Drive Revocation List are the first entries in the MKB to make it easy for drives to extract the data. The most recent version of the HRL and DRL will be stored in non-volatile memory of the device.

In case an attack occurred, future discs will contain a new MKB excluding the compromised Device Keys, so that the Media Key calculations will not result in the key needed to decrypt the content of the disc. This mechanism ensures that every disc released after the attack will revoke the compromised device and prevent playback.

Sequence Keys

One of the problems with revocation is that it is based on knowing the Device Key of this particular device. Without knowing them, nothing can be revoked. In order to identify the device keys, Sequence Keys are used. The Sequence Key approach will provide different alternate versions of the movie to be played. However, the alternate versions don't differ in terms of the content or the story line, but the forensic mark that is embedded into the video. Meaning, every user will see the exact same movie, but analyzing the forensic mark of every player will have different results enabling the content owner to identify which player a pirated movie was coming from. Granted, analyzing the forensic mark of one movie may only point to a group of players, but analyzing the data of a few movies will identify an individual player. The way it works is illustrated in Figure 17.2.

Figure 17.2 Application Image of Sequence Key Section

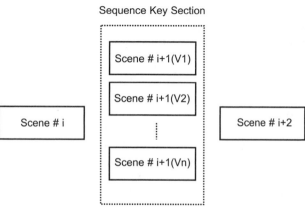

17

Basically, the feature on a disc contains multiple Sequence Key Sections. Each of these sections contains multiple versions of the same scene with different forensic marks. Having multiple Sequence Key Sections, each player will calculate its own way through the movie, making it possible to find the player producing the pirated movie.

The implementation for HD DVD and Blu-ray format are slightly different. HD DVD uses Interleaved Blocks similarly to Multi-Angle sequences. This is a well-known implementation and the same restrictions as for Multi-Angle scenes also apply for Sequence Key Sections. Looking at the disc, there are a few additional files inside the AACS directory. The SKB.AACS file contains six Sequence Key Blocks and one Volume Variant Unique Key per Sequence Key Block. There is also a SKF.AACS file accompanying the Sequence Key Block file. The SKF.AACS file contains six Segment Key Group fields with 1024 Segment Key Units each. One Segment Key Unit out of one Segment Key Group will be decrypted with the Volume Variant Unique Key. One Segment Key Group corresponds to one Sequence Key Range, which is a set of P-EVOBS on the disc. There is a maximum of 32 Sequence Key Sections per Sequence Key Range. Before playing back Sequence Key Sections, the SKB.AACS and SKF.AACS file will be processed and result in 6 Sequence Key Tables with 32 pairs of Segment Key Numbers each. These numbers will then determine which path through the movie the player will take.

> **WARNING**
>
> ⊗ Since Sequence Key Sections are using Interleaved Blocks, there are no other Multi-Angle features allowed during this sequence.

In Blu-ray this is implemented using multiple Playlists. Each disc can contain six Sequence Key Blocks, with every Sequence Key Block containing 256 PlayLists. Each PlayList contains a set of PlayItems for Sequence Key segments and non-Sequence Key segments. Every Sequence Key Variant (PlayItem) is stored as an individual Clip AV Stream file on the disc and encrypted with a different Segment Key. Also, each Segment Key is encrypted with the Volume Variant Unique Key. The Segment Key File is recorded in the Segment_Key.inf in the AACS directory of the disc together with the files for each Sequence Key Block, SKB1.AACS – SKB6.AACS.

> **NOTE**
>
> At least one non-Sequence Key segment has to be located between two Sequence Key segments.

Analog Sunset

A very big challenge that AACS had to master is the so-called analog sunset. This means that all products manufactured and sold after December 31, 2010, are only allowed to pass decrypted AACS content through an analog output with Standard Definition Interlace Mode

(i.e. Composite, S-Video, 480i component). Decrypted high definition content will not be allowed through analog outputs. Products manufactured and sold after December 31, 2013 are prohibited to pass any content (protected or unprotected) through analog outputs.

However, what happens until this time comes? There are a lot of analog display devices in the market. Most of the early high definition displays don't have any digital interfaces. Since the CE manufacturers rushed them into the market, most of them only provide analog inputs. The AACS specifications define a number of settings that can be applied to AACS protected content. The two most important settings are the Image Constrained Token (ICT) and Digital Only Token (DOT). The Image Constrained Token is a flag specifying that the content is allowed to pass analog outputs, but the maximum image is restricted to 520,000 pixels per frame. This translates to a resolution of 960 x 540 pixels for a 16:9 aspect ratio. The constrained image can be displayed using scaling algorithms, line doubling, or other video processing techniques. The Digital Only Token on the other hand is a flag preventing decrypted AACS content to be passed through analog outputs. If this flag is set, only digital outputs are allowed to pass the decrypted content.

Products manufactured and sold prior to December 31, 2010 are allowed to output decrypted AACS content through specified analog outputs. The below Table 17.2 describes all of them.

Table 17.2 Allowed Analog Output Devices for AACS Protected Content

Computer Monitors	● ICT must be set
Component Video Outputs (SD and HD)	● ICT must be set for HD
	● Macrovision APS1[a] setting has to be applied
	● CGMS-A[b] has to be applied with RC[c] optional
Composite Video Output (SD only)	● Macrovision APS1[a] setting has to be applied
	● CGMS-A[b] has to be applied with RC[c] optional

[a]Macrovision is providing content protection for analog output. There are 3 different options for the analog protection system (APS). APS 1 is the most common and widely supported implementation.

[b]The analog copy generation management system (CGMS-A) conveys basic copy protection information over an analog video interface. It describes in 2-bit information whether:

    00: Unlimited copies allowed

    01: One generation of copies already been made, no more copying

    10: One generation of copies may be made

    11: No copies may be made

[c]The Redistribution Control (RC) is an optional bit as part of the CGMS-A. It conveys information relating to control over unauthorized consumer redistribution.

For digital interfaces, the outputs have to be protected by DTCP (Digital Transmission Content Protection), HDCP (High-bandwidth Digital Content Protection) or WMDRM (Microsoft's Windows Media Digital Rights Management).

HD DVD Implementation

As mentioned above, although both formats are using the same underlying technology and features of AACS, the implementation for HD DVD and Blu-ray is different. Every format is responsible for their own implementation to make it work within their specific framework.

For HD DVD this means that first of all, AACS is optional. So the content owner has the choice whether they want to protect their content with AACS or not. Since a licensing fee is involved in using AACS protection, a content owner, particularly smaller studios, may decide that they don't want to spend their money on this and go without content protection. However, non-AACS protected discs are somewhat restricted in functionality, i.e. persistent storage and network access is not permitted. Since these two features are able to use parts of the player that are usually protected, for instance shared storage, discs have to be AACS encrypted to authenticate themselves for usage.

In addition to that certain APIs, mainly related to file access, but also some others, may behave differently than on AACS protected discs. So it is important for the content author to familiarize him/herself with the details of these restrictions in order to ensure proper playback of the disc.

So the first check a player has to do is to look for either the PMSN (Prerecorded Media Serial Number) or the Volume ID on the disc. The Volume ID is divided into two parts. One part is stored in the BCA area; the other part is stored in the Lead-in area of the disc.

If it can read one of these numbers, it will start the playback in AACS mode. Generally speaking, the content of an AACS protected HD DVD has two forms of encryption. The encryption of all audio-visual material contained in the Primary and Secondary EVOBs (P/S-EVOB), and the encryption of all advanced resource elements used for navigation material (i.e. menu elements). For the encryption, Title Keys will be used and there are 64 of them to choose from for each HD DVD disc.

 TIP   One Title Key can be used to encrypt multiple Titles, so if there is a disc with more than 64 Titles, they can still all be encrypted using the same Key on multiple Titles. For instance, all Bonus Pieces can use the same Title Key.

The Title keys get encrypted with the Volume Unique Key using AES-128 encryption. The name for the Title Key File for a Standard Content disc is VTKF.AACS. Due to the structure of Advanced Content discs, there will be multiple Title Key Files, one for each Playlist on the disc, the filenames are VTKF###.AACS ("###" being the number of the Playlist ranging from 000 to 999). There's also an accompanying Title Usage File for each Title Key File describing the usage rules for each particular EVOB. The usage rules also contain the CGMS information such as Copy Control Information, whether the Analog Protection System shall be used, and whether the Image Constrained Token or Digital Only Token shall be set. The name for the Title Usage File is VTUF###.AACS, with the "###" being the number of the Playlist. The specific VTUF number defines the rules of the same VTKF number. For Standard Content discs, the file name for the Title Usage File is VTUF.AACS.

Each disc also contains two Content Hash Tables. CONTENT_HASH_TABLE1.AACS listing all the hashes used to encrypt all the P/S-EVOBs on the disc. The other hash table, CONTENT_HASH_TABLE2.AACS, contains the hashes of the following files:

- DISCID.DAT
- Directory Key File (DKF.AACS)
- Managed Copy Manifest File (MNGCPY_MANIFEST.XML) – if applicable
- Title Usage Files on the disc
- All XML document files on the disc
- All ECMAScript files on the disc

To protect the directory name of a content provider in persistent Storage, the DISCID.DAT and the Directory Key File (DKF.AACS) are used. The DISCID.DAT is the configuration file containing the Provider ID, which is used for the segmentation of the Persistent Storage inside the player. This way it can be ensured that the content inside the Persistent Storage is only accessible from discs created by a particular Content Provider. The DISCID.DAT file is stored in the ADV_OBJ directory on the disc, whereas the DKF.AACS file is stored inside the AACS directory on the disc.

The Advanced Elements on the disc used for the Menu Navigation are protected by encapsulation. There are five formats for encapsulation as per Table 17.3 below. The first two methods in the table require an AACS signature, so they guarantee a higher integrity.

Table 17.3 Encapsulation Formats for Advanced Elements on HD DVD

Encapsulation Format for Hash	Content will be protected by Content Hashing and Content Certificate.
Encapsulation Format for Encryption and Hash	Data will be encrypted and protected by Content Hashing.
Encapsulation Format for MAC	The MAC value calculated using one of the Title Keys.
Encapsulation Format for Encryption	Data will be protected by encryption.
Encapsulation Format for non-protected Advanced Elements	Only filenames will be encrypted. Nothing else will be protected. This method prevents substituting a protected file with an unprotected file.

Basically, almost all Advanced Elements are protected by Encapsulation in some shape or form. All images, animations and fonts shall be encapsulated using method 3 or 4 in the above table. Even non-protected elements should be encapsulated using the Encapsulation method 5 in the table 17.3. Also, all ECMAScript and XML documents are to be encapsulated using the encapsulation format for hash. If the author chooses to also encrypt the ECMAScript files, encapsulation format for encryption and hash can be used. Since most of these Advanced Resource Files may be stored in an archive file (ACA), the ACA itself will not be encapsulated, but all the contents within will be.

17

There are a few exceptions to the general rules. For instance, the XML file of the Managed Copy Manifest shall not be encapsulated, but will be protected by content hashing. Also, the encapsulation of Advanced Subtitle Files is left for the author to decide – it is optional. As for Effect Sound Files (.wav), they should not be encapsulated at all.

WARNING	Images not used for Advanced Navigation of the Disc shall not be encapsulated. For instance, images used for Player Setup Displays (i.e. Provider Directory) or as PC ROM data (i.e. Wallpapers) should not be encapsulated.

Also, the protection methods for files in Persistent Storage differ slightly. For instance, there are no content hash tables or content certificates for files stored in persistent storage. The files are mainly protected by MAC verification. Since the verification of these files happens in the presentation engine immediately before playback, parsing or interpretation, it can be ensured that no data manipulation is possible. Also, since it will only be accessed while an authenticated disc is played back, the data stored in persistent storage doesn't affect the integrity of the data on the disc.

In addition to that, the Media Key Block for the data on the HD DVD, a content certificate and a content revocation list are stored on the disc as well. The Media Key Block (MKB.AACS) is stored in the data area of the disc, but a partial MKB containing Type, Version Record and Host Revocation List are stored in the access restricted lead-in area of the disc. The content certificate contains information such as number of layers and both content hash tables (P/S-EVOBs and Advanced Elements) and will be signed by the AACS LA. The file name is CONTENT_CERT.AACS. The content revocation list is just another means to ensure renewability in case of an attack and follows the same principle as the Host Revocation List or the Drive Revocation List. It allows storing a list of disc IDs in the player that it will refuse to play. The filename is CONTENT_REVOCATION_LIST.AACS.

All the above-mentioned files, Title Key Files, Title Usage Files, Content Hash Tables, Directory Key Files, Media Key Blocks, Content Certificate and Content Revocation List are all stored in the AACS directory on the disc as shown in Figure 17.3 below. A backup for each of these files is stored in the AACS_BAK directory on the disc.

The security flow for disc applications is the following. AACS verifies the Content Certificates, checks the hashes of Content Hash Table 1 and 2, and checks the Title Usage Files. The Content Hash Table 1 is also constantly being checked during P/S-EVOB playback. Also, before any XML or ECMAScript files are processed, the hashes for all referenced files will be verified with the values in Content Hash Table 2. For applications in Persistent Storage, the security flow differs since there are no Content Hash Tables available. So the Playlist resides in Persistent Storage with its associated Title Key File and Title Usage File. There is also a Content Certificate File associated with this Playlist in the same directory. The Content Certificate contains the hash of the Title Usage File, which would be verified by the AACS module.

Figure 17.3 Structure of AACS directory for HD DVD

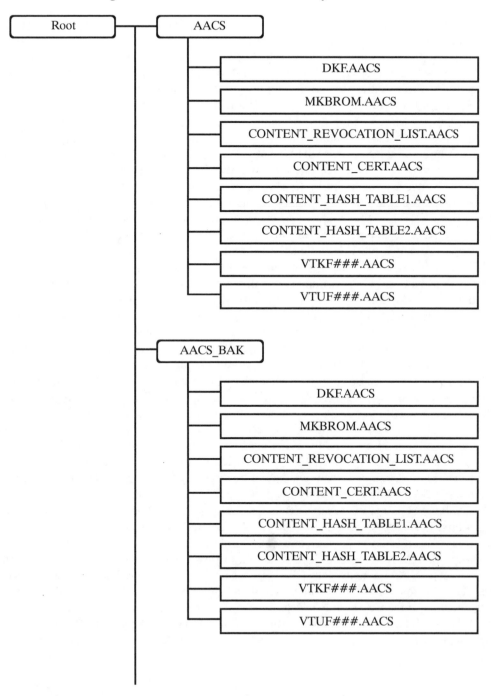

17

Blu-ray Implementation

Contrary to HD DVD, the Blu-ray Disc specifications define AACS as mandatory for all BD-ROM discs. It isn't the content provider's choice whether to protect the disc with AACS. If the content provider wants to release titles in this format, AACS has to be applied. Due to this requirement, every Blu-ray disc contains an AACS folder in the Root directory as shown in Figure 17.4 . In case the original file cannot be read, a backup copy for all the files in this AACS directory is stored in the DUPLICATE folder inside the AACS directory.

Basically, every layer of the disc contains its own Content Certificate, which has to be signed by the AACS LA. It contains the hash values of all other AACS files (and the BDMV directory), and also the calculated SHA-1 hash value for the BD-J Root Certificate for application security. In order to use certain functionality of the BD-J specification (i.e. Persistent Storage or Network access), applications have to be signed. For that, there are various signature files stored within the signed JAR file and a Root Certificate will reside on the disc (inside the CERTIFICATE folder) verifying the authenticity of these signed applications using the certificate chain. The Hash value of the Root Certificate is stored in the Content Certificate for AACS protection, and the Content Certificates are named Content000.cer and Content001.cer for a dual layer disc and are stored on Layer 0 and Layer 1 respectively.

There is also a Content Revocation List (ContentRevocation.lst) and Content Hash Tables stored in the AACS directory. The Content Hash Table contains a SHA-1 hash for every Clip AV Stream file and a SHA-1 value for the Managed Copy Manifest File if applicable. There is a separate Hash Table for each layer and the filenames are Content Hash000.tbl and Content_Hash001.tbl – they are stored on Layer 0 and Layer 1 respectively. The player will verify the hash values and constantly check them during playback of the disc.

Each BD-ROM Disc also contains a Media Key Block named MKB_RO.inf that is stored in the AACS directory. Also, a Partial Media Key Block containing the Type, Version Number and Host Revocation List is stored in the Lead-in area of the disc. In order to decrypt the disc content, the Media Key is retrieved from the calculations of the Media Key Block and the Device Key. In addition to the Media Key, the Volume Identifier is needed to calculate the Volume Unique Key. The Volume Identifier is stored in the ROM-Mark of the disc. ROM-Mark is in a place where it cannot be duplicated, adding another level of security to the Blu-ray format, more details are described later in this book.

The Volume Unique Key is then used to decrypt the so-called CPS Units. A CPS Unit contains a First Playback, Top Menu, and/or Titles encrypted with the same Unit Key. All Clip AV Stream files referenced by the same CPS Unit have to be encrypted with the exact same CPS Key. All CPS Keys on one disc are stored in the Unit_Key-RO.inf file located in the AACS directory. Additionally, each CPS Unit has its own CPS Usage File defining the usage rules. The filename for the Usage File is CPSUnitXXXXX.cci, with "XXXXX" defining the 5-digit number of the CPS Unit to which the Copy Control Information File (.cci file) is associated. The CCI file contains all the information related to which copy protection systems are to be applied, such as Analog Copy protection, Image Constrained Token, Digital Only Token, etc.

Figure 17.4 Structure of AACS directory for Blu-ray Disc

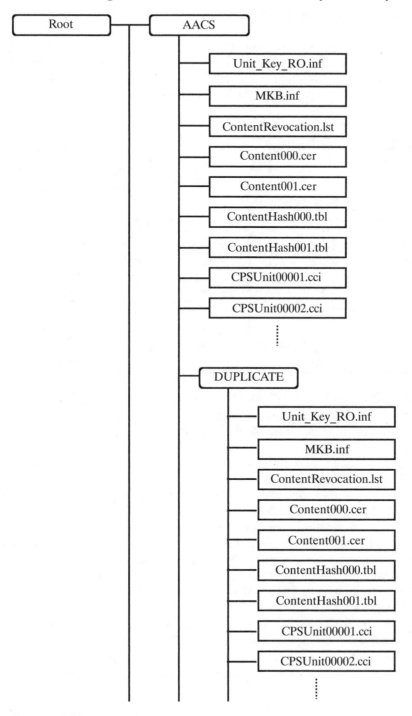

17

AACS Production Workflow

Understanding the technology details is certainly very helpful, but what exactly does that mean from a workflow perspective you may wonder? Which of these steps happen where in the production process? How complicated is it in reality to implement AACS? The below Figure 17.5 will give an overview about the production realities.

Figure 17.5 AACS Workflow

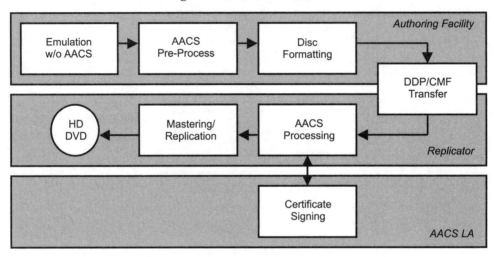

There are slight differences between HD DVD and Blu-ray in terms of the actual technology implementation the workflow is about the same. Basically, there are three main entities involved in the production process - the Authoring Facility, the Replicator, and AACS LA. The Authoring Facility usually programs the full disc and simulates and/or emulates the disc as much as possible making sure everything functions as expected. Since there are usually various rounds for emulation, this step is done without any AACS being applied; otherwise it would be to time consuming and expensive. Only once a version is approved for a check disc run, the AACS Pre-Process begins. This is the stage at which the Authoring Facility would prepare the content for a later encryption. General information required for AACS encryption would be entered such as Provider ID, Content ID, etc. Then some files may be encapsulated, signed or hashed. Basically, everything that will be needed for an encryption at a later stage will be flagged and prepared in some way. The Authoring Tool used does the pre-processing. The final disc will then be formatted and the DDP or CMF transferred to the Replicator.

The actual encryption of the content will then take place at the Replicator. There is a very strong connection between the Replicator and AACS LA. For instance, the Replicator is licensed by AACS LA and will receive the AACS keys needed for encryption. Since the content of disc is already pre-processed from the Authoring Facility, it is fairly easy for the

Mastering equipment to understand where to apply the encryption itself. After all the content has been encrypted with the AACS keys and the hashes for everything have been created, the resulting Content Certificate containing all the content hashes will be sent to AACS LA. Once the Content Certificate is received, it will be signed by AACS LA and sent back to the Replicator. As soon as it is back at the Replicator, the manufacturing process can begin.

As one can imagine, this is a rather complicated process that may take quite some time as well. At the time of writing this book, the interaction between a Replicator and AACS LA is still not fully automated. Sending the Content Certificates back and forth takes at least one day. However, in the future this will most likely be automated and a central server somewhere should do most of the work. This would simplify this process and improve the turnaround times for a check disc dramatically.

Online Connections and Network Content

For an AACS protected next generation disc, there is no online connection necessary to play back discs. However, both formats advertise and implement broadband connectivity their players to provide a new user experience and additional features. Network related content could be divided into at least four groups of use cases as outlined in Table 17.4.

Table 17.4 Network Related Content Use Cases

AACS Network Download Content	The AACS protected content is downloaded and stored in persistent storage or a recordable disc. An online transaction binds the content to this particular media.
AACS Online Enabled Content	The AACS protected content is already on the disc, but an online transaction is required to "unlock" or "enable" the content.
AACS Streamed Content	Content associated to a particular AACS protected disc would be streamed. It will be delivered live on demand.
AACS Managed Copy	The content on the AACS protected disc contains at least one offer for a copy to another device (i.e. Home Media Server). The device performing this copy process has to be authorized by a server before allowing this operation.

Use case 1 and 2 will have content either already stored on the disc or downloaded to persistent storage or a recordable media. But in order to play back the content, an online connection and possibly a money transaction will be necessary to retrieve playback permissions. Without connecting to the server, this content could not be played back. Use case 3 does not have the content present on the disc or in persistent storage. Instead, it will provide a live data stream on demand after the connection and transaction has been successful. Use case 4 is the same principle as use case 2. The content is already on the disc, but an online connection is necessary to perform a transaction and allow the managed copy to be made. In addition to these four use cases, there may be other scenarios possible utilizing AACS protected content.

Now how would this look in terms of disc playback? There are again a few different scenarios as outlined in Figure 17.6 below. There could be multiple titles on a disc. Title 1 for instance would linearly play back the feature. The content is stored on the disc and since there are no additional permissions needed, accessing the network is not necessary. However, Title 2 would provide additional scenes to the movie (i.e. Directors Cut) that could be purchased separately. The user would have to connect to a server and perform a transaction to unlock these scenes. The additional scenes could be integrated seamlessly into the existing movie on the disc. Or, the user could unlock additional features in Title 3 such as bonus content like deleted scenes that are not tied into the feature playback. In any case, a transaction would be necessary to unlock this content as well.

Figure 17.6 Example Title Configuration for Online Content

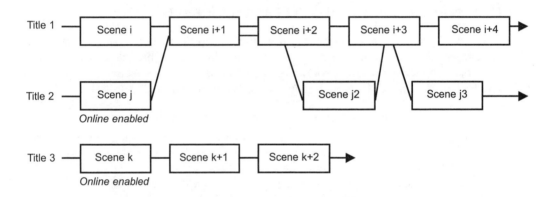

Title 2 and Title 3 in Figure 17.6 are both called Enhanced Titles, as they require a connection to a remote server to obtain playback permissions. Without a connection to the server, content playback would not be allowed. Now the question becomes how all this ties into the AACS protection flow? Well, the content on a disc and on the remote server has to be AACS encrypted and will be bound to a specific media. Of course, the remote server doesn't have a Device Key, but that is not required since the playback device will have device keys to decrypt the content. Also, the server receives certain information from the AACS media (i.e. Media Key Block) to encrypt the Title Keys for the Network Content and make it playable in this particular device. This step is called Media Binding as it cryptographically binds the network content to the media.

As a result, the Title Keys itself can be stored either on the disc that this content is bound to or they can be downloaded as part of the transaction process. However, an online connection is required in either case to encrypt the Title Keys and eventually grant access. With every Title having an entry in the Title Usage File, this is also the place where the information is stored whether or not a title requires external permissions and if so, the remote server address will be contained in the Title Usage File as well. Basically, before a player can play back a title, it has to check the Title Usage File to find out whether or not this particular title is an Enhanced Title and requires external permissions.

Managed Copy

One of the new concepts with AACS is the option to allow a "Managed Copy" of a disc. It is a way to offer more value to the end user by enabling an authorized movement of the content to a home media server or a portable device. This also opens the door for new business models related to electronic distribution of content. At the same time, the content owner doesn't lose control of the content and is able to define the rules for the allowed copies on a title-by-title basis. For instance, on one title, the content owner might allow one copy of this movie to the home media server, and on a different title, there might be additional bonus material available for download. The possibilities with Managed Copy are very diverse making it a powerful feature for the content owner.

Basically, an AACS protected disc may provide Managed Copy offers, in which case the PMSN (Prerecorded Media Serial Number) would be required to authenticate an individual disc. Additionally, a disc using Managed Copy will contain a Managed Copy Manifest file defining the URL and some other information. Also, the player will need to have a Managed Copy machine built in to manage this process. Figure 17.7 outlines the process of a Managed Copy transaction.

Figure 17.7 Managed Copy Process

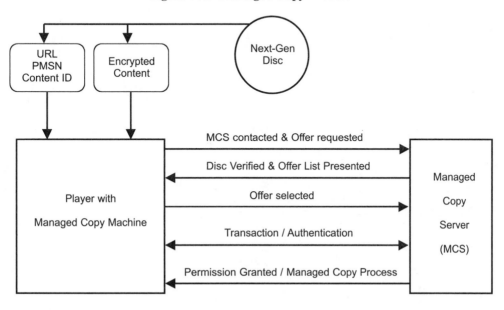

17

In case a user purchased a disc containing a Managed Copy offer, the content on the disc would be encrypted and the disc would contain a URL for the Managed Copy Server, the Content ID and a unique PMSN for an individual disc. The Managed Copy Machine inside the player would read the Manifest file and contact the URL of the Managed Copy Server as specified in the Manifest File to request the offer. The Managed Copy Server would verify the disc based on the PMSN and the Content ID, which would be provided by the Managed Copy Machine. Once the verification was successful, an Offer List would be presented to the user. Now the user can browse through the various offers and choose one, at which point the transaction and authentication process between the Player and the Managed Copy Server would occur. Once the transaction is completed, permissions are granted and the actual copy process can begin. This process could either include a monetary transaction or could be free. This is left to the content owner and the defined rules for the particular title.

Chapter 18
Blu-ray Disc's BD+

Although AACS provides an improved copy protection compared to CSS, some content providers were not satisfied. They were scared that history may repeat itself and a major attack may occur making it impossible or impractical to be fixed with the revocation mechanisms provided by AACS. Suddenly all the titles, already released and also future titles, would be exposed and unprotected. This scenario is certainly the nightmare of Hollywood and the reason for the strong argument to provide additional security measures. But only the Blu-ray Disc™ Association incorporated a solution for the problem on the Blu-ray Disc. They provided an additional layer of content protection on top of AACS including two technologies – BD-ROM Mark and SPDC (Self-Protecting Digital Content), also called BD+. Both technologies provide additional security on top of AACS as described in Figure 18.1. HD DVD on the other hand does not include any of these technologies and only relies on AACS as a copy protection mechanism.

Figure 18.1 Overview of copy protection layers for Blu-ray

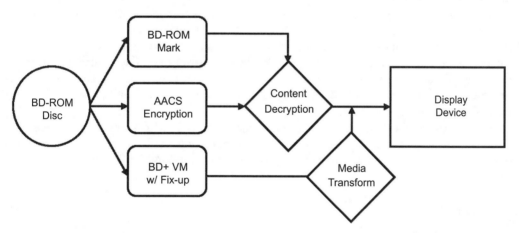

As shown in Figure 18.1, AACS encryption is only one part of the disc security. But in order to actually decrypt the content of a disc, a key that is stored in the BD-ROM Mark is necessary. Otherwise, the disc could not be played back. The other protection layer is SPDC. It offers a variety of additional protection measures such as content code, or counter measures, and in the picture above, the Fix-up Tables are specifically pointed out. On a very high level this means that the Transport Stream on the disc is stored in a non-format, so that even if the content has been successfully decrypted, it would not be viewable. Only some special code that is running while the disc plays back knows how to fix the stream to make it viewable. While BD-ROM Mark is mandatory on all Blu-ray discs, SPDC is optional and hence the content providers choice whether to use it or not. However, every player has to support SPDC.

BD-ROM Mark

BD-ROM Mark is a physical mark on the disc that is intended to prevent piracy on a consumer and professional level. The concept of a physical mark was first used with Super Audio CD and proved to be very useful. So it doesn't come as a surprise that the Blu-ray Disc Association considered this technology, because it provides a very effective copy protection solution for the movie industry in combination with the encryption technology AACS.

Basically, this physical mark contains a 128-bit key as payload. This key is called the Volume ID and it is required to decrypt the AACS protected content. Without this Volume ID, the disc content could not be played back. In addition to this, it can also be used for tracing purposes in case pirated discs would be discovered. Because the Volume ID binds a manufactured disc to the mastering facility, it would be used to trace where the disc was replicated.

The effectiveness of BD-ROM Mark is the fact that all BD Prerecorded discs will need to contain this mark. Otherwise they could not be played back. At the same time, disc recorders cannot reproduce this mark. This makes it a great solution against piracy because bit-to-bit copies of a disc would not work. On top of this, only licensed BD Players are able to read the BD-ROM Mark. And since the mark is never transmitted outside of the player, a manipulation would be very hard.

On the disc mastering side, a special piece of hardware, called BD-ROM Mark Inserter, is required. It is a secure encoder that will insert the key (Volume ID) and format the mark. Additionally, it will analyze the inserted data to verify readability of the key to ensure that it can be used for AACS decryption. The BD-ROM Mark Inserter is only available to licensed BD-ROM Disc Manufacturers and the license has to be renewed on a regular basis. This provides a higher level of security, because stolen equipment could not be used for a long time to pirate content.

Bottom-line, both the Mastering equipment and the BD-ROM player have to be licensed. The mastering equipment has to be licensed in order to apply the BD-ROM Mark and without the mark, a disc would not play in a licensed player. This combination is adding another security level to AACS.

SPDC

SPDC (Self-Protecting Digital Content) was developed by Cryptography Research Inc. and provides a renewable security mechanism. For that, every BD-ROM player has a so-called Security Virtual Machine (SVM) built in. With SPDC being optional, this SVM will not do anything unless the content owner decides to use this technology. However, whenever the content provider decides to apply SPDC, a BD-ROM disc will contain special "security code" that would be loaded into the SVM. This "security code" is also referred to as "content code" as it secures the content of the disc. The files for the content code are stored in a BDSVM directory on the disc (with additional backup files in the BACKUP directory within the same folder) as pictured in Figure 18.2.

Figure 18.2 Directory Structure for SPDC Content Code

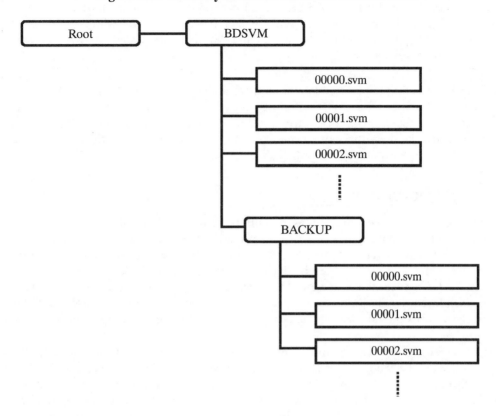

When the disc is started, the content code (*.svm files) is loaded from the BDSVM directory into the Security Virtual Machine. During disc playback the content code will be executed. Since the content owner defines the behavior in such cases, it may differ on each individual disc.

The SVM is also able to receive Interrupts from the player, such as notifications on Title or PlayList changes, general status changes or other player commands. And it provides a set of commands for the content code. For instance, three Player Status Registers (PSRs) are dedicated to the communication between HDMV and content code. Additionally, there is a set of BD-J APIs that can be used for the communication between content code and BD-J navigation layer. This way, parameters could be passed back and forth between playback and security environment allowing the content code to examine the playback environment, perform some calculations and based on the outcome determine where to jump next. In case the player was compromised, the playback behavior of the disc could differ.

> **NOTE** PSR 102, 103, and 104 are dedicated to the communication between content code and HDMV. PSR 102 allows content code to write data and the HDMV layer is allowed to read the data. It is the other way with PSR 103; the HDMV layer is allowed to write data. PSR 104 grants read and write access to both sides.

As explained above, content code can do arbitrary things. It can check results of some APIs, interface with BD-J and HDMV, store notes to itself, read the PMSN (Prerecorded Media Serial Number), or perform various other operations. In addition to these general abilities of code, there are three other features that can be controlled by the content code - Counter Measure, Native Code, and Media Transform.

Counter Measures

The content code is able to read areas of the BD-ROM player's memory space. This mechanism is used to compare values of known areas in the player with the values expected. In case they don't match, a security problem (player attack) has been encountered. If that would be the case, counter measures could be deployed. The content provider or player manufacturer would define the actual behavior of the counter measure. Some counter measures may change the playback behavior and may only be loaded as long as the disc is inserted. Others may fix the security problem by making persistent changes to the player applying a firmware update.

As one can imagine, it is very important to make sure that the detected security problem is indeed a player hack and not just a false alarm or wrong calculation. To ensure this is the case, each individual player manufacturers encrypts the counter measure code with a model-specific key. This ensures that the code executed is specific for that particular player. This type of player specific content code will be developed by the player manufacturer to ensure that only hacked players will be affected by the counter measure.

Native Code

If there is no way to fix the security problem from the content side (i.e. modifying some playback parameters), the SVM can request that player specific native code will be executed. The intention of this mechanism is to recover the player from a security problem. Since this kind of operation has a very serious impact on the player, an advanced mechanism is in place to ensure that only authenticated applications will be run. The native code is usually developed and tested by the player manufacturer and jointly integrated with the content code developer. The authentication process would require the native code to be signed with the player manufacturer's private key, and the SVM would verify the signature, before execution, with the player's public keys. Only if the authentication were successful, the native code would be run.

Media Transform

Content code has the ability to apply transformations to the disc content. This mechanism is called Media Transform and is illustrated in Figure 18.3. This transformation of the content ensures that only secure players will perform this operation correctly, hacked players on the other hand will not perform this operation, which may result in an unpleasant user experience.

Figure 18.3 Principle of Media Transform

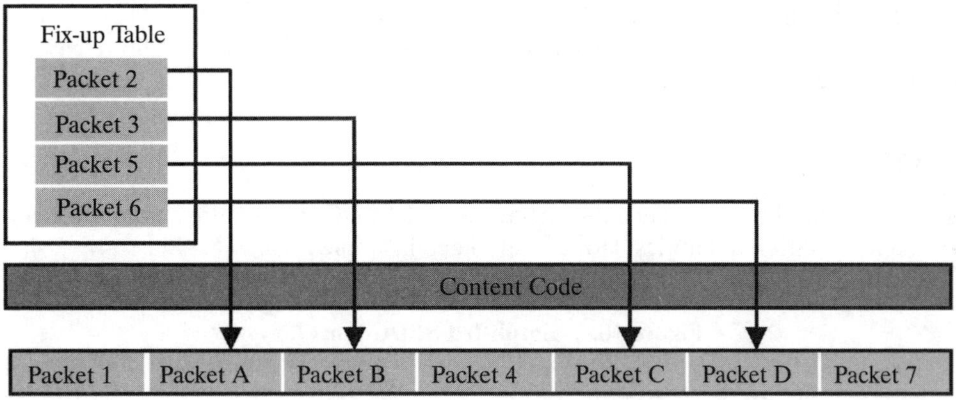

Fix-up Table

| Packet 2 |
| Packet 3 |
| Packet 5 |
| Packet 6 |

Content Code

| Packet 1 | Packet A | Packet B | Packet 4 | Packet C | Packet D | Packet 7 |

During disc production, specific sections of the fully functional video will be chosen. The packets in the transport stream of the chosen sections will then be modified (obscured) and the correct values of these packets are stored in a Fix-up Table. As a result, the transport stream on the disc will be obscured. However, while the disc plays back, the content code accesses the Fix-up table and replaces the obscured values with the correct values in the transport stream to make the video viewable. Since content code has to run to decode video pirates cannot simply post decryption keys on the Internet to circumvent security. The content code would notice this hack and prevent the Media Transform security. The content video output.

The Fix-up Table is stored in two places of the disc. One version is stored as a separate file on the disc; another version is multiplexed into the transport stream itself. The reason for this is that the separate file would have to be loaded into player memory at once, which would take up a lot of space. Another method was added allowing to only load the relevant part of the Fix-up table from within the multiplexed file. Finally, it is the player manufacturer's choice, which file to read.

Since the Fix-up Tables are multiplexed into the transport stream, this means that the m2ts file (transport stream file) will be modified after it was multiplexed. For that reason, it is necessary that the authoring application leave padding within the stream that would later be replaced with the actual Fix-up table avoiding the file structure to change.

NOTE The Fix-Up Table is usually masked or otherwise obscured to add another level of security. This way it is not easy to read and reconstruct the content some other way.

While the general Media Transform is necessary to "fix" the obscured video, this also provides a mechanism for yet another optional feature — forensic marking. A player specific "payload" could be inserted into the video as a watermark making it easy to identify the player that was hacked. For example, the player brand, model and serial number could be inserted into the watermark, so that an unauthorized video on the Internet could be traced back to the player that it was coming from. And with the next disc the same studio releases, this particular player could be recovered using a countermeasure.

Workflow

The actual workflow to create a disc protected with SPDC can be rather complex as it may involve a lot of different entities. However, to keep it simple Figure 18.4 illustrates a simplified workflow for SPDC.

Figure 18.4 Simplified SPDC Workflow

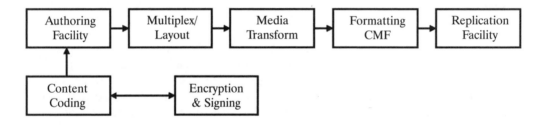

The content code is usually created by a separate entity. It requires a very specialized skill set writing secure code, which is very different from general software programming. Additionally, to make sure the content code is authorized and tested, it is necessary to get the code signed and encrypted by the Key Issuance Center (KIC). Once this is done, the code package will be provided to the Authoring Facility.

After the contents are encoded and the disc is fully programmed by the Authoring Facility, the content code will be incorporated. Depending on the content code implementation, there may be special requirements for authoring as well. For instance, it may be necessary that specific methods are called from within the BD-J environment or that certain values have to be written into some of the PSR's necessary for the content code. In any case, once this has been fully integrated and tested, the streams will be multiplexed and the disc layout will be created. Once this is finished, the Media Transform can be applied. This is the step where the transport stream would be obscured and the Fix-up table generated. These new files will then be incorporated into the existing layout before the final formatting can take place. Once the CMF has been created, it can be transferred to the Replicator for Mastering.

Before the disc is shipped, it will be properly tested by a Playability Test Center to make sure it functions as desired and doesn't cause any problems during playback in secure players.

When to use SPDC

With SPDC being optional on a disc, the question becomes when does it make sense for a content owner to implement this technology? Generally speaking, SPDC can be used for every disc. It is not dependant on the content at all. However, since an additional per-disc licensing fee is applicable for this technology, content providers may focus on high profile titles first.

Also, it is expected that as long as AACS provides a good enough security and hasn't been hacked, there might not be a very big demand for applying SPDC, despite the fact that it would secure the disc against future attacks. But as soon as AACS is exposed, SPDC offers a great way to renew the security. By that time, it is expected that most studios will apply SPDC on their discs since it will be the best alternative to protect the content. In the meantime, the Media Transform and possibly the Forensic Marking capabilities of SPDC provide interesting features for studios. Particularly the fact that a forensic mark can be applied at the end of the production chain will make it very easy to apply different marks for different versions of a disc without having to create separate masters and encodes for each version. This will make the production process easier and offer a great flexibility and added value.

Appendix A
About The Disc

The accompanying disc is provided as an Appendix to the book. This appendix document is duplicated on the disc in a file named **AboutThisDisc.pdf**.

There are multiple folders and files on the disc that provide additional tools and assets for the reader, aka the enduser, of our book. We have drawn these resources from a variety of suppliers in the hope that they will serve to augment and to broaden the reader's understanding of and enthusiasm for learning about the next generation HD disc formats.

To display the various folders and files on the disc, please use the traditional method of clicking on the folder and file names presented within the Windows Explorer directory structure. There is also a web page called **ReadMe.htm** on the disc that contains links to the disc items. Double click to open the file.

Some of the files may contain links to websites and, as such, will require an Internet connection. And, some of the files may contain executable files that will open and, possibly, install elements and/or files to your computer. Although we did exercise our best efforts at vetting the data and information furnished to us for inclusion on the disc, the reader is advised that the authors of this book and the disc make no guarantees about the reliability of said executable files or links, and do not accept responsibility for any untoward event that may result from activating said files or links.

Having expressed that caveat emptor, we hope that the reader will peruse the assorted data provided on the disc because we did go to great lengths to gather the materials and create the disc. There are some really cool tools, shortcuts, calculators and editors provided herein, along with the latest information from some of the leading suppliers of tools and equipment for producing next generation discs!

Please be aware that elements on this disc may require unique minimum system capabilities. Information about system requirements is provided in a later section of this appendix.

Assets on This Disc

The disc is organized by named folders. By opening these folders you will find the following —

Audio Volume Calculator

When programming HD DVD Advanced Content, this calculator will help an author to set appropriate values for mixing main, sub (secondary), and effects audio levels. This folder includes a Microsoft Excel calculator and a PDF tutorial.

A

The calculator is a Microsoft Excel spreadsheet that can be manipulated by changing values within the sheet. The output can help an HD DVD Advanced Content author stay within the limitations of the HD DVD specification. Please read the PDF document accompanying the spreadsheet for more details.

Cyberlink

Cyberlink has provided a software tool that will convert a variety of video and audio file formats, including high definition video, and output the assets in a variety of optical disc formats, such as BDAV on BD-RE, DVD-Video on standard DVD, and more. Create your own high definition discs at home.

The Cyberlink folder includes an installation executable, a PDF tutorial, and a software key text file. Click the **PowerProducer.exe** file to install.

Dolby

The Dolby folder includes demonstration software (for Macintosh) and a specification sheet for Dolby Media Producer, which is capable of producing Dolby high definition audio for next generation optical discs.

Please read the PDF document included in the folder to install the demonstration Dolby encoder, decoder, and tools for the Dolby high definition audio software.

DTS

DTS has provided documents for their Master Audio Suite software, encoding tips, and their DTS logo sheet for high definition audio. Sample DTS audio and MPEG-4 video assets have also been provided.

There are two example DTS HD (.dtshd) assets and one MPEG-4 (.mp4) asset for demonstration purposes only. These assets can be used for initial testing purposes of the new file types.

HD DVD Archive Editor 1.0

This folder includes a PC software tool created by the book authors for editing HD DVD Archive files (.aca files). This file type is defined in the HD DVD specification for Advanced Content programming.

On a PC, install the software with the **setup.exe** file in the Install directory. By adding files intended for HD DVD Advanced Applications, this software will build a properly formatted archive file (.aca) for inclusion in the ADV_OBJ folder. Using the extract functions, this tool can also be used to extract files within an .aca file.

interVideo / Ulead

A PC software package named *ULead DVD MovieFactory 5 Plus* allows a user to import video, and write to popular optical disc formats such as CD, DVD, and HD DVD. Software and product documentation are provided here.

WinDVD for HD/BD is an HD DVD and Blu-ray software player that may be used to enjoy these new disc formats on your PC, but it has not been included on this disc. To learn more about this player, please go to the interVideo / Ulead website at — http://www.intervideo.com/WinDVD .

Reference Tables

This disc includes several of the reference tables and figures from the book that outline scheduling, workflows, disc budgeting for small and large scale operations, and a quality control checklist sample. Click to open the individual PDF files. Microsoft Excel documents have also been provided for mutable operational needs, wherein you may enter the data that is applicable to your task.

Sonic Solutions

Sonic Solutions has provided PDF brochures for their CineVision high definition encoding software, ScenaristHD Standard Content Authoring software, and DVDit Pro HD software. Click to open the files.

Sony

Sony has provided PDF documentation about their Blu-print Blu-ray HDMV authoring solution and high definition products. Click to open the files.

Sounds

A series of AIF one second and 3/4 second audio files are provided on the disc that may be used as button sounds on Blu-ray and HD DVD disc menus. The Willie Tones and Button Sounds have been provided free for your use but you must reference Willie Chu NYC with disc and/or jacket credit in a form similar to the following — *Interactive audio sound effects courtesy of Willie Chu NYC*.

To use these AIF sound files, they must first be converted into a file format defined in the HD DVD or Blu-ray specifications, like the WAV file format. Tools to do this are not included on this disc.

A   Minimum System Requirements

The following folders and files have minimum system requirements —

Cyberlink Power Producer

- OS   Windows® 2000 Home Edition or higher
- CPU   Intel® Pentium® 4.2.2 GHz (Pentium 4.2.4 GHz recommended) or AMD Athlon® XP 2000+
- Memory   512 MB
- Hard drive space   25 GB
- DVD/HDV   OHCI IEEE 1394 compliant device
- Analog video capture   USB or PCI capture devices compliant to WDM standards
- DV video capture   IEEE 1394 I/O devices compliant with OHCI standards
- DSC photo import   USB interface
- Graphics card   nVidia® 6200, or ATI X700 graphics card

Dolby Media Producer

- OS   Mac OS X Tiger™ version 10.4.2 or higher
- CPU   Apple Power Mac G5® dual 2.3 GHz or faster Power PC® G5 processor
- Memory   512 MB
- Hard drive space   100 MB
- I/O requirements   USB to 9-pin adaptor for RS-422 control
- I/O audio requirements   eight channel (or more) low-latency audio device with Cor Audio/ASIO support

DTS

The example data files included with DTS will require other high definition software that is not included on this disc. The new flavors of high definition software will have their own system requirements. To play the DTS examples provided on this disc —

- BT-1618-48kHz-24bit-1509kbps-5_1.dtshd — requires high definition authoring tools, or DTS encoding/decoding tools will be needed to make use of this example DTSHD audio file.
- BT-1618-48kHz-24bit-LL-7_1.dtshd — requires high definition authoring tools, or DTS encoding/decoding tools will be needed to make use of this example DTSHD audio file.
- BT-1618_720p.mp4 — to view this MPEG-4 video/audio file, high definition software playback tools will be required.

Excel Calculators/Spreadsheets

To open the spreadsheet calculators on this disc, Microsoft® Excel® 2007 or higher.

Ulead DVD MovieFactory®5 Plus

- OS   Microsoft® Windows® 2000 with SP4 or higher
- CPU   Intel® Pentium® III 800 MHz, AMD Athlon® XP 1800+
- Memory   512 MB (1 GB or above recommended)
- Hard drive space   30 GB or higher for 3 hours of HD quality video capture
- DVD/HDV   OHCI IEEE 1394 compliant device
- Analog video capture   PCI, TV tuner, USB capture device for analog capture (WDM support)
- DV video capture   IEEE 1394 I/O devices compliant with OHCI standards cards for use withDV/HDV/D8 camcorders
- DSC photo import   USB 1.0/2.0 or PCI capture devices compliant with WDM standards and PC cameras
- Graphics card   Windows® compatible AGP or PCI graphics card (overlay support is recommended)
- Display   Windows® compatible display with 1024×768 resolution or above
- Other software support DirectX® 9.0, Windows Media Format 9, Macromedia Flash Player 7, Microsoft Internet Explorer 6 or later

PDF Documentation

To view the PDF files on this disc, Adobe® Reader® version 7 or later is needed.

Glossary

1080i 1080 lines of interlaced video (540 lines per field). This usually refers to a 1920×1080 resolution in a 1.78 aspect ratio.

1080p 1080 lines of progressive video (1080 lines per frame). This usually refers to a 1920×1080 resolution in a 1.78 aspect ratio.

2:3 pulldown The process of converting 24-frame-per-second film to video by repeating one film frame as two fields, and then the next film frame as three fields.

3:2 pulldown An uncommon variation of 2:3 pulldown, where the first film frame is repeated for three fields instead of two. Most people mean 2:3 pulldown when they say 3:2 pulldown.

4:1:1 A component digital video format with one C_b sample and one C_r sample for every four Y' samples. This uses 4:1 horizontal downsampling with no vertical downsampling. Chroma is sampled on every line, but only for every four luma pixels (one pixel in a 1×4 grid). This amounts to a subsampling of chroma by a factor of two compared to luma (and by a factor of four for a single C_b or C_r component).

4:2:0 A component digital video format, with one C_b sample and one C_r sample for every four Y' samples (one pixel in a 2×2 grid). This uses 2:1 horizontal downsampling and 2:1 vertical downsampling. C_b and C_r are sampled on every other line, with one set of chroma samples for each two luma samples on a line. This amounts to a subsampling of chroma by a factor of two, compared to luma (and by a factor of four for a single C_b or C_r component). DVD, HD DVD, and BD use 4:2:0 sampling.

4:2:2 A component digital video format, with one C_b sample and one C_r sample for every two Y' samples (one pixel in a 1×2 grid). This uses 2:1 horizontal downsampling with no vertical downsampling. This allocates the same number of samples to the chroma signal as to the luma signal.

4:4:4 A component digital video format, where Y', C_b, and C_r are sampled equally.

525/60 The scanning system of 525 lines per frame and 60 interlaced fields (30 frames) per second. This is used by the NTSC television standard.

625/50 The scanning system of 625 lines per frame and 50 interlaced fields (25 frames) per second. This is used by PAL and SECAM television standards.

720p 720 lines of progressive video (720 lines per frame). 720p60 refers to 60 frames per second, 720p30 refers to 30 frames per second, and 720p24 refers to 24 frames per second (film source). This usually refers to a 1280×720 resolution in a 1.78 aspect ratio.

A

AAC Advanced Audio Coder. An audio-encoding standard for MPEG-2 that is not backward-compatible with MPEG-1 audio.

AC Alternating Current. An electric current that regularly reverses direction. It has been adopted as a video term for a signal of non-zero frequency. Compare this to DC.

Glossary

AC-3 The former name of the Dolby Digital audio-coding system, which is still technically referred to as AC-3 in standards documents. AC-3 is the successor to Dolby's AC-1 and AC-2 audio coding techniques.

access time The time it takes for a drive to access a data track and begin transferring data. In an optical jukebox, the time it takes to locate a specific disc, insert it in an optical drive, and begin transferring data to the host system.

aacPlus v1 (Advanced Audio Coding) A new audio compression technology which supports 5.1 channel audio based on the MPEG-4 Audio technology.

aacPlus v2 An enhancement of aacPlus v1, with new, more powerful encoding techniques with Parametric Stereo in MPEG.

AACS The security mechanism that has been adopted for use by Blu-ray Disc and HD DVD high definition video formats. This protection scheme is better than CSS which protects DVD discs. AACS is more robust and has a mechanism (MKB) to stop players that have been compromised to play discs.

AACS LA The organization that grants AACS licensing is the Advanced Access Content System Licensing Administrator (AACS LA).

ActiveMovie The former name for Microsoft's DirectShow technology.

ADA (Application Data Area) In Blu-ray, the ADA is a storage area required to be on all Blu-ray players. Typical items to place in this smaller storage area are small files, bookmark files, disc profile files.

Adobe After Effects An Adobe software application that is intended for 2D graphics animation and for compositing motion images. Adobe After Effects is a good application to model interactive menus before they are developed for Blu-ray and HD DVD menus.

Adobe Illustrator An Adobe software application that is intended for 2D text and image creation. It is a vector based program that creates images that are smaller in file size.

Adobe Photoshop An Adobe software application that is intended for 2D still graphics creation and alteration. Photoshop is a tool commonly used to create still graphics for menus.

Advanced Application HD DVD provides a platform for more programming possibilities known as Advanced Applications. Using JavaScript and XML an author can program moving images, picture in picture video, persistent storage, networking functions, etc. Typical file types found in an Advanced Application include .js, .xmf, .xmu, .png, and .ttf. These files are archived (zipped) together into an Advanced Content Archive (.aca). The archive file is referenced in the Playlist file (.xpl).

Advanced Content The HD DVD Specification is divided into two sections: Standard Content, and Advanced Content. Advanced Content is generally comprised of Playlist, Primary Video Set, Secondary Video Set, Advanced Application, and Advanced Subtitles.

Advanced Subtitle In HD DVD, Advanced Subtitles are delivered in the Advanced Content Stream rather than the Subtitle Stream. Advanced Subtitles have the ability to be downloaded, while subtitles presented through the Subtitle Stream must be multiplexed into the stream and cannot be downloaded.

Advanced VTS As defined in the HD DVD Specification, Advanced VTS have eliminated the layered structure of Standard VTS such as Title, PGC, PTT and Cell, and introduced Time Map Information. An Advanced VTS is comprised of the following — one Main Video

Stream, eight Main Audio Streams, one Sub Video Stream, eight Sub Audio Streams, 32 Subtitle Streams, and one Advanced Stream.

ADV_OBJ The root of every HD DVD video disc must have two folders. The advanced content (i.e., an interactive menu) is stored in a folder labeled ADV_OBJ, and the video is stored in a folder labeled HVDVD_TS. An ADV_OBJ folder typically contains the following — DISCID.DAT configuration file, .xpl playlist files, .aca application files (which contain (manifest files, mark-up files, and script files), advanced subtitles, and non-video assets (graphics, audio files, and fonts).

AGC Automatic gain control. A circuit designed to boost the amplitude of a signal to provide adequate levels for recording. See Macrovision.

aliasing A distortion (artifact) in the reproduction of digital audio or video that results when the signal frequency is more than twice the sampling frequency. The resolution is insufficient to distinguish between alternate reconstructions of the waveform, thus admitting additional noise that was not be present in the original signal.

analog A signal of (theoretically) infinitely variable levels. Compare this to digital.

angle In DVD-Video this is a specific view of a scene, usually recorded from a certain camera angle. Different angles can be chosen while viewing the scene.

anamorphic A widescreen cinema theatrical release film format.

ANSI American National Standards Institute.

API An application programming interface (API) is an access point into another system, often as a set of functions that can be called by a programmer. An example of an API in HD DVD is the JavaScript Player API, which grants programmers access to the HD DVD player functions such as — changing audio streams, switching video streams, turning on picture in picture video, etc.

APL Average Picture Level is the average signal level with respect to blanking during the active picture time. APL represents the video signal level, during the active picture part of each horizontal line, and is mathematically averaged over the period of a frame to come up with APL.

APLSTnnn.XPL audio-only playlist file A playlist file that describes Audio Titles on a disc.

APM The APM (Auto Play Menu) is the menu on auto-play discs. These discs simply play the feature when you put them in a DVD player. At the end of the feature, the APM is seen, which simply allows the user to play the feature again.

application clock One of three tick-based clocks. The Application clock starts upon initialization of an Advanced Application.

application format A specification for storing information in a particular way to enable a particular use.

Application Segment Titles defined in an HD DVD playlist may have Application Segments defined within them, which contain Advanced Applications such as menus, pop-up graphics, advanced subtitles, etc. When the title is active, the Application Segment can be accessed.

Application Tick For communication between applications and the presentation engine, application ticks are defined by the TickBase and TickBaseDivisor in the <TitleSet> and <Title> elements in the playlist. Each tick carries time values for all three Clocks — Activation Clock, Page Clock, and Title Clock.

APS Auto Play and Stop. An APS disc is a technique to produce a title that has limited or no interactivity and is slated for a quick release after the theatrical run.

artifact Any distortion caused during any part of the digitization process of video or audio.

aspect ratio The width-to-height ratio of an image. A 4:3 aspect ratio means the horizontal size is a third wider than the vertical size. The standard television ratio is 4:3 (or 1.33:1). The widescreen DVD and HTDV aspect ratio is 16:9 (or 1.78:1). Common film aspect ratios are 1.85:1 and 2.35:1. Aspect ratios normalized to a height of one are abbreviated by leaving off the final :1, such as 1.85 or 2.35.

ASVOBS Audio Still Video Set.

ATAPI Advanced Technology Attachment (ATA) Packet Interface. An interface between a computer and its internal peripherals such as DVD-ROM drives. ATAPI provides the command set for controlling devices connected via an IDE interface. ATAPI is part of the Enhanced IDE (E-IDE) interface, also known as ATA-2. ATAPI was extended for use in DVD-ROM drives by the SFF 8090 specification.

ATSC The Advanced Television Systems Committee. In 1978, the Federal Communications Commission (FCC) empanelled the Advisory Committee on Advanced Television Service (ACATS) as an investigatory and advisory committee to develop information that would assist the FCC in establishing an advanced broadcast television (ATV) standard for the U.S. This committee created a subcommittee, the ATSC, to explore the need for and to coordinate development of the documentation of Advanced Television Systems. In 1993, the ATSC recommended that efforts be limited to a digital television system (DTV), and in September 1995 issued its recommendation for a DTV standard, which was approved with the exclusion of compression format constraints (picture resolution, frame rate, and frame sequence).

authoring Authoring refers to the process of designing, creating, collecting, formatting, and encoding material for an optical disc project.

autoplay or **automatic playback** A feature of players that automatically begins playback of a disc if so encoded.

AVC Advanced Video Coding (AVC) or H.264, MPEG-4 Part 10, is a digital video codec that is one of three codecs approved for the Blu-ray Discand HD DVD technologies. The other two video codecs are VC1, and MPEG-2.

B

bandwidth Strictly speaking, this is the range of frequencies (or the difference between the highest and the lowest frequency) carried by a circuit or signal. Loosely speaking, this is the amount of information carried in a signal. Technically, bandwidth does not apply to digital information; the term data rate is more accurate.

BCA Burst cutting area. A circular section near the center of a disc where ID codes and manufacturing information may be inscribed in barcode format.

BD+ An additional component beyond the AACS protection mechanism that adds a higher level of protection to Blu-ray Discs.

BD 25 A single layer Blu-ray Disc that can contain up to 25 gigabytes of data.

BD 50 A dual layer Blu-ray Disc that can contain up to 50 gigabytes of data.

BD-J Blu-ray Disc Java is a Java-based interactivity standard that is a part of every Blu-ray player. BD-J is compatible with GEM and, therefore, is based on MHP.

BD-R/RE BD-R and BD-RE are two recordable Blu-ray Disc formats. A BD-R disc can be burned (written to) once, and BD-RE can be burned multiple times.

bit A binary digit. The smallest representation of digital data: zero/one, off/on, no/yes. Eight bits make one byte.

bitmap An image made of a two-dimensional grid of pixels. Each frame of digital video can be considered a bitmap, although some color information is usually shared by more than one pixel.

bitrate The quantity of data per second that can be streamed through a system. The maximum stream bitrate of DVD is 11 Mbps (Megabits per second) before buffering. For increased information in high definition video and audio, the maximums for the high definition formats increased the necessary limits. The maximum stream bitrate of HD DVD is over 30 Mbps. The maximum stream bitrate of Blu-ray Disc is over 40 Mbps.

bits per pixel The number of bits used to represent the color or intensity of each pixel in a bitmap. One bit enables only two values (black and white), two bits enable four values, and so on. Bits per pixel is also referred to as color depth or bit depth.

Bit shaving When done intentionally, bit shaving is a technique that simply removes bit depth from an audio stream with the goal of achieving a smaller file that still sounds good.

bitstream Digital data, usually encoded, that is designed to be processed sequentially and continuously.

blocked-based video compression A video compression technique that defines blocks of pixels grouped together, that are similar between frames, and repeats them to save space. Common video artifacts seen from this technique are mosquito wings and blocking.

blockiness A term referring to the occasional blocky appearance of compressed video (an artifact). Blockiness is caused when the compression ratio is high enough that the averaging of pixels in 8×8 blocks becomes visible.

Blue Book The document that specifies the CD Extra interactive music CD format. The original CDV specification was also in a blue book. See Enhanced CD.

Blu-ray Disc (BD) Blu-ray Disc, often said just Blu-ray, is one of two competing video high definition optical disc formats. The technology uses a blue purple 405 nanometer laser. The other competing format is HD DVD.

Blu-ray Disc Founders (BDF) Blu-ray Disc Founders industry group developed the use of blue lasers for reading and writing discs. This Group is led by Sony, and the following companies are among the Board of Directors — Apple, Inc., Dell, Inc., Hewlett-Packard Company, Hitachi, Ltd., LG Electronics Inc., Matsushita Electric Industrial Co., Ltd., Mitsubishi Electric Corporation, Pioneer Corporation, Royal Philips Electronics, Samsung Electronics Co., Ltd; Sharp Corporation, Sun Microsystems, Inc., TDK Corporation, Thomson, 20th Century Fox, Walt Disney Company, and Warner Bros. Entertainment Inc.

Blu-ray Java (BD-J) In the Blu-ray format there are two technologies that enable the interactivity of the disc. One is called HDMV, and the other is called BD-J (Blu-ray Disc Java). The more powerful of the two is BD-J because it is based on the Sun Microsystems Java Programming Language.

BUDA Binding Unit Data Area In Blu-ray, this is an additional large storage space outside of the required player storage area (i.e., a USB drive). Typical files on BUDA would be, downloaded video, audio, or subtitle files.

.bup (Time Map backup) This is a file type in the HD DVD folder HVDVD_TS. A .bup file is an exact copy of the .map file which describes the time information of a video object.

burst A short segment of the color subcarrier in a composite signal that is inserted to help the composite video decoder regenerate the color subcarrier.

button A rectangular area appearing on the screen of a title, used to initiate navigation commands.

byte A unit of data or data storage space consisting of eight bits, commonly representing a single character. Digital data storage is usually measured in bytes, kilobytes, megabytes, etc.

C

CABAC (Context-Adaptive Binary Arithmetic Coding) A coding mechanism used in the AVC (H.264) video compression standard. This mechanism requires more processing than other mechanisms like CAVLC, but gives a better compression ratio.

capture In video, this is the process of changing a video analog signal into a digital file. This can be done with a video capture card and software.

CAV Constant angular velocity. Refers to rotating disc systems in which the rotation speed is kept constant, where the pickup head travels over a longer surface as it moves away from the center of the disc. The advantage of CAV is that the same amount of information is provided in one rotation of the disc. Contrast with CLV and ZCLV.

CAVLC (Context Adaptive Variable Length Coding) Many video coding compression standards use the CAVLC coding method. This coding technique varies based on the context of the picture.

CBR Constant bit rate. Data compressed into a stream with a fixed data rate. The amount of compression (such as quantization) is varied to match the allocated data rate, but as a result, quality may suffer during high-compression periods. In other words, the data rate is held constant, while quality is allowed to vary. Compare this to VBR.

CCA Copy Control Authority. System specifying if the content is allowed to be copied.

CCIR Rec. 601 A standard for digital video. The CCIR changed its name to ITU-R, and the standard is now properly called ITU-R BT.601.

CDV A combination of laserdisc and CD that places a section of CD-format audio on the beginning of the disc and a section of laserdisc-format video on the remainder of the disc.

CE (Consumer Electronics) The popular electricity-based products that are a means of entertainment for humans. Popular consumer electronics are — the television, a video game system, a stereo, a VCR, or a digital camera. Items that do not fit into this category are — the toaster, a blender, or an electric toothbrush.

cell In DVD-Video, a unit of video with a duration that is anywhere from a fraction of a second to several hours long. Cells enable the video to be grouped for sharing content among titles, interleaving for multiple angles, etc.

cell block A group of cells.

CEA The Consumer Electronics Association. A subsidiary of the Electronics Industry Association (EIA).

CGMS The Copy Guard Management System. A method of preventing copies or controlling the number of sequential copies allowed. CGMS/A is added to an analog signal (such as line 21 of NTSC). CGMS/D is added to a digital signal, such as IEEE 1394.

challenge key Data used in the authentication key exchange process between a ROM drive and a host computer, where one side determines if the other side contains the necessary authorized keys and algorithms for passing encrypted (scrambled) data.

channel A part of an audio track. Typically, one channel is allocated for each loudspeaker.

channel bit The bits stored on the disc after being modulated.

channel data The bits physically recorded on an optical disc after error-correction encoding and modulation. Because of the extra information and processing, channel data is larger than the user data contained within it.

Chapter list A list of timecodes that are marker points within a video piece that allow ease of navigation for a user.

chroma (C′) The nonlinear color component of a video signal, independent of the luma. It is identified by 'C prime' (C′), where prime indicates nonlinearity, but it is usually written as C because it's never linear in practice.

chrominance (C) The color component (hue and saturation) of light, independent of luminance. Technically, chrominance refers to the linear component of video, as opposed to the transformed nonlinear chroma component.

Cinevision A software package made by Sonic Solutions that is designed to encode video for the next generation HD formats Blu-ray and HD DVD. It will encode all three video codecs H.264 (AVC), VC-1, and MPEG-2.

CIRC Cross-interleaved Reed Solomon code. An error-correction coding method that overlaps small frames of data.

clamping area The area near the inner hole of a disc where the drive grips the disc in order to spin it.

closed captions Textual video overlays that are not normally visible, as opposed to open captions, which are a permanent part of the picture. Captions are usually a textual representation of the spoken audio. In the US, the official NTSC Closed Caption standard requires that all TVs larger than 13 inches include circuitry to decode and display caption information stored on line 21 of the video signal.

CLV Constant linear velocity. This refers to a rotating disc system in which the head moves over the disc surface at a constant velocity, requiring that the motor vary the rotation speed as the head travels in and out. The further the head is from the center of the disc, the slower the rotation. The advantage of CLV is that data density remains constant, optimizing the use of the surface area. Contrast this with CAV and ZCLV.

CMF (Cutting Master Format) A standard to describe the contents of an optical disc. It is generally used between an authoring facility and a replication facility. CMF is a subset of DDP® 2.0.

codec Coder/decoder. The circuitry or computer software that encodes and decodes a signal.

Coherent Acoustics The full name for the multi-channel audio format called DTS.

color banding A problem related to graphics created with computers. This problem occurs most visibly in slow gradients like the fade of the blue sky. It appears as hard lines (bands) as the shade of the color changes.

colorburst See burst.

color depth The number of levels of color (usually including luma and chroma) that can be represented by a pixel. It is generally expressed as a number of bits or a number of colors. The color depth of MPEG video in DVD is 24 bits, although the chroma component is shared across four pixels (averaging 12 actual bits per pixel).

color difference A pair of video signals that contain the color components minus the brightness component, usually B'-Y' and R'-Y' (G'-Y' is not used, since it generally carries less information). The color-difference signals for a black-and-white picture are zero. The advantage of color-difference signals is that the color component can be reduced more than the brightness (luma) component without being visually perceptible.

colorist Someone who operates a telecine machine to transfer film to video.

Common Area In the context of persistent storage on a high definition player such as an HD DVD or Blu-ray player, the Common Area is a place to store data that can be shared between multiple discs.

companded In digital data, first compressed and then expanded, resulting in nonlinear code representation.

component video A video system containing three separate color component signals, either red/green/blue (RGB) or chroma/color difference ($Y'C_bC_r$, $Y'P_bP_r$, YUV), in analog or digital form. The MPEG-2 encoding system used by DVD is based on color-difference component digital video.

composite video An analog video signal in which the luma and chroma components are combined (by frequency multiplexing), along with sync and burst. This is also called CVBS. Most televisions and VCRs have composite video connectors, which are usually colored yellow.

compression rate A measurement of how much something has been compressed in percent. Usually, the higher the percentage is, the better the quality of the image, and the bigger the file size.

configuration In HD DVD, the Playlist file located in the ADV_OBJ folder has a section tagged <Configuration>. This area defines initial player settings such as video aperture size, network content, and streaming buffer size.

configuration file In HD DVD, the configuration file is referring to the DISCID.DAT which is located in the ADV_OBJ folder. This file contains GUID's that identify the disc, the provider (studio), and network identification.

constant data rate or **constant bit rate** See CBR.

content code BD+ security software is also known as content code. This content code is included as a small piece of software that is included on the Blu-ray Disc, and is run by the Java Security Virtual Machine on the player.

Content ID In HD DVD, the Content ID is one of the three GUID's that identify a disc. It is stored on the DISCID.DAT file in the ADV_OBJ folder of every HD DVD video disc. One of its purposes is to uniquely identify the disc among many discs within a studio.

control area A part of the lead-in area on a disc containing one ECC block (16 sectors) repeated 192 times. The repeated ECC block holds information about the disc.

core substream In context of DTS-HD audio, a core substream is defined with 4 extension substreams. The core substream carries the signal that is backwards compatibility with legacy DTS equipment.

CPPM Content Protection for Prerecorded Media. Copy protection for DVD-Audio.

CPRM Content Protection for Recordable Media. Copy protection for writable DVD formats.

CPSA Content Protection System Architecture. An overall copy protection design for DVD.

CPTWG Copy Protection Technical Working Group. The industry body responsible for developing or approving DVD copy protection systems.

CPU Central processing unit. The integrated circuit chip that forms the brain of a computer or other electronic device.

crop To trim and remove a section of the video picture in order to make it conform to a different shape. Cropping is used in the pan and scan process, but not in the letterbox process.

CRT monitor Since televisions were invented, the moving images were viewed on CRT (cathode-ray tubes) monitors. CRT works by quickly shooting an electron gun in the back of the TV in horizontal lines down the back of the screen which is filled with glowing phosphor material. This form of monitor tends to be a big box so the gun can cover the screen in front of it.

CVBS Composite video baseband signal. This is a standard single-wire video, mixing luma and chroma signals together.

D

D5 D5 is a digital video format by Panasonic. D5 can be used to store 1080p, 1080i, 720p, and standard definition resolutions.

DAC Digital-to-analog converter. Circuitry that converts digital data (such as audio or video) to analog data.

DAE Digital audio extraction. Reading digital audio data directly from a disc.

DAT Digital audio tape. A magnetic audio tape format that uses PCM to store digitized audio or digital data.

Data Access Manager As outlined in the HD DVD Specification, the Data Access Manager is a defined component within an HD DVD player that manages access to the disc, to persistent storage, and to the network.

data rate The volume of data measured over time. The rate at which digital information can be conveyed. This is usually expressed as bits per second with notations of kbps (thousand/sec), Mbps (million/sec), and Gbps (billion/sec). Digital audio data rate is generally computed as the number of samples per second times the bit size of the sample. For example, the data rate of uncompressed 16-bit, 48-kHz, two-channel audio is 1536 kbps. The digital video bitrate is generally computed as the number of bits per pixel times the number of pixels per line times the number of lines per frame times the number of frames per second. Compression reduces the data rate. Digital data rate is sometimes inaccurately equated with bandwidth.

dB See decibel.

DBS Digital (or Direct) Broadcast Satellite The general term for 18-inch digital satellite systems.

DC Direct Current. The electrical current flowing in one direction only. Adopted in the video world to refer to a signal with zero frequency. Compare this to AC.

DCC Digital Compact Cassette. A digital audio tape format based on the popular compact cassette that was abandoned by Philips in 1996.

DCT Discrete Cosine Transform. An invertible, discrete, orthogonal transformation. A mathematical process used in MPEG video encoding to transform blocks of pixel values into blocks of spatial frequency values with lower-frequency components organized into the upper-left corner, allowing the high-frequency components in the lower-right corner to be discounted or discarded. DCT also stands for digital component technology, a videotape format.

DD Dolby Digital.

DDWG Digital Display Working Group. See DVI.

deblocking filter A deblocking filter is a tool in video compression to reduce blocking artifacts of video compression software. Blocking artifacts are a result of compression tools that run compression algorithms by slicing up an image in blocks to help in the compression process. The edges of these blocks sometimes become visible as a blocking artifact.

decibel (dB) Used commonly as a measurement of sound. 140 decibels (a gunshot) is considered dangerously loud and can damage hearing. 45 decibels (a refrigerator) is considered quiet.

Decklink A high definition video capture card, made by Blackmagic Design Pty. Ltd. that supports 10 bit SDI or HDMI video. In early 2007, this card cost approximately $1000.00 USD.

decimation A form of subsampling that discards existing samples (pixels, in the case of spatial decimation, or pictures, in the case of temporal decimation). The resulting information is reduced in size but may suffer from aliasing.

Declarative In respect to HD DVD, declarative programming references XML .xmu markup in HD DVD advanced application programming.

decode To reverse the transformation process of an encoding method. The process of converting a digitally encoded signal into its original form. Decoding processes are usually deterministic.

decoder A circuit that decodes compressed audio or video, taking an encoded input stream and producing output such as audio or video. Players use the decoders to recreate information that was compressed by systems such as MPEG-2 and Dolby Digital.

delta picture or **delta frame** A video picture based on the changes from the picture before (or after) it. MPEG P pictures and B pictures are examples. Contrast this with key picture.

de-packetizer To send video through a system like the internet, the data must be sent in chunks called packets. When the packets get to the other side (your computer) they are reassembled in the original order. A function or piece of software that performs this function can be called a de-packetizer.

deterministic A process or model in which the outcome does not depend upon chance, and a given input always produces the same output. Audio and video decoding processes are mostly deterministic.

Device ID A Device ID is a unique identifier for a component connected to a system. In this way, the system can communicate with the device. An example of a device that has an ID is a printer that is connected to a computer.

digital Expressed in digits. A set of discrete numeric values, as used by a computer. Analog information can be digitized by sampling.

digital signal processor (DSP) A digital circuit that can be programmed to perform digital data manipulation tasks such as decoding or audio effects.

digital video noise reduction (DVNR) Digitally removing noise from video by comparing frames in sequence to spot temporal aberrations.

directory The part of a disc that indicates which files are stored on the disc and where they are located.

DirectShow A software standard developed by Microsoft for the playback of digital video and audio in the Windows operating system. This has replaced the older MCI and Video for Windows software.

DIN The German Institute for Standardization.

directory The section of a disk that contains information about which files are stored and where.

DISCID.DAT disc configuration file In HD DVD, the DISCID.DAT file is found in the ADV_OBJ file and contains GUID values for the Provider ID, Content ID, and Disc ID. Using these values, data specific to the disc can be stored on the player like downloaded material, or bookmarks the user has made.

disc key A value used to encrypt and decrypt (scramble) a title key on discs.

disc menu The main menu of a disc from which titles are selected. This is also called the system menu or title selection menu.

discrete cosine transform See DCT.

discrete surround sound Audio in which each channel is stored and transmitted separate from and independent of other channels. Multiple independent channels, directed to loudspeakers in front of and behind the listener, enable precise control of the soundfield in order to generate localized sounds and simulate moving sound sources.

display rate The number of times per second the image in a video system is refreshed. Progressive scan systems such as film change the image once per frame. Interlace scan systems such as television change the image twice per frame, with two fields in each frame. Film has a frame rate of 24 fps, but each frame is shown twice by the projector for a display rate of 48 fps. 525/60 (NTSC) television has a rate of 29.97 frames per second (59.94 fields per second). 625/50 (PAL/SECAM) television has a rate of 25 frames per second (50 fields per second).

DivX Digital Video Express. A short-lived pay-per-viewing-period variation of DVD. The term has been resurrected for DivX, which is a proprietary encode/decode scheme for audio/video content and presentation.

DL (Dual Layer) To get more physical data on an optical disc, a second layer of data can be added. To read the second layer, the laser refocuses through the top layer, to the bottom layer data.

DLT Digital linear tape. A digital archive standard using half-inch tapes, may be used for submitting a premastered disc image to a replication service.

Dolby Digital (AC-3) Dolby Digital (AC-3) is a digital audio codec by Dolby that first appeared in the consumer market in Laserdiscs. The typical file extension for these audio files is .ac3.

Dolby® Digital Plus (DD +) Dolby® has entered the high definition consumer market with higher definition audio based on the AC-3 codec. It offers higher bitrates and more audio channels. It is important to note that DD+ is a lossy audio codec. The typical file extension for these audio files is .ec3. This technology is directed toward many HD delivery technologies like optical media, online content, satellite content, terrestrial broadcast and more.

Dolby Digital EX Using matrix encoding the Left and Right surround signal, Dolby Digital EX extends the Dolby Digital 5.1 surround capabilities to up to 7.1 channels intended for a fuller surround experience similar to the theatre experience. The typical file extension for these audio files is .ac3.

Dolby® Lossless MLP Lossless is the another name for this Dolby® technology. It offers up to six channels of lossless 96 kHz/24 bit audio, which means that it will sound exactly as it does in the studio. This technology has been approved for use in Blu-ray and HD DVD discs. The typical file extension for these audio files is .mlp.

Dolby Pro Logic The technique (or the circuit that applies the technique) of extracting surround audio channels from a matrix-encoded audio signal. Dolby Pro Logic is a decoding technique only, but it is often mistakenly used to refer to Dolby Surround audio encoding.

Dolby Surround The standard for matrix encoding surround-sound channels in a stereo signal by applying a set of defined mathematical functions when combining center and surround channels with left and right channels. The center and surround channels can then be extracted by a decoder such as a Dolby Pro Logic circuit that applies the inverse of the mathematical functions. A Dolby Surround decoder extracts surround channels, while a Dolby Pro Logic decoder uses additional processing to create a center channel. The process is essentially independent of the recording or transmission format. Both Dolby Digital and MPEG audio compression systems are compatible with Dolby Surround audio.

Dolby® True HD Another approved Dolby technology for high definition media is Dolby® TrueHD. This lossless technology supports up to 8.1 channels and can reach up to 18 Mbps variable bitrate. To enjoy all 8.1 channels at home, the user must have HDMI enabled equipment. This product is targeted towards HD optical media.

domain A domain is a large area that contains several groups of video. While in a domain, there are certain rules that are forced. In DVD, for example, there are three domains — the menu domain, the title domain, and the video manager domain.

downmix To convert a multichannel audio track into a two-channel stereo track by combining the channels with the Dolby Surround process. Players are required to provide downmixed audio output from Dolby Digital audio tracks.

downsampling See subsampling.

double-sided disc A type of disc on which data is recorded on both sides.

drop frame timecode The method of timecode computation that accounts for the reality of there being only 29.97 frames of video per second. The 0.03 frame is visually insignificant, but mathematically very significant. A one-hour video program will have 107,892 frames of video (29.97 frames per second x 60 seconds x 60 minutes). The drop frame timecode method of accommodating reality was developed where 2 frames are dropped from the numerical count for every minute in an hour except for every tenth minute, when no frames are dropped. See also non-drop frame timecode and timecode.

DRC See dynamic range compression.

DSD Direct Stream Digital. An uncompressed audio bitstream coding method developed by Sony. It is used as an alternative to PCM.

DSI Data search information. Navigation and search information contained in the DVD-Video data stream. DSI and PCI together make up an overhead of about one Mbps.

DSP Digital signal processor (or processing).

DTS Digital Theater Sound. A perceptual audio-coding system developed for theaters. A competitor to Dolby Digital and an optional audio track format for DVD-Video and DVD-Audio.

DTS 96/24 A typical DTS-HD stream reference where 96 represents the sample rate (96 kHz) and 24 represents the bit depth (24 bit).

DTS-ES A version of DTS decoding that is compatible with 6.1-channel Dolby Surround EX. DTS-ES Discrete is a variation of DTS encoding and decoding that carries a discrete rear center channel instead of a matrixed channel.

DTS-HD DTS (Digital Theater System) has built onto their core DTS codec and extended it to support up to 8.1 channels for the new HD formats. This is a lossy format. The typical file extension for DTS-HD is .dtshd.

DTS-HD LBR DTS-HD LBR (Low Bit Rate) This DTS format is intended for high quality DTS audio for the secondary audio streams defined in Blu-ray and HD DVD optical discs.

DTS-HD Lossless DTS-HD Lossless audio was the original name for DTS-HD Master Audio.

DTS-HD Master Audio DTS-HD Master Audio is intended for Blu-ray and HD DVD and is an exact lossless recreation of the sound mastered in the studio. It offers up to 7.1 channels of audio. It offers up to 24.5 Mbps for Blu-ray and 18.0 Mbps for HD DVD. This format can support an unlimited number of channels.

DTV Digital television. In general, any system that encodes video and audio in digital form. Specifically, the Digital Television System proposed by the ATSC or the digital TV standard proposed by the Digital TV Team founded by Microsoft, Intel, and Compaq.

DV Digital Video. This usually refers to the digital videocassette standard developed by Sony and JVC.

DVB Digital video broadcast. A European standard for broadcast, cable, and digital satellite video transmission.

DVC Digital video cassette. The early name for DV.

DVCAM Sony's proprietary version of DV.

DVCD Double Video Compact Disc. A long-playing (100-minute) variation of VCD.

DVCPRO Matsushita's proprietary version of DV.

DVD An acronym that officially stands for nothing but is often expanded as Digital Video Disc or Digital Versatile Disc. The audio/video/data storage system based on 12- and 8-cm optical discs.

DVD-Audio (DVD-A) The audio-only format of DVD that primarily uses PCM audio with MLP encoding, along with an optional subset of DVD-Video features.

DVD-R One of three recordable DVD formats approved by the DVD forum. It has two sister technologies: **DVD-RW** rewritable, **DVD-R DL** Dual Layer (8.74 GB of data)

DVD+R One of three recordable DVD formats approved by the DVD Forum, developed by the DVD+RW Alliance. DVD+R has two sister technologies: **DVD+RW** rewritable, **DVD+R DL** Dual Layer (8.74 GB of data)

DVD-RAM One of three recordable DVD formats approved by the DVD Forum. DVD-RAM is used mostly for data storage and is not compatible with most DVD video players.

DVD-ROM The base format of DVD-ROM stands for read-only memory, referring to the fact that standard DVD-ROM and DVD-Video discs can't be recorded on. A DVD-ROM can store essentially any form of digital data.

DVD-Video (DVD-V) A standard for storing and reproducing audio and video on DVD-ROM discs, based on MPEG video, Dolby Digital and MPEG audio, and other proprietary data formats.

DVI Digital Visual Interface. The digital video interface standard developed by the Digital Display Working Group (DDWG). A replacement for analog VGA monitor interface.

DVNR See digital video noise reduction.

dye polymer The chemical used in DVD-R and CD-R media that darkens when heated by a high-power laser.

dye-sublimation An optical disc recording technology that uses a high-powered laser to burn readable marks into a layer of organic dye. Other recording formats include magneto-optical and phase-change.

dynamic range The difference between the loudest and softest sound in an audio signal. The dynamic range of digital audio is determined by the sample size. Increasing the sample size does not allow louder sounds; it increases the resolution of the signal, thus allowing softer sounds to be separated from the noise floor (and allowing more amplification with less distortion). Therefore, the dynamic range refers to the difference between the maximum level of distortion-free signal and the minimum limit reproducible by the equipment.

dynamic range compression A technique of reducing the range between loud and soft sounds in order to make dialog more audible, especially when listening at low volume levels. It is used in the downmix process of multichannel Dolby Digital sound tracks.

E

ECC See error-correction code.

ECD Error-detection and correction code. See error-correction code.

ECMA European Computer Manufacturers Association.

ECMAScript ECMAScript is the programming language used for dynamic content in HD DVD, Sony UMD, and the World Wide Web.

edge enhancement When films are transferred to video in preparation for DVD encoding, they are commonly run through digital processes that attempt to clean up the picture. These processes include noise reduction (DVNR) and image enhancement. Enhancement increases the contrast (similar to the effect of the sharpen or unsharp mask filters in Photoshop), but it can tend to overdo areas of transition between light and dark or different colors. This causes a chiseled look or a ringing effect like the haloes you see around streetlights when driving in the rain. Video noise reduction is a good thing when done well, because it can remove scratches, spots, and other defects from the original film. Enhancement, which is rarely done well, is a bad thing. The video may look sharper and clearer to the casual observer, but fine tonal details of the original picture are altered and lost.

EDTV Enhanced-definition television. A system that uses existing transmission equipment to send an enhanced signal that looks the same on existing receivers, but carries additional information to improve the picture quality on enhanced receivers. PALPlus is an example of EDTV. Contrast this with HDTV and IDTV.

EFM Eight-to-14 modulation. A modulation method used by CD. The 8/16 modulation used by DVD is sometimes called EFM plus.

EIA Electronics Industry Association.

E-IDE Enhanced Integrated Drive Electronics. These are extensions to the IDE standard that provide faster data transfers and enable access to larger drives, including CD-ROM and tape drives, using ATAPI. E-IDE was adopted as a standard by ANSI in 1994. ANSI calls it Advanced Technology Attachment-2 (ATA-2) or Fast ATA.

elementary stream A general term for a coded bitstream such as audio or video. Elementary streams are made up of packs of packets.

emulate To test the function of a disc on a computer after formatting a complete disc image.

ENAV Enhanced DVD Specifications, Enhanced Navigation After the success of DVD, another book was added to the DVD spec called DVD ENAV. It was not widely adopted.

encode To transform data for storage or transmission, usually in such a way that redundancies are eliminated or complexity is reduced. Most compression is based on one or more encoding methods. Data such as audio or video is encoded for efficient storage or transmission and is decoded for access or display.

encoder 1) A circuit or a program that encodes audio or video.

Enhanced CD A music CD that has additional computer software and can be played in a music player or read by a computer. Also called CD Extra, CD Plus, hybrid CD, interactive music CD, mixed-mode CD, pre-gap CD, or track-zero CD.

entropy coding Variable-length, lossless coding of a digital signal to reduce redundancy. MPEG-2, DTS, and Dolby Digital apply entropy coding after the quantization step. MLP also uses entropy coding.

error-correction code Additional information added to data to enable errors to be detected and possibly corrected.

Ethernet An ethernet is a typical way to connect computers together in a LAN (Local Area

Network) to share files. HD DVD players and Blu-ray players have ethernet ports that will enable next-generation player access to the Internet. DVD players do not have this capability.

ETSI European Telecommunications Standards Institute.

EVO enhanced video object file In HD DVD, this is the video file type found in the HVDVD_TS folder. An .evo file contains video, audio, sub-video, sub-audio, and subtitle data.

EVOB (Enhanced Video Object) A program stream made up of elementary streams. Examples of elementary streams are video, audio, subtitles, etc.

EVOBS enhanced video object set A group of EVOBs (the s is for plurality). There are three types of EVOBS — Video Manager EVOBS, Menu Space EVOBS, and Title Space EVOBS.

EVOBU enhanced video object units A part of an EVOB that ranges between .4 seconds and 1.001 seconds.

Extension substream In context of DTS-HD audio, a core substream is defined with 4 extension substreams. The core substream carries the legacy DTS audio asset, while the extension substreams can contain up to 8 audio assets.

F

father The metal master disc formed by electroplating the glass master. The father disc is used to make mother discs from which multiple stampers (sons) can be made.

field A set of alternating scan lines in an interlaced video picture. A frame is made of a top (odd) field and a bottom (even) field.

FIFO Buffer FIFO stands for "First In/First Out". A buffer is a place to store data before processing. A FIFO buffer is a good mechanism for collecting information before it is presented. An example of this is in the process of reading data from an optical disc, it will be placed in a buffer before it is presented to the screen.

File Cache This is a very fast, but usually small area of computer memory. In the context of Blu-ray, Java .jar menu files are stored into File Cache. In the context of HD DVD, .aca menu files are stored in File Cache.

File Cache Manager A File Cache Manager is responsible for storing and discarding data in the file cache.

file system A defined way of storing files, directories, and information about such files and directories on a data storage device.

Film Grain Technology A technology invented by Thomson that is employed in HD content to reduce the bitrate of a feature. Thomson has created a technology to add grain after the film has been compressed, and this will achieve a much lower bitrate.

FireWire A standard for the transmission of digital data between external peripherals, including consumer audio and video devices. The official name is IEEE 1394, based on the original FireWire design by Apple Computer.

First Play title This is the in-point of control when a disc is initialized in a player.

fixed rate Information flow at a constant volume over time. See CBR.

Fixed Storage A data storage unit that is meant to stay in one place. It is usually accessible through a network.

forced display A feature that enables subpictures to be displayed even if the player's sub-picture display mode is turned off. It is also designed to show subtitles in a scene where the language is different from the native language of the film.

formatting The process of preparing a data storage medium to store data in a particular way.

FP, First Play First Play discs generally have no menus and usually no extra content on them. The user simply put the disc in a player, and it immediately plays through the feature with no main menu or sub menus.

fps Frames per second. A measure of the rate at which pictures are shown to create a motion video image. In NTSC and PAL video, each frame is made up of two interlaced fields.

frame The piece of a video signal containing the spatial detail of one complete image, or the entire set of scan lines. In an interlaced system, a frame contains two fields.

frame doubler A video processor that increases the frame rate (display rate) in order to create a smoother-looking video display. Compare this to line doubler.

frame rate The frequency of discrete images. This is usually measured in frames per second (fps). Film has a rate of 24 frames per second, but it usually must be adjusted to match the display rate of a video system.

frequency The number of repetitions of a phenomenon in a given amount of time. The number of complete cycles of a periodic process occurring per unit of time.

FRExt (Fidelity Range Extensions) The Fidelity Range Extensions outline further capabilities of H.264/AVC over the older less superior MPEG-2 coding scheme. These extensions explain Fidelity increase by as much as 3:1 compared to MPEG-2.

FTP (File Transfer Protocol) One of the several Internet protocols used to connect two computers over the Internet using TCP/IP so that the user of one computer can transfer files and perform file commands on the other computer.

G

G or Giga An SI prefix for denominations of one billion (10^9).

G byte One billion (10^9) bytes. Not to be confused with GB or gigabyte (2^{30} bytes).

GB Gigabyte.

Gbps Gigabits/second. Billions (10^9) of bits per second.

gigabyte 1,073,741,824 (2^{30}) bytes. (See Introduction, for more information.)

GEM Globally Executable MHP is a specification that extracts the interoperable "common core" from the MHP specification, in order to create a transport-neutral platform for applications that can run globally, over any network transport or on optical discs. See MHP.

GOF Group of Audio Frames, a data area containing 20 audio frames of Linear PCM audio.

GOP Group of pictures. In MPEG video, one or more I pictures followed by P and B pictures. A GOP is the atomic unit of MPEG video access. GOPs are limited in DVD-Video to maximums of 18 frames for 525/60 and 15 frames for 625/50.

GPRM (General Parameters) In DVD, HD DVD, and Blu-ray, GPRM's are 16 to 32 bit registers that store integer values in the player memory. These are intended as variables for authors to store information like, audio setting, game values, disc flow, etc. DVD has 16 GPRM's for use, Blu-ray has 4096 for use, and HD DVD has 64 for use.

gray market Dealers and distributors who sell equipment without proper authorization from the manufacturer.

Green Book The document developed in 1987 by Philips and Sony as an extension to CD-ROM XA for the CD-i system.

GUIDs (Globally Unique Identifiers) A number intended to be unique across space and time. These numbers are used in HD DVD and Blu-ray to uniquely identify discs, providers, and other information. Microsoft supports tools that can generate these numbers which have the following signature — 88888888-4444-4444- 4444-121212121212

H

Hard Sync In HD DVD, two types of sync values that an Advanced Application can have — Hard Sync, and Soft Sync. Hard Sync requires the application to be loaded into the file cache before the video begins playback. If the application is loaded in the middle of the Title Time Line, then a Hard Sync application will cause an audio and video pause.

HDAC High Definition Advanced Content.

HDCAM SR Sony's professional digital video format. HDCAM SR is a 10-bit 4:4:4 RGB up to 1080p format that can be used both as a storage format and recording format.

HDCD High-definition Compatible Digital. A proprietary method of enhancing audio on CDs.

HDD HDD (Hard Disk Drive) is more commonly referred to as hard disk. There is typically at least one hard disk in a Personal Computer. Some digital video cameras now record directly to hard disk.

HD DVD HD DVD is a new high definition video format intended for use in a home theater system. It offers high definition picture quality up to six times better than DVD, and also offers new high definition audio capabilities that offer up to 8.1 discrete channels of audio. To fit more data on a disc, a smaller blue laser is used for this format. Each layer of an HD DVD disc can store 15 GB of data. As of this writing, HD DVD is in a format war with Blu-ray Disc.

HD DVD 15 A single layer HD DVD disc with a data capacity of 15 GB.

HD DVD 30 A dual layer HD DVD disc with a data capacity of 30 GB.

HD DVD_TS A folder in the root of an HD DVD disc that contains the video objects.

HD DVD-Video A format supported by the HD DVD Forum intended for high definition video.

HD DVD-VR (Video Recording) A format supported by the DVD Forum that allows video recording onto an HD DVD disc.

HDMI HDMI is a format for a new type of cable that can carry both video and audio signals that will support the latest high definition signals. HDMI 1.3 supports the most advanced audio and video that Blu-ray and HD DVD can deliver.

HD Movie Mode (HDMV) HDMV is one of two methods to deliver interactive menus in the Blu-ray format. BD-J is the second method, which is more advanced than HDMV.

HD SDI HD SDI (High Definition Serial Digital Interface) is a standard intended for high definition video with a data rate of 1.485 Gbit/s.

HDTV HDTV (High Definition Television) is a digital format for broadcast television. It supports up to 1080i, and all television in the United States will transmit HDTV in the future. An HD tuner is needed in the home to receive this signal.

H/DTV High-definition/digital television. A combination of acronyms that refers to both HDTV and DTV systems.

HFS Hierarchical file system. A file system used by Apple Computer's Mac OS operating system.

highlight A method of display that emphasizes a selected item on a menu screen by increasing the brightness level to show which function is executed.

High Sierra The original file system standard developed for CD-ROM, later modified and adopted as ISO 9660.

HLI highlight information In DVD and Standard Content HD DVD, highlight information contains information such as button number, highlight timing, color palette, etc. This method of enabling buttons is not used in Advanced Content HD DVD, and has been replaced with a more advanced Web-like model using XML and JavaScript.

HSF See High Sierra.

HTML Hypertext Markup Language. This is a tagging specification based on the standard generalized markup language (SGML) for formatting text to be transmitted over the Internet and displayed by client software.

hue The color of light or a pixel. The property of color determined by the dominant wavelength of light.

Huffman coding A lossless compression technique of assigning variable-length codes to a known set of values. Values that occur most frequently are assigned the shortest codes. MPEG uses a variation of Huffman coding with fixed code tables, often called variable-length coding (VLC).

HVA0001.VTI Advanced VTS information file In HD DVD video, One of several information files that describe the video objects. It is located in the HVDVD_TS folder.

HVDVD_TS One of two folders in the root of a HD DVD video disc. This folder contains the video objects, while the ADV_OBJ folder contains the Advanced Objects such as the interactive menus.

I

I picture or **I frame** In MPEG video, this is an intra picture that is encoded independent of other pictures (see intraframe). Transform coding (DCT, quantization, and VLC) is used with no motion compensation, resulting in only moderate compression. I pictures provide a reference point for dependent P pictures and B pictures and enable random access into the compressed video stream.

i.Link Trademarked Sony name for IEEE 1394.

IDE Integrated Drive Electronics. An internal bus or standard electronic interface between a computer and internal block storage devices. IDE was adopted as a standard by ANSI in November 1990. ANSI calls it Advanced Technology Attachment (ATA). See E-IDE and ATAPI.

IDTV Improved-definition television. A television receiver that improves the apparent quality of the picture from a standard video signal by using techniques such as frame doubling, line doubling, and digital signal processing.

IEC International Electrotechnical Commission.

IEEE Institute of Electrical and Electronics Engineers, an electronics standards body.

IEEE 1394 A standard for the transmission of digital data between external peripherals, including consumer audio and video devices. Also known as FireWire.

IFE In-flight entertainment.

iHD Name historically given to specification during development discussion in the DVD forum. Not to be confused with HDi which is Microsoft's trademark name for their specific implemention, or iHD.org, Internet High Definition, a standards organization for internet video.

immediate value In programming, immediate is a value. In the following code example, 3 is the immediate: X = 3

I-MPEG Intraframe MPEG. An unofficial variation of MPEG video encoding that uses only intraframe compression.

inter Spanning more than one thing.

interactive The capability to respond to commands issued by a user and to prompt a user to issue commands.

inter-coded pictures Coding that happens across more than one frame. For example, if you have a picture of a newscaster on a black background, coding the black part of the picture over multiple frames (inter-coding) helps achieve a good data rate.

interlace A video scanning system in which alternating lines are transmitted so that half a picture is displayed each time the scanning beam moves down the screen. An interlaced frame comprises two fields.

interleave To arrange data in alternating chunks so that selected parts can be extracted while other parts are skipped, or so that each chunk carries a piece of a different data stream.

interpolate To increase the pixels, scan lines, or pictures when scaling an image or a video stream by averaging adjacent pixels, lines, or frames to create additional inserted pixels or frames. This generally causes a softening of still images and a blurring of motion images because no new information is created. Compare this to filter.

intra Within one thing.

intraframe Something that occurs within a single frame of video. Intraframe compression does not reduce temporal redundancy but enables each frame to be independently manipulated or accessed. See I picture. Compare this to interframe.

intra-picture pulsing An I-Frame (intra-picture frame) is the beginning of a GOP (Group of Pictures) in an MPEG encoded stream. The I-Frame marks the beginning of an encode chunk. Inferior encoding tools were sometimes not efficient when encoding between the last frame of a GOP and the first I-Frame of the next GOP causing a visible pulsing of the video image.

inverse telecine The reverse of 2:3 pulldown, where the frames that were duplicated to create 60-fields/second video from 24-frames/second film source are removed.

iso isolated camera or recorder.

ISO International Organization for Standardization.

ISO-8959-1 This is the nomenclature for the Unicode Character Code UTF-8. This Character Set holds most of the world's common characters (Tens of thousands of characters), and is used extensively in the definition of many XML documents.

ISO 9660 The international standard for the file system used by CD-ROM. ISO 9660 allows filenames of only eight characters plus a three-character extension.

ISO/IEC 13818-2 The ISO/IEC nomenclature for H.262 MPEG 2 encoding scheme. This encoding scheme is part of the DVD Specification for video.

ISO/IEC MPEG-4 Part 10 (ISO/IEC 14496-10) This is the nomenclature for one of the video encoding schemes for the high definition technologies including HD DVD and Blu-ray.

ISRC International Standard Recording Code.

IT (Information Technology) A broad term covering all aspects of digital information in use today. This term is also used as a department name in many organizations, whose job is caretaker of the computer network, software, and databases within the company. When your computer is broken, call IT.

ITU International Telecommunication Union.

ITU-T H.264 Together with ISO/IEC MPEG group, a very powerful encoding scheme (H.264/MPEG-4 AVC) has been developed that is a part of HD DVD and Blu-ray specifications.

ITU-T Video Coding Experts Group (VCEG) This group has been very instrumental in developing video coding mechanisms in use today.

ITU-R BT.601 The international standard specifying the format of digital component video.

J

Java A programming language with specific features designed for use with the Internet and HTML.

JavaScript A programming language developed by Netscape. This language has been incorporated into some video platforms to enable interactive content like dynamic menus. Versions of JavaScript can be found in HD DVD and PSP UMD discs.

.jar A file format used for Java programming applications.

jewel box The plastic clamshell case that holds a disc.

jitter A temporal variation in a signal from an ideal reference clock. Many kinds of jitter can occur, including sample jitter, channel jitter, and interface jitter.

JPEG Joint Photographic Experts Group. The international committee that created its namesake standard for compressing still images.

JS files JS files are files with JavaScript code in them.

JVT (Joint Video Team) The ISO/IEC MPEG Group and ITU-T Video Coding Experts Group joined together as JVT to build a very powerful coding scheme called MPEG-4.

K

k or Kilo An SI prefix for denominations of one thousand (10^3). Also used, in capital form, for 1024 bytes of computer data (see kilobyte).

k byte One thousand (10^3) bytes. Not to be confused with KB or kilobyte (2^{10} bytes). Note the small "k."

karaoke Literally, "empty orchestra". The social sensation from Japan where sufficiently inebriated people embarrass themselves in public by singing along to a music track. Karaoke was largely responsible for the success of laserdisc in Japan, thus supporting it elsewhere.

KB Kilobyte.

kbps Kilobits/second. Thousands (10^3) of bits per second.

key picture or key frame A video picture containing the entire content of the image (intraframe encoding), rather than the difference between it and another image (interframe encoding). MPEG I pictures are key pictures. Contrast this with delta picture.

kHz Kilohertz. A unit of frequency measurement. It is one thousand cycles (repetitions) per second or 1,000 Hertz.

kilobyte 1024 (2^{10}) bytes. (See Introduction, for more information.)

L

land The raised area of an optical disc.

Language Unit In DVD programming, a Language Unit is in menu space that is associated with a language. If a user were to press menu on their remote control, they would be taken to the Language Unit that their player was set to (English, French, etc.).

laserdisc A 12-inch (or 8-inch) optical disc that holds analog video (using an FM signal) and both analog and digital (PCM) audio. Laserdisc was a precursor to DVD.

latency The time it takes to get a packet through a system. If there are problems in the system causing it to slow down, this is known as latency.

layer The plane of a disc where information is recorded in a pattern of microscopic pits. Each substrate of a disc can contain one or two layers. The first layer, closest to the readout surface, is layer 0; the second is layer 1.

layout In the authoring process, a layout occurs after multiplexing is complete and a fully qualified root structure of a disc is made. This layout can be played as a disc would play, but it is not yet an image intended for an optical disc.

LBR (Low Bit rate extension) DTS-HD LBR (Low Bit Rate) This DTS format is intended for high quality DTS audio for the secondary audio defined in Blu-ray and HD DVD discs.

LCD monitor LCD (Liquid Crystal Display) monitors are a flat panel, energy efficient display that are made of thousands of tiny crystal coils that control the amount of light passing through them. They were used extensively on laptops and are a popular technology for flat panel widescreen high definition television sets.

lead in The physical area that is 1.2 mm or wider preceding the data area on a disc. The lead in contains sync sectors and control data including disc keys and other information.

lead out On a single-layer disc or PTP dual-layer disc, this is the physical area 1.0 mm or wider toward the outside of the disc following the data area. On an OTP dual-layer disc, this is the physical area 1.2 mm or wider at the inside of the disc following the recorded data area (which is read from the outside toward the inside on the second layer).

letterbox The process or form of video where black horizontal mattes are added to the top and bottom of the display area to create a frame in which to display video using an aspect ratio different than that of the display. The letterbox method preserves the entire video picture, as opposed to pan and scan. Players can automatically letterbox an anamorphic widescreen picture for display on a standard 4:3 TV.

letterbox filter The circuitry in a player that reduces the vertical size of anamorphic widescreen video (combining every four lines into three) and adds black mattes at the top and bottom.

level In MPEG-2, levels specify parameters such as resolution, bitrate, and frame rate. Compare this to profile.

Line 21 The specific part of the NTSC video signal that carries the teletext information for closed captioning.

linear PCM A coded representation of digital data that is not compressed. Linear PCM spreads values evenly across the range from highest to lowest, as opposed to nonlinear (that which is first compressed then expanded, see companded) PCM that allocates more values to more important frequency ranges.

line doubler A video processor that doubles the number of lines in the scanning system in order to create a display with scan lines that are less visible. Some line doublers convert from an interlaced to a progressive scan.

lines of horizontal resolution Sometimes abbreviated as TVL (TV lines) or LoHR, this is a common but subjective measurement of the visually resolvable horizontal detail of an analog video system, measured in half-cycles per picture height. Each cycle is a pair of vertical lines, one black and one white. The measurement is usually made by viewing a test pattern to determine where the black and white lines blur into gray. The resolution of VHS video is commonly gauged at 240 lines of horizontal resolution, broadcast video at 330, laserdisc at 425, and DVD at 500 to 540. Because the measurement is relative to picture height, the aspect ratio must be taken into account when determining the number of vertical units (roughly equivalent to pixels) that can be displayed across the width of the display. For example, an aspect ratio of 1.33 multiplied by 540 gives 720 pixels.

locale See regional code.

localization The process of translating something into another language. In terms of optical disc creation, menus must be prepared to be viewed in other countries. The process of preparing a disc to be readable and usable in a foreign language is called localization.

logical An artificial structure or organization of information created for convenience of access or reference, usually different from the physical structure or organization. For example, the application specifications of DVD (the way information is organized and stored) are logical formats.

logical unit A physical or virtual peripheral device, such as a DVD-ROM drive.

lossless compression Compression techniques that enable the original data to be recreated without loss. Contrast with lossy compression.

lossy compression Compression techniques that achieve very high compression ratios by permanently removing data while preserving as much significant information as possible. Lossy compression includes perceptual coding techniques that attempt to limit the data loss so that it is least likely to be noticed by human perception.

lossless extension (XLL) Increase the dimension of the JPEG image in one or more dimensions. The size of the extension in each direction must be a multiple of the MCU (usually $\times 8$, but $\times 16$ if chroma subsampling is used). For example, one can place a 16 pixel border around a digital photo, leaving the original content intact perfectly. In JPEG club terminology, this operation is still called "lossless crop", even though it is a crop with negative dimensions. The JPEGclub page doesn't make this feature particularly obvious.

lower third In television, this refers to the lower third of the screen that contains text. For example, when a person appears on the screen, their name and title will appear in the lower third. CNN runs news text in the lower third.

LP Long-playing record. An audio recording on a plastic platter turning at 33 1/3 rpm and read by a stylus.

LPCM See linear PCM.

L_t/R_t Left total/right total. Four surround channels matrixed into two channels. The mandatory downmixing method in Dolby Digital decoders.

luma (Y′) The brightness component of a color video image (also called the grayscale, monochrome, or black-and-white component) with nonlinear luminance. The standard luma signal is computed from nonlinear RGB as $Y' = 0.299 \, R' + 0.587 \, G' + 0.114 \, B'$.

luminance (Y) Loosely, the sum of RGB tristimulus values corresponding to brightness. This may refer to a linear signal or (incorrectly) a nonlinear signal.

M

M or **Mega** An SI prefix for denominations of one million (10^6).

Mac OS The operating system used by Apple Macintosh computers.

macroblock In MPEG MP@ML, the four 8×8 blocks of luma information and two 8×8 blocks of chroma information that form a 16×16 area of a video frame.

macroblocking An MPEG artifact. See blocking.

Macrovision An anti-taping process that modifies a signal so that it appears unchanged on most televisions but is distorted and unwatchable when played back from a videotape recording. Macrovision takes advantage of the characteristics of AGC circuits and burst decoder circuits in VCRs to interfere with the recording process.

magneto-optical A recordable disc technology using a laser to heat spots that are altered by a magnetic field. Other formats include dye-sublimation and phase-change.

manifest In HD DVD, this file (.xmf) describes the assets that will be used in an Advanced Application. It is generally included into the archive file (.aca)

.MAP time map In HD DVD, this file accompanies an .EVO file and describes its time characteristics.

markup This is a shortened term for Markup Language. In HD DVD there is a file with a file extension .xmu which defines the layout of objects on the screen. This file is sometimes called Markup. Generally, it is any tagged (ex: <Tag>) file that describes the layout of things on a page. Web pages are created with Markup called HTML.

master The metal disc used to stamp replicas of optical discs, or the tape used to make additional recordings.

mastering The process of replicating optical discs by injecting liquid plastic into a mold containing a master. This is often used inaccurately to refer to premastering.

matrix encoding The technique of combining additional surround-sound channels into a conventional stereo signal. See Dolby Surround.

matte An area of a video display or motion picture that is covered (usually in black) or omitted in order to create a differently shaped area within the picture frame.

MB Megabyte.

Mbps Megabits/second. Millions (10^6) of bits per second.

M byte One million (10^6) bytes. Not to be confused with MB or megabyte (2^{20} bytes).

Media Attributes List In HD DVD this is found in the Playlist file (.xpl) which describes a list of video and audio files that will be on the disc. The tag as is follows: <MediaAttributeList>

megabyte 1,048,576 (2^{20}) bytes. (See Introduction, for more information.)

megapixel An image or display format with a resolution of approximately one million pixels.

memory Data storage used by computers or other digital electronics systems. Read-only memory (ROM) permanently stores data or software program instructions. New data cannot be written to ROM. Random-access memory (RAM) temporarily stores data, including digital audio and video, while it is being manipulated and holds software application programs while they are being executed. Data can be read from and written to RAM. Other long-term memory includes hard disks, floppy disks, digital CD formats (CD-ROM, CD-R, and CD-RW), and DVD formats (DVD-ROM, DVD-R, DVD+R, DVD±RW and DVD-RAM).

menu A graphic image provided in a title to assist in picture, audio, subpicture and multi-angle selections recorded on a DVD Video disc.

menu state The condition of a player when a menu is presented.

metadata Metadata is information about data. For example — Author, Publisher, Copyright are metadata of a book.

MHP Multimedia Home Platform is a Java-based specification for interactivity and television. MHP and MHP-based systems are used for cable, satellite, terrestrial TV and IPTV systems, as well as BD-J.

MHz One million (10^6) Hertz.

Microsoft Windows The leading operating system for Intel CPU-based computers developed by Microsoft.

middle area On a dual-layer OTP disc, the physical area 1.0 mm or wider on both layers, adjacent to the outside of the data area.

MKB (Media Key Block) A set of keys used in CPPM and CPRM to authenticate players.

MLP (Meridian Lossless Packing) A lossless compression technique (used by DVD-Audio) that removes redundancy from PCM audio signals to achieve a compression ratio of about

2:1 while allowing the signal to be perfectly recreated by the MLP decoder.

MO Magneto-optical rewritable discs.

modulation Replacing patterns of bits with different (usually larger) patterns designed to control the characteristics of the data signal. DVD uses 8/16 modulation, where each set of eight bits is replaced by 16 bits before being written onto the disc.

mosquitoes A term referring to the fuzzy dots that can appear around sharp edges (high spatial frequencies) after video compression. Also known as the Gibbs Effect.

mother The metal discs produced from mirror images of the father disc in the replication process. Mothers are used to make stampers, often called sons.

motion compensation In video decoding, the application of motion vectors to already-decoded blocks in order to construct a new picture.

motion estimation In video encoding, the process of analyzing previous or future frames to identify blocks that have not changed or have changed only their location. Motion vectors are then stored in place of the blocks. This is very computation-intensive and can cause visual artifacts when subject to errors.

motion vector A two-dimensional spatial displacement vector used for MPEG motion compensation to provide an offset from the encoded position of a block in a reference (I or P) picture to the predicted position (in a P or B picture).

MPEG-2 A popular video encoding standard for standard and high definition content. It is not as efficient as MPEG-4.

MPEG-4 AVC A very efficient video codec approved in the Blu-ray and HD DVD Specifications.

MPEG-4 HE AAC v2 MPEG-4 HE (High Efficiency) is a lossy encoding scheme intended for streaming audio. Some examples of use are in XM Satellite Radio, HD Radio, and Digital Radio Mondiale.

MP@ML Main profile at main level. The common MPEG-2 format used by DVD (along with SP@SL).

MP3 MPEG-1 Layer III audio. A perceptual audio coding algorithm. Not supported in DVD-Video or DVD-Audio formats.

MPEG Moving Picture Experts Group. An international committee that developed the MPEG family of audio and video compression systems.

MPEG audio Audio compressed according to the MPEG perceptual encoding system. MPEG-1 audio provides two channels, which can be in Dolby Surround format. MPEG-2 audio adds data to provide discrete multichannel audio. Stereo MPEG audio is one of two mandatory audio compression systems for 625/50 (PAL/SECAM) DVD-Video.

MPEG video Video compressed according to the MPEG encoding system. MPEG-1 is typically used for low data rate video, as on a Video CD. MPEG-2 is used for higher-quality video, especially interlaced video, such as on DVD or HDTV.

MTBF Mean time between failure. A measure of reliability for electronic equipment, usually determined in benchmark testing. The higher the MTBF, the more reliable the hardware.

Mt. Fuji See SFF 8090.

multichannel Multiple channels of audio, usually containing different signals for different speakers to create a surround-sound effect.

multimedia Information in more than one form, such as text, still images, sound, animation, and video. Usually implies that the information is presented by a computer.

multiplexing Combining multiple signals or data streams into a single signal or stream. This is usually achieved by interleaving at a low level.

MultiRead A standard developed by the Yokohama group, a consortium of companies attempting to ensure that CD and DVD hardware can read all CD formats.

multisession A technique in write-once recording technology that enables additional data to be appended after data is written in an earlier session.

multistory A configuration in which one title contains more than one story.

mux Short for multiplex.

mux_rate In MPEG, the combined rate of all packetized elementary streams (PES) of one program. For example, the mux_rate of DVD is 10.08 Mbps.

N

NAB National Association of Broadcasters.

NAS network attached storage A NAS is a hard disk that is attached to a network to perform one task — store data. This is a place in a business to store large amounts of data like raw video data, audio data, etc., that must be shared among the users on a network.

navigation The process of manipulating a title toward a particular destination, allowing the user access to the contents and operation functions of a title. This process is defined and programmed during the authoring process.

Navigation Manager In HD DVD, a module in between the user (remote control) the disc and the presentation engine (the video displaying on the TV). The Navigation Manager handles information going between these units.

Navigation Pack (NV_PCK) A Navigation Pack contains a PCI (presentation control information) and a DSI (data search information). The Navigation Pack is important for functions such as fast forward and rewind.

Nero An advance software program that specializes in audio/video multimedia data file conversion, burning, ripping, viewing, creating, etc.

Network Manager In HD DVD, the network manager is a module inside the Data Access Manager, and it handles transactions between the server, and Navigation Manager. If a user is watching streaming content, and hits pause on the remote, then this information will be passed to the network manager to handle correctly.

noise Noise is any interference observable in a signal that is caused by another interfering signal. On a television, one can observe noise in a broadcast if someone is using a blow dryer next to the TV. The noise may appear as "snow" on the TV screen.

Non-drop frame timecode The method of timecode computation where there are 30 numerical frames per second of video. "There are 30 frames of video per second," you say. Wrong. There are only 29.97 frames of video per second. In a mathematical hour, there would be 108,000 frames (30 frames per second $\times$ 60 seconds $\times$ 60 minutes). So, a mathematical hour is 108 frames longer than an hour of reality. See also drop frame timecode and timecode.

NRZI Non-return to zero, inverted. A method of coding binary data as waveform pulses. Each transition represents a one, while a lack of a transition represents a run of zeros.

NTSC National Television Systems Committee. A committee organized by the Electronic Industries Association (EIA) that developed commercial television broadcast standards for the US. The group first established black-and-white TV standards in 1941, using a scanning system of 525 lines at 60 fields per second. A second committee standardized color enhancements using 525 lines at 59.94 fields per second. The NTSC standard is also used in Canada, Japan, and other parts of the world. NTSC is facetiously referred to as meaning "never the same color" because of the system's difficulty in maintaining color consistency.

NTSC-4.43 A variation of NTSC in which a 525/59.94 signal is encoded using the PAL sub-carrier frequency and chroma modulation. Also called 60-Hz PAL.

Numerical aperture Numerical Aperture is a measurement of how accurate a lens can resolve the light passing through it. In the context of high definition lasers, the numerical aperture of HD DVD is .65, while the higher numerical aperture of Blu-ray is .85.

NV-RAM (non-volatile random memory) NVRAM is static memory that does not require power. It is used in devices like — thumb drives, digital camera memory cards, video game memory cards, etc.

O

operating system The primary software in a computer, containing general instructions for managing applications, communications, input/output, memory, and other low level tasks. DOS, Windows, Mac OS, Linux, and Unix are examples of operating systems.

opposite path See OTP.

option card In DVD, and in any environment, an option card is a menu that presents a user with several languages they may choose to enjoy the disc. Often, discs are created with several menu languages to suit many regions the disc is produced for. An option card will allow the user to choose which language they will see the menus appear in. This is an alternative to using built in methods such as Language Unit as described in the DVD Specification.

Orange Book The document begun in 1990 that specifies the format of recordable CD. Its three parts define magneto-optical erasable (MO) and write-once (WO) discs, dye-sublimation write-once (CD-R) discs, and phase-change rewritable (CD-RW) discs. Orange Book also added multisession capabilities to the CD-ROM XA format.

OS Operating system.

OSTA Optical Storage Technology Association.

OTP Opposite track path. A variation of DVD dual-layer disc layout where readout begins at the center of the disc on the first layer, travels to the outer edge of the disc, then switches to the second layer and travels back toward the center. Designed for long, continuous-play programs. Also called RSDL. Contrast this with PTP.

overscan The area at the edges of a television tube that is covered to hide possible video distortion. Overscan typically covers about four or five percent of the picture.

P

P or Peta A prefix that defines quantity of a thing such as a byte or bit. Peta means 10^{15} or 2^{50}. The progression of byte measurement is as follows — kilobyte, megabyte, gigabyte, terabyte, petabyte. If you owned more than 100,000 DVDs, the data added up on all of them might be close to a petabyte.

pack A group of MPEG packets in a DVD-Video program stream. Each DVD sector (2,048 bytes) contains one pack.

packet A low level unit of data storage containing contiguous bytes of data belonging to a single elementary stream such as video, audio, control, et cetera. Packets are grouped into packs.

packetized elementary stream (PES) The low level stream of MPEG packets containing an elementary stream, such as audio or video.

page clock In HD DVD this clock is independent of the title time and begins when the active markup page initializes.

PAL Phase alternate line. A video standard used in Europe and other parts of the world for composite color encoding. Various versions of PAL use different scanning systems and color subcarrier frequencies (identified with letters B, D, G, H, I, M, and N), the most common being 625 lines at 50 fields per second, with a color subcarrier of 4.43 MHz. PAL is also said to mean "picture always lousy" or "perfect at last," depending on which side of the Atlantic ocean the speaker comes from.

palette A table of colors that identifies a subset from a larger range of colors. The small number of colors in the palette enables fewer bits to be used for each pixel. Also called a color look-up table (CLUT).

pan and scan The technique of reframing a picture to conform to a different aspect ratio by cropping parts of the picture. Players can automatically create a 4:3 pan and scan version from widescreen anamorphic video by using a horizontal offset encoded with the video.

parallel path See PTP.

Parametric Stereo (PS) this is an added function of aacPlus audio compression that separates aacPlus v2 from aacPlus v1. It is a good technique to get lower bandwidth in stereo audio.

parental control function This function automatically compares the upper limit level of permissible Parental Level playback preset by the users in the player with the Parental Level contained in the disc.

Parental ID Instructions contained in a DVD-Video player that identify the Parental Level of a title.

Parental Level A permissible level to be observed on screen depending on the age of viewers and the nature of the content of a Title.

parental management An optional feature of DVD-Video that prohibits programs from being viewed or substitutes different scenes within a program depending on the parental level set in the player. Parental control requires that Parental Levels and additional material (if necessary) be encoded on the disc.

part of title In DVD-Video, a division of a title representing a scene. Also called a chapter. Parts of titles are numbered 1 to 99.

patchiness A term referring to the occasional patchy appearance of compressed video.

PC (personal computer) a microcomputer designed for personal use. Originally referred to an IBM PC compatible computer, currently refers to any machine that runs on Microsoft Windows operating systems.

PCI Presentation control information. A DVD-Video data stream containing details of the timing and presentation of a program (aspect ratio, angle change, menu highlight and selection information, etc). PCI and DSI comprise an overhead of about one Mbps.

PCM An uncompressed, digitally coded representation of an analog signal. The waveform is sampled at regular intervals and a series of pulses in coded form (usually quantized) are generated to represent the amplitude.

pel See pixel.

perceived resolution The apparent resolution of a display from the observer's point of view, based on viewing distance, viewing conditions, and physical resolution of the display.

perceptual coding Lossy compression techniques based on the study of human perception. Perceptual coding systems identify and remove information that is least likely to be missed by the average human observer.

Persistent Storage In Blu-ray and HD DVD, persistent storage is an area to store data on a player. An example of data that would go onto persistent storage is bookmarks. This way, if a user were to save bookmarks for a particular disc, when they put it back into the player later, the bookmarks will be available as they are stored in persistent storage.

PGC Program Chain, a group of cells linked together creating a program sequence.

PGCI Program chain information. Data describing a chain of cells (grouped into programs) and their sector locations that compose a sequential program. PGCI data is contained in the PCI stream.

PG programs In DVD, a Program is a unit of time in a piece of video. It is typical that the feature divided into PG's (i.e. chapters) which the user can skip back and forth between.

phase-change A technology for rewritable optical discs using a physical effect in which a laser beam heats a recording material to reversibly change an area from an amorphous state to a crystalline state, or vice versa. Continuous heat just above the melting point creates the crystalline state (an erasure), while high heat followed by rapid cooling creates the amorphous state (a mark). Other recording technologies include dye-sublimation and magneto-optical.

physical format The low-level characteristics of the DVD-ROM and DVD-Video standards, including pits on the disc, the location of data, and the organization of data according to physical position.

picture In video terms, a single still image or a sequence of moving images. Picture generally refers to a frame, but for interlaced frames it may refer instead to a field of the frame. In a more general sense, picture refers to the entire image shown on a video display.

picture stop A function of DVD-Video where a code indicates that video playback should stop and a still picture be displayed.

PIP Picture in picture. A feature of some televisions that shows another channel or video source in a small window superimposed in an area of the screen.

pit A microscopic depression in the recording layer of an optical disc. Pits are usually 1/4 of the laser wavelength to cause cancellation of the beam by diffraction.

pit art A pattern of pits stamped onto a disc to provide visual art rather than data. An alternative to a printed label.

pixel The smallest picture element of an image (one sample of each color component). A single dot in the array of dots that comprise a picture, sometimes abbreviated to pel. The resolution of a digital display is typically specified in terms of pixels (width by height) and color depth (the number of bits required to represent each pixel).

pixel aspect ratio The ratio of width to height of a single pixel. This often means the sample pitch aspect ratio (when referring to sampled digital video). Pixel aspect ratio for a given raster can be calculated as y/x × w/h (where x and y are the raster horizontal pixel count and vertical pixel count, and w and h are the display aspect ratio width and height). Pixel aspect ratios are also confusingly calculated as x/y × w/h, giving a height-to-width ratio.

pixel depth See color depth.

platform A platform is an environment in which applications can operate. Examples of a platform are Microsoft XP, Macintosh OSX, the Sony PS3, the Nintendo WII, etc.

Playlist application In HD DVD, a playlist application, defined in the Playlist file (.xpl), that can run in each title by spanning all titles.

PMMA Polymethylmethacrylate. A clear acrylic compound used in laserdiscs and as an intermediary in the surface transfer process (STP) for dual-layer DVDs. PMMA is sometimes used for DVD substrates.

PNG Portable Network Graphic A Portable Network Graphics (PNG) is a compressed image file that is popular in HD DVD, Blu-ray, and the World Wide Web. This format was not used in DVD.

POP Picture outside picture. A feature of some widescreen displays that uses the unused area around a 4:3 picture to show additional pictures.

Postcommand A navigation command executed after a PGC has been read by a player.

Precommand A navigation command executed before a PGC is read by a player.

premastering The process of preparing data in the final format to create a DVD disc image for mastering. This includes creating DVD control and navigation data, multiplexing data streams together, generating error-correction codes, and performing channel modulation. It also often includes the process of encoding video, audio, and subpictures.

presentation data DVD-Video information such as video, menus, and audio that is presented to the viewer. See PCI.

primary audio/video clips In HD DVD, primary audio/video clips are intended for presentation on the main video plain. In contrast Secondary audio/video clips are pieces that might belong in Picture in Picture.

procedural In the context of a programming language, this is a step by step method of programming a computer.

program In a general sense, a sequence of audio or video. In a technical sense for DVD-Video, a group of cells within a program chain (PGC) containing up to 999 Cells.

program chain In DVD-Video, a collection of programs, or groups of cells, linked together to create a sequential presentation.

program streams (PS) A single file that contains audio, video, and subtitle files. It is a file that is a combination of all of its elementary streams.

progressive scan A video scanning system that displays all lines of a frame in one pass. Contrast this with interlaced scan.

Provider ID In HD DVD, this is a GUID that is stored in the ADV_OBJ folder in a file called DISCID.DAT. It represents a unique number that correlates to a studio. If used as intended, all discs of a particular studio will store common media in a folder labeled with this GUID.

psychoacoustic See perceptual coding.

PTP Parallel track path. A variation of dual-layer disc layout where readout begins at the center of the disc for both layers. This is designed for separate programs (such as a widescreen and a pan and scan version on the same disc side) or programs with a variation on the second layer. PTP is most efficient for random-access application. Contrast this with OTP.

PTT A PTT (Part of Title) as defined in DVD is a segment of video that can be referenced, and jumped to. An example is a chapter within a movie. Typically, the Scene Selection menus use a JumpPTT command to jump directly to the chapter from that menu.

Q

quantization levels The predetermined levels at which an analog signal can be sampled, as determined by the resolution of the analog-to-digital converter (in bits per sample) or the number of bits stored for the sampled signal.

quantize To convert a value or range of values into a smaller value or smaller range by integer division. Quantized values are converted back (by multiplying) to a value that is close to the original but may not be exactly the same. Quantization is a primary technique of lossless encoding.

R

RAM Random-access memory. This generally refers to solid-state chips. In the case of DVD-RAM, the term was borrowed to indicate the capability to read and write at any point on the disc.

random access The capability to jump to a point on a storage medium.

raster The pattern of parallel horizontal scan lines that comprise a video picture.

raw Uncompressed data.

read-modify-write An operation used in writing to DVD-RAM discs. Because data can be written by the host computer in blocks as small as two KB, but the DVD format uses ECC blocks of 32 KB, an entire ECC block is read from the data buffer or disc, modified to include the new data and new ECC data, and then written back to the data buffer and disc.

Red Book The document first published in 1982 that specifies the original compact disc digital audio format developed by Philips and Sony.

Reed-Solomon An error-correction encoding system that cycles data multiple times through a mathematical transformation to increase the effectiveness of the error correction, especially for burst errors (errors concentrated closely together, as from a scratch or physical defect). DVD uses rows and columns of Reed-Solomon encoding in a two-dimensional lattice called Reed-Solomon product code (RS-PC).

re-encode Often, if a video has been encoded, but the encode is not sufficient, then it will need a re-encode before it would be released to the replication.

reference picture or **reference frame** An encoded frame that is used as a reference point from which to build dependent frames. In MPEG-2, I pictures and P pictures are used as references.

regional management A mandatory feature of DVD-Video to restrict the playback of a disc to a specific geographical region. Each player and DVD-ROM drive includes a single regional code, and each disc side can specify in which regions it is permitted to be played. Regional coding is optional; a disc without regional codes will play in all players in all regions.

regional playback control function A system to control the playable geographic region of any disc.

render farms A dedicated set of processors, data storage, and a software manager dedicated to processing video. In the context of HD DVD video, a render farm is for encoding video. The processors can be thought of as the farm, the software manager is the farmer, and the video encode is the crop.

replication 1) The reproduction of media such as optical discs by stamping (contrast this with duplication); 2) A process used to increase the size of an image by repeating pixels (to increase the horizontal size) and/or lines (to increase the vertical size) or to increase the display rate of a video stream by repeating frames. For example, a 360×240 pixel image can be displayed at 720×480 size by duplicating each pixel on each line and then duplicating each line. In this case, the resulting image contains blocks of four identical pixels. Obviously, image replication can cause blockiness. A 24-fps video signal can be displayed at 72 fps by repeating each frame three times. Frame replication can cause jerkiness of motion. Contrast this with decimation. See interpolate.

resampling The process of converting between different spatial resolutions or different temporal resolutions. This can be based on a sample of the source information at a higher or lower resolution or it can include interpolation to correct for the differences in pixel aspect ratios or to adjust for differences in display rates.

resolution 1) A measurement of the relative detail of a digital display, typically given in pixels of width and height; 2) The capability of an imaging system to make the details of an image clearly distinguishable or resolvable. This includes spatial resolution (the clarity of a single image), temporal resolution (the clarity of a moving image or moving object), and perceived resolution (the apparent resolution of a display from the observer's point of view). Analog video is often measured as a number of lines of horizontal resolution over the number of scan lines. Digital video is typically measured as a number of horizontal pixels by vertical pixels. Film is typically measured as a number of line pairs per millimeter; 3) The relative detail of any signal, such as an audio or video signal. See lines of horizontal resolution.

Resource Management API In HD DVD, this API gives the programmer a way to manage the file cache to ensure that the file cache is at full potential.

resources a source of supply or support, an available means — usually used in plural.

resume A function used when a player returns from a menu state.

RGB Video information in the form of red, green, and blue tristimulus values. The combination of three values representing the intensity of each of the three colors can represent the entire range of visible light.

ROM Read-only memory.

Roxio Roxio is a division of Sonic Solutions that provides many DVD and other media software solutions.

rpm Revolutions per minute. A measure of rotational speed.

Root menu The lowest level menu contained in a VTS, from which other menus may be accessed.

RS Reed-Solomon. An error-correction encoding system that cycles data multiple times through a mathematical transformation in order to increase the effectiveness of the error correction. DVD uses rows and columns of Reed-Solomon encoding in a two-dimensional lattice, called Reed-Solomon product code (RS-PC).

RS-CIRC See CIRC.

RSDL Reverse-spiral dual-layer. See OTP.

RS-PC Reed-Solomon product code. An error-correction encoding system used by DVD employing rows and columns of Reed-Solomon encoding to increase error-correction effectiveness.

run length coding Method of lossless compression that codes by analyzing adjacent samples with the same values.

S

sample A single digital measurement of analog information or a snapshot in time of a continuous analog waveform. See sampling.

sample rate The number of times a digital sample is taken, measured in samples per second, or Hertz. The more often samples are taken, the better a digital signal can represent the original analog signal. The sampling theory states that the sampling frequency must be more than twice the signal frequency in order to reproduce the signal without aliasing.

sample size The number of bits used to store a sample. Also called resolution. In general, the more bits allocated per sample, the better the reproduction of the original analog information. The audio sample size determines the dynamic range.

sampling Converting analog information into a digital representation by measuring the value of the analog signal at regular intervals, called samples, and encoding these numerical values in digital form. Sampling is often based on specified quantization levels. Sampling can also be used to adjust for differences between different digital systems. See resampling and subsampling.

sampling frequency The frequency used to convert an analog signal into digital data.

saturation The intensity or vividness of a color.

SBR (Spectral Band Replication) A coding technology included in MPEG-4 audio technologies aacPlus V1 and aacPlus v2. It specializes in low bandwidth encoding.

scaling Altering the spatial resolution of a single image to increase or reduce the size, or altering the temporal resolution of an image sequence to increase or decrease the rate of display. Techniques include decimation, interpolation, motion compensation, replication, resampling, and subsampling. Most scaling methods introduce artifacts.

scan line A single horizontal line traced out by the scanning system of a video display unit. 525/60 (NTSC) video has 525 scan lines, about 480 of which contain the actual picture. 625/50 (PAL/SECAM) video has 625 scan lines, about 576 of which contain the actual picture.

scanning velocity The speed at which the laser pickup head travels along the spiral track of a disc.

Scenarist SCA Sonic Scenarist has created a software tool that specializes in HD DVD Standard Content Authoring (SCA). At format launch, this is a quick way to author HD DVD discs.

SCMS The serial copy management system used by DAT, MiniDisc, and other digital recording systems to control copying and limit the number of copies that can be made from copies.

screen grab this is a quick way to turn the full image of a computer screen into an image file. In the Microsoft Windows OS, this is done by pressing [ALT]+[Print Scrn], saving the image into the clip board.

script In HD DVD, this is a short form for JavaScript. Also referenced as the entire collection of script files.

SCSI Small Computer Systems Interface. An electronic interface and command set for attaching and controlling internal or external peripherals, such as a DVD-ROM drive, to a computer. The command set of SCSI was extended for DVD-ROM devices by the SFF 8090 specification.

SD Card The SD (Secure Digital) card is a small flash memory card that is about the size of a US quarter. It is the storage device of many consumer electronic products including digital cameras, video cameras, cell phones, etc.

SDI See Serial Digital Interface. Also Strategic Defense Initiative, aka Star Wars, which was finally released on DVD so fans can replace their bootleg copies.

SDDI Serial Digital Data Interface. A digital video interconnect designed for serial digital information to be carried over a standard SDI connection.

SDDS Sony Dynamic Digital Sound. A perceptual audio-coding system developed by Sony for multichannel audio in theaters. A competitor to Dolby Digital and an optional audio track format for DVD.

SDMI Secure Digital Music Initiative. Efforts and specifications for protecting digital music.

SDTV Standard-definition television. A term applied to 4:3 television (in digital or analog form) with a resolution of about 700×480. Contrast this with HDTV.

SECAM Sequential couleur avec mémoire/sequential color with memory. A composite color standard similar to PAL but currently used only as a transmission standard in France and a few other countries. Video is produced using the 625/50 PAL standard and is then transcoded to SECAM by the player or transmitter.

secondary audio In HD DVD, secondary audio will often accompany secondary video in picture in picture. Secondary audio does not need to have accompanying video and might be applied as an audio commentary track by the movie director.

secondary video plane In the Blu-ray and HD DVD specifications, this is a plane that is dedicated to the picture in picture video which is above the main video plane. This plane can be programmatically turned on and off.

Secondary Video Player The component dedicated to playing secondary video in an HD DVD or Blu-ray player.

Secondary Video Sets The HD DVD spec defines three different kinds of Secondary Video Sets: 1) Substitute Audio 2) Substitute Audio Video 3) Secondary Audio Video.

sector A logical or physical group of bytes recorded on the disc, the smallest addressable unit. A DVD sector contains 38,688 bits of channel data and 2,048 bytes of user data.

seek time The time it takes for the head in a drive to move to a data track.

SEI (Supplemental Enhancement Information) Information in an MPEG-4 bitstream that can be used for any variety of reasons to communicate information to a playback device.

Serial Digital Interface (SDI) The digital video connection format using a 270-Mbps transfer rate. A 10-bit, scrambled, polarity-independent interface, with common scrambling for both component ITU-R 601 and composite digital video and four groups each of four channels of embedded digital audio. SDI uses standard 75-ohm BNC connectors and coax cable.

sequential presentation Playback of all Programs in a Title in a specified order determined in the authoring process.

setup The black level of a video signal.

SFF 8090 The specification number 8090 of the Small Form Factor Committee, an ad hoc group formed to promptly address disk industry needs and to develop recommendations to be passed on to standards organizations. SFF 8090 (also known as the Mt. Fuji specification) defines a command set for CD-ROM- and DVD-ROM-type devices, including implementation notes for ATAPI and SCSI.

shuffle presentation A function of a DVD-Video player enabling programs in a title to play in an order determined at random.

SI Système International (d'Unités)/International System (of Units). A complete system of standardized units and prefixes for fundamental quantities of length, time, volume, mass, et cetera.

signal-to-noise ratio The ratio of pure signal to extraneous noise, such as tape hiss or video interference. Signal-to-noise ratio is measured in decibels (dB). Analog recordings almost always have noise. Digital recordings, when properly prefiltered and not compressed, have no noise.

simulate To test the function of a disc in the authoring system without actually formatting an image.

single-sided disc A type of disc on which data is recorded on one side only.

skin a custom and interchangeable graphic element for a user interface with computing software.

SL (Single Layer) In the context of optical media discs, Single Layer means that there is data on only one layer. For example, a single layer single sided disc is known as a DVD 5.

SMPTE The Society of Motion Picture and Television Engineers. An international research and standards organization. This group developed the SMPTE timecode, used for marking the position of audio or video in time.

SMPTE VC-1 VC-1 is a video standard accepted by Blu-ray and HD DVD standards as a codec that all players must support. It was developed by SMPTE and the Microsoft Corporation.

S/N Signal-to-noise ratio. Also called SNR.

Soft Sync In HD DVD, Soft Sync gives preference to seamlessness. If a Soft Sync application is loaded in the middle of a Title timeline, then there will not be audio or video drop outs.

son The metal discs produced from mother discs in the replication process. Fathers or sons are used in molds to stamp discs.

Sonic Solution Scenarist BD A software solution created by Sonic Solutions that specializes in the authoring of Blu-ray Discs.

Sony Blu-Print A software solution created by Sony that specializes in the authoring of Blu-ray Discs.

space The reflective area of a writable optical disc. Equivalent to a land.

spatial Relating to space, usually two-dimensional. Video can be defined by its spatial characteristics (information from the horizontal plane and vertical plane) and its temporal characteristics (information at different instances in time).

spatial resolution The clarity of a single image or the measure of detail in an image. See resolution.

SPDC (Self Protecting Digital Content) An architecture that applies content protection mechanisms (code) on the content (disc) rather than on the playback device. It is employed in AACS.

S/PDIF Sony/Philips digital interface. A consumer version of the AES/EBU digital audio transmission standard. Most DVD players include S/P DIF coaxial digital audio connectors providing PCM and encoded digital audio output.

specification verification The process of ensuring that something built is in compliance with a specification. One popular verification tool is called the MEI Verifier. This tool will report errors on a disc build that can be addressed before replication.

SPRM system parameters Similar to GPRMs, SPRMs are registers that describe features of the player. They are read only. Some SPRMs have player information like — menu language, region, default audio setting, default subtitle setting, etc.

SRP Sustained read performance The highest read/write speed at which an optical disc can be operated on. Often read as a multiple — 1X, 2X, 4X, etc.

stamping The process of replicating optical discs by injecting liquid plastic into a mold containing a stamper (father or son). Also (inaccurately) called mastering.

Standard Content In HD DVD, Standard Content is a less functional disc that does not take advantage of higher Advanced Content capabilities. Standard Content supports new audio/video high definition, but does not have full functionality like networking capabilities, among others. This term should not be confused with standard definition, a lower resolution audio/video format as seen in DVD.

statistically lossless A statistically lossless compression algorithm was developed for CR and DR that makes it possible to increase compression ratios by approximately three-fold over current lossless entropy coding schemes. The algorithm cascades a quantization companding function with a new lossless image compression technique called JPEG-LS. The companding function is derived from the inherent noise properties of the image acquisition system using a theory, which describes the modification of statistical moments in images due to quantization. In addition, image distortion effects are quantified that are associated with the use of a unique feature of JPEG-LS, which allows the user to trade constrained reconstruction errors for increased amounts of compression.

stream A continuous flow of data, usually digitally encoded, that is designed to be processed sequentially. Also called a bitstream.

streaming buffer In HD DVD, this component handles incoming audio/video streams from the Internet, and is configured in the Playlist and DISCID.DAT.

subpicture Graphic bitmap overlays used in DVD-Video to create subtitles, captions, karaoke lyrics, menu highlighting effects, etc.

subsampling The process of reducing spatial resolution by taking samples that cover areas larger than the original samples, or the process of reducing temporal resolutions by taking samples that cover more time than the original samples. This is also called downsampling. See chroma subsampling.

substrate The clear polycarbonate disc onto which data layers are stamped or deposited.

subtitle A textual representation of the spoken audio in a video program. Subtitles are often used with foreign languages and do not serve the same purpose as captions for the hearing impaired. See subpicture.

surround sound A multichannel audio system with speakers in front of and behind the listener to create a surrounding envelope of sound and to simulate directional audio sources.

surround An audio system that produces a vivid stereophonic sound environment by arranging two or more speakers around the listener.

SVCD Super Video Compact Disc. MPEG-2 video on CD. Used primarily in Asia.

S-VHS Super VHS (Video Home System). An enhancement of the VHS videotape standard using better recording techniques and Y/C signals. The term S-VHS is often used incorrectly to refer to s-video signals and connectors.

s-video A video interface standard that carries separate luma and chroma signals, usually on a four-pin mini-DIN connector. Also called Y/C. The quality of s-video is significantly better than composite video because it does not require a comb filter to separate the signals, but it's not quite as good as component video. Most high-end televisions have s-video inputs. S-video is often erroneously called S-VHS.

SXGA A video graphics resolution of 1280×1024 pixels.

sync A video signal (or component of a video signal) containing information necessary to synchronize the picture horizontally and vertically. Also, sync is specially formatted data on a disc that helps the readout system identify location and specific data structures.

syntax The rules governing the construction or formation of an orderly system of information. For example, the syntax of the MPEG video encoding specification defines how data and associated instructions are used by a decoder to create video pictures.

system menu The main menu of a disc, from which titles are selected. Also called the title selection menu or disc menu.

system parameters A set of conditions defined in the authoring process used to control basic player functions.

System Space In DVD, System Space refers to PGC Units within the Video Manager Domain. The command JumpSS is a call to get into System Space.

T

T or Tera An SI prefix for denominations of one trillion (10^{12}), that defines quantity of a thing such as a byte or bit. The progression of byte measurement is as follows — kilobyte, megabyte, gigabyte, terabyte, petabyte. If a 4 Gigabyte iPod can hold 1000 songs, a 4 Terabyte iPod could hold 1,000,000 songs. As of the writing of this book, a 4 Terabyte iPod does not exist.

TEG2 (Technical Experts Group 2) One of five technical groups in the Joint Technical Committee in the (BDA) Blu-ray Disc Association. TEG2 is responsible for Audio/Video applications.

telecine The process (and the equipment) used to transfer film to video. The telecine machine performs 2:3 pulldown by projecting film frames in the proper sequence to be captured by a video camera.

telecine artist The operator of a telecine machine. Also called a colorist.

temporal Relating to time. The temporal component of motion video is broken into individual still pictures. Because motion video can contain images (such as backgrounds) that do not change much over time, typical video has large amounts of temporal redundancy.

temporal resolution The clarity of a moving image or moving object, or the measurement of the rate of information change in motion video. See resolution.

timecode Information recorded with audio or video to indicate a position in time. This usually consists of values for hours, minutes, seconds, and frames. It is also called SMPTE timecode. Some DVD-Video material includes information to enable the player to search to a specific timecode position. There are two types of timecode — non-drop frame and drop frame. Non-drop frame timecode is based on 30 frames of video per second. Drop frame timecode is based on 29.97 frames of video per second. Truth be told, there are only 29.97 frames of video per second. For short amounts of time, this discrepancy is inconsequential. For longer periods of time, however, it is important. One hour of non-drop frame timecode will be 108 frames longer than one hour of real time. See non-drop frame timecode and drop frame timecode.

title The largest unit of a DVD-Video disc (other than the entire volume or side). A title is usually a movie, TV program, music album, or so on. A disc can hold up to 99 titles, which can be selected from the disc menu. Entire DVD volumes are also commonly called Titles.

title key A value used to encrypt and decrypt (scramble) user data on DVD-Video discs.

title menu A graphic image presented by the VMG for a user to select a title.

Title Set In DVD, a group of titles that have the same aspect ratio, and combined with menu language units comprise a Title Set.

Title Space In DVD, title space is the area where the major video content (feature video) of a disc will reside. Menus typically reside in menu space.

track 1) A distinct element of audiovisual information, such as the picture, a sound track for a specific language, or the like; 2) One revolution of the continuous spiral channel of information recorded on a disc.

transfer rate The speed at which a certain volume of data is transferred from a device such as a DVD-ROM drive to a host such as a personal computer. This is usually measured in bits per second or bytes per second. It is sometimes confusingly used to refer to the data rate, which is independent of the actual transfer system.

trick play modes In digital video mediums, trick play is a term that covers types of play other than normal play like fast forward, rewind, slow motion, etc.

trim See crop.

tristimulus A three-valued signal that can match nearly all the colors of visible light in human vision. This is possible because of the three types of photoreceptors in the eye. RGB, $Y'C_bC_r$, and similar signals are tristimulus and can be interchanged by using mathematical transformations (subject to a possible loss of information).

TT Title In DVD, a VTS is made of Title Space (VTS-TT) and Menu Space (VTS-M). TT Title is a way to refer to a Title, and DVD commands often end with TT in a title call like JumpTT.

TVL Television line. See lines of horizontal resolution.

TWG Technical Working Group. A general term for an industry working group. Specifically, the predecessor to the CPTWG. It is usually an ad hoc group of representatives working together for a period of time to make recommendations or define standards.

U

UDF Universal Disc Format. A standard developed by the Optical Storage Technology Association designed to create a practical and usable subset of the ISO/IEC 13346 recordable, random-access file system and volume structure format.

UDF Bridge A combination of UDF and ISO 9660 file system formats that provides backward-compatibility with ISO 9660 readers while allowing the full use of the UDF standard.

UOP A UOP (User Operation) is function that a user can perform generally with a remote control. If a UOP is permitted, the user can use it. Examples of UOPs are — stop, pause, fast forward, etc. A title may have all UOPs locked during the Warnings as studios desire that the user fully views them.

URI A Uniform Resource Identifier (URI) is a string of characters usually delimited with several slashes (/) to identify the location of a file. URI's are used in the World Wide Web (ex: http://www.xxxx.com/folder/file.htm) URI's are also used in HD DVD to identify assets on the disc, player and network.

USB flash drive A small portable data storage device connects to a computer or other device through USB.

user The person operating the player. Sometimes referred to as wetware, as in, the operation experienced a wetware breakdown.

user data The data recorded on a disc independent of formatting and error-correction over-head. Each DVD sector contains 2,048 bytes of user data.

UXGA A video graphics resolution of 1600 x 1200.

V

VBR Variable bit rate. Data that can be read and processed at a volume that varies over time. A data compression technique that produces a data stream between a fixed minimum and maximum rate. A compression range is generally maintained, with the required bandwidth increasing or decreasing depending on the complexity (the amount of spatial and temporal energy) of the data being encoded. In other words, the data rate fluctuates while quality is maintained. Compare this to CBR.

VC-1 VC-1 is a video standard accepted by Blu-ray and HD DVD standards as a codec that all players must support. It was developed by SMPTE and the Microsoft Corporation.

VCD Video Compact Disc. Near-VHS-quality MPEG-1 video on CD. Used primarily in Asia.

VfW See Video for Windows. Not to be confused with Veterans of Foreign Wars, a fine organization composed of persons who occasionally experience wetware problems when operating a VCR, DVD player, or computer.

VGA (Video Graphics Array) A standard analog monitor interface for computers. It is also a video graphics resolution of 640 × 480 pixels.

VHS Video Home System. The most popular system of videotape for home use. Developed by JCV.

Video CD A CD extension based on MPEG-1 video and audio that enables the playback of near-VHS-quality video on a Video CD player, CD-i player, or computer with MPEG decoding capability.

Video for Windows The system software additions used for motion video playback in Microsoft Windows. Replaced in newer versions of Windows by DirectShow (formerly called ActiveMovie).

Video manager (VMG) The disc menu. Also called the title selection menu.

Video title set (VTS) A set of one to ten files holding the contents of a title.

Video_TS In DVD, this is the folder on the root of a DVD contains all data for a DVD video disc. There is another folder on the root of a DVD video disc labeled Audio_TS.

videophile Someone with an avid interest in watching videos or in making video recordings. Videophiles are often very particular about audio quality, picture quality, and aspect ratio to the point of snobbishness. Videophiles never admit to a wetware failure.

VLC Variable length coding. See Huffman coding.

VMG Video Manager Domain In DVD, this domain is generally the area where programmatic traffic travels from VTS to VTS. Sometimes, video or menu data can be placed inside the Video Manager. The Title button on a DVD remote control is associated with the Title Entry PGC in this domain.

VOB Video object. A small physical unit of DVD-Video data storage, usually a GOP.

VOBS Video object set In DVD, A group of VOBs (the s is for plurality). There are three types of VOBS: Video Manager VOBS, Menu Space VOBS, and Title Space VOBS.

volume A logical unit representing all the data on one side of a disc.

VPLStnnn.XPL video playlist file In HD DVD, a playlist is an XML document that describes video objects in the HVDVD_TS folder. It outlines time codes, chapters, title numbers, and incorporates Advanced Applications in relation to the Titles.

VSDA Video Software Dealers Association.

VTS Video Title Set, containing up to ten files with all of the contents of a Title.

VTSM video title set menu In DVD the VTSM is the root flagged menu of a Video Title set. If playing a video in a title, if the user were to push the Menu button, then they would be directed to this VTSM PGC.

W

WAM!NET WAM!NET is a company that provides a solution for transferring large data files. WAM!NET is an alternative to physically sending a driver with tapes to a replication facility.

watermark Information hidden as invisible noise or inaudible noise in a video or audio signal.

wetware see user.

WG1 (Working Group 1) A group formed from members of the DVD Forum that concentrates on technical DVD-Video issues.

White Book The document from Sony, Philips, and JVC begun in 1993 that extended the Red Book CD format to include digital video in MPEG-1 format. It is commonly called Video CD.

widescreen A video image wider than the standard 1.33 (4:3) aspect ratio. When referring to DVD or HDTV, widescreen usually indicates a 1.78 (16:9) aspect ratio.

window A usually rectangular section within an entire screen or picture.

Windows See Microsoft Windows.

Windows Media Audio (WMA) WMA is a Windows audio compression format that is supported on many devices.

Windows Media Audio Professional (WMAPro) A high resolution Windows audio compression format that supports 96 kHz / 24 bit 5.1 channel sound.

X

XBR (High-Bit Rate Extension) A DTS-HD extension for higher constant bit rates intended for Blu-ray (6 Mbps) and HD DVD (3Mbps).

XGA A video graphics resolution of 1024 × 768 pixels.

XHTML extensible hypertext markup language HTML was strictly defined, but browsers became more lenient to bad HTML coders to make web page creation easier. This eventually led to sloppy html code all over the world. When XML was designed, this flaw was taken into account and XML was made very strict. XHTML is a strict form of HTML that follows the strict XML guideline.

XLL A DTS-HD extension for Lossless encoding intended for Blu-ray (up to 24.5 Mbps) and HD DVD (up to 18Mbps)

XML Extensible markup language XML (Extensible Markup Language) is a text based container for information. It can really be used for anything (extensible). Because of this openness, it is used for many things in the computing world now and has been quickly adopted for many Web needs. In HD DVD there are 3 types of XML documents designed that have different purposes.

XXCH A DTS-HD extension that goes beyond 6.1 channels.

XVCD A non-standard variation of VCD.

Y

Y The luma or luminance component of video, which is the brightness independent of color.

Y/C A video signal in which the brightness (luma, Y) and color (chroma, C) signals are separated. This is also called s-video.

$Y'C_bC_r$ A component digital video signal containing one luma and two chroma components. The chroma components are usually adjusted for digital transmission according to ITU-R BT.601. $Y'C_bC_r$ applies only to digital video, but it is often incorrectly used in reference to the $Y'P_bP_r$ analog component outputs of DVD players.

Yellow Book The document produced in 1985 by Sony and Philips that extended the Red Book CD format to include digital data for use by a computer. It is commonly called CD-ROM.

$Y'P_bP_r$ A component analog video signal containing one luma and two chroma components. It is often referred to loosely as YUV or Y', B'-Y', R'-Y'.

YUV In the general sense, any form of color-difference video signal containing one luma and two chroma components. Technically, YUV is applicable only to the process of encoding component video into composite video. See $Y'C_bC_r$ and $Y'P_bP_r$.

Z

ZCLV Zoned constant linear velocity. This consists of concentric rings on a disc within which all sectors are the same size. It is a combination of CLV and CAV.

Index

A

C

D

H

I

J

L

M

P

Q

T

U

V

W

X